THERMAL ELECTROCYCLIC REACTIONS

This is Volume 43 of
ORGANIC CHEMISTRY
A series of monographs
Editor: HARRY H. WASSERMAN

A complete list of the books in this series appears at the end of the volume.

THERMAL ELECTROCYCLIC REACTIONS

Elliot N. Marvell

Department of Chemistry
Oregon State University
Corvallis, Oregon

1980

ACADEMIC PRESS

A Subsidiary of Harcourt Brace Jovanovich, Publishers
New York   London   Toronto   Sydney   San Francisco

ACADEMIC PRESS, INC.
111 Fifth Avenue, New York, New York 10003

United Kingdom Edition published by
ACADEMIC PRESS, INC. (LONDON) LTD.
24/28 Oval Road, London NW1   7DX

Library of Congress Cataloging in Publication Data

Marvell, Elliot Nelson, Date
    Thermal electrocyclic reactions.

    (Organic chemistry series)
    Includes bibliographies and index.
    1. Conservation of orbital symmetry. I. Title.
II. Series: Organic chemistry series (New York)
QD476.M29   547'.2   80–761
ISBN 0–12–476250–6

PRINTED IN THE UNITED STATES OF AMERICA

80 81 82 83   9 8 7 6 5 4 3 2 1

CONTENTS

v

9   Odd Electron Systems

PREFACE

Woodward and Hoffmann defined, in their original set of communications on the conservation of orbital symmetry theory, three specific types of possible concerted reactions: i.e. electrocyclizations, cycloadditions, and sigmatropic shifts. The latter pair have received considerable attention in reviews and monographs, but the electrocyclic reaction has been completely neglected except for those books which treat pericyclic reactions in general. This result may pertain because of the impression that the electrocyclic process is very limited and thus has been viewed as important only in tests of the orbital symmetry theory. I hope that this impression will be corrected by this review, and that the electrocyclic reactions will be seen to have an importance in synthesis as well as theory. I trust also that the total picture will lead to wider usage of the reaction and to the development of new and valuable adaptations to stereospecific syntheses.

I have made an attempt to cover the literature through mid 1979, but as anyone who has attempted to penetrate that mass will understand the coverage is certainly incomplete. My hope is that the majority of the most important results have been uncovered. Perhaps this book may best be viewed as a tribute to the ingenuity of those researchers who have had the imagination to see, within the apparently limited scope of the definition of an electrocyclic reaction, a multitude of ideas and applications which have lengthened this manuscript beyond my original anticipations.

Beyond the words and the pages there always exists a special group of people who make the writing possible. I extend my thanks to my wife for her constant encouragement, to my students who have added to the vision, and to Mrs. Alma Rogers who, without previous experience with organic chemists, tackled the unenviable task of translating my handwritten manuscript into a typed copy.

Elliot N. Marvell

CHAPTER 1
INTRODUCTION

Reaction classification is based in its most primitive sense on structural relations, as for example substitution, elimination, cyclization, etc. More sophisticated classification adds an element of mechanism, i.e., S_N1, E_2, etc. For many years the electrocyclic reaction remained hidden among reactions having a primitive label as cyclization or ring scission, their mechanistic interrelations unrecognized for about a century. In 1934 Baker[1] placed the reaction as a valence isomerization and that term has been often applied since, but it is mechanistically too diffuse and includes too many diverse reaction types. Thus it often disguises rather than illuminates the modes of interconversion of these isomers. In 1965 Woodward and Hoffmann in their initial set of papers on the conservation of orbital symmetry theory grouped the reactions to which the theory applied into three classes, electrocyclic reactions,[2] sigmatropic shifts,[3] and cycloadditions.[4] The results have been mechanistically enlightening and exceedingly helpful in synthesis.

Electrocyclic reactions were defined by Woodward and Hoffmann as "the formation of a single bond between the termini of a linear system containing $k\,\pi$ electrons, and the converse." This definition is not sufficiently precise, but it was diagramatically refined by showing that the single bond was formed at the expense of two of the π electrons.[5] Furthermore the mechanistic content must be assumed from the context in which the definition was made, i.e., conservation of orbital symmetry must apply. The $k\,\pi$ electrons must therefore constitute a conjugated system, and formation of the new single bond and rebonding in the resultant π system of k-2 electrons must occur in "concerted" fashion, since "orbital symmetry is conserved in concerted reactions."[6]

To permit experimental tests of the validity of the theory it is desirable to have a definition of the structural requirements of the reaction, a means of testing whether the theory should apply to a given experimental example and finally a way to verify whether the predictions of the theory are born out. In this sense the electrocyclic reaction is defined as an uncatalyzed intramolecular process forming a new single bond between the terminal atoms of a linear conjugated system with n π electrons giving a cyclic system with n-2 π electrons. According to Woodward and Hoffmann if this process occurs in "concerted" fashion the theory should apply. Objections have been raised to this one-to-one correspondence between "concerted" and orbital symmetry conservation.[7,8] Experimental tests of whether the theory should apply to a given example have been difficult to develop. In terms of the theory "concerted" means that the electronic system of the k electrons remains sufficiently interconnected throughout the reaction as to permit orbital phasing information to be relayed to all parts of the system. One interpretation is that this demands that all bonding reaffil-iations occur simultaneously, not consecutively. Thus any experimental consequences of simultaneity or lack thereof can be used as a test of concertedness.[9] Another test for "concertedness" is based on the pre-diction of the orbital symmetry theory that a "concerted allowed" process should require a lower energy than reaction via an intermediate. Thus a significant difference between the measured activation energy (or entropy) and those calculated for possible intermediates by thermochemical means

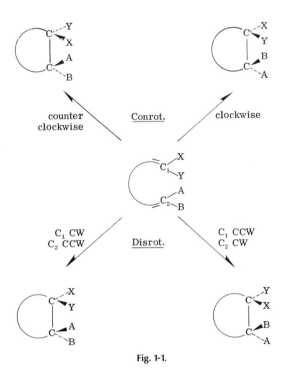

Fig. 1-1.

is a measure of concert. Benson[10] and Doering[11] have been proponents of this test.

Quite obviously the electrocyclic reaction is intended to apply to a rather restricted group of processes. Unfortunately two events have combined to encourage a misuse of the term. Orbital symmetry theory applies generally

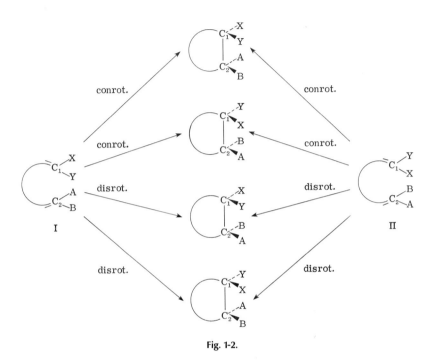

Fig. 1-2.

to reactions having transition states with closed cycles of electrons, hence in the absence of any other general term for all such reactions, electrocyclic appears to be a natural. Later pericyclic was coined to fill this gap,[7] as the term applied generally to all reactions whose bonding relations change in

Suprafacial   Antarafacial

Fig. 1-3.

concert via a closed cycle. The conservation of orbital symmetry theory was introduced to chemists via electrocyclic processes,[2] and so the term is still applied erroneously instead of pericyclic.

Electrocyclic reactions generally cause a stereochemical alteration at the terminal atoms via a rotational process. This converts the trigonal termini of the linear system to tetrahedral atoms in the cyclic system. Four possible rotational routes can exist. Starting with one geometric isomer rotation of both terminal atoms in the same direction (conrotation) can lead to two enantiomers of the cyclic form. Similarly rotation in opposing directions (disrotation) leads to the enantiomers of a second cyclic isomer. Considering terminal atoms only there exist four geometric isomers for the open chain molecules and four optical isomers for the cyclic molecules. These can be interconverted, as is illustrated here for two of the geometric isomers, by two conrotatory or two disrotatory routes. The preference for conrotatory or disrotatory routes will be determined generally by orbital symmetry, while the preference for one or the other of the two conrotatory (or two disrotatory) paths will be determined by steric or electronic interactions.

The comparative terminology of con and disrotatory is not always convenient, and a more general description has been developed in terms of the

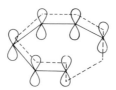

A connectivity cycle

Fig. 1-4.

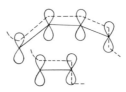

Broken into two
components

Fig. 1-5.

reconnection of orbitals occurring during reaction.[3,12] Thus for a π bond a suprafacial process forms new connections on the same side of the nodal plane, whereas formation of the new bonds on opposite sides is an antarafacial process. A more precise topological definition of the terms has been provided by Day.[13] He defines a *connectivity cycle* as a continuous closed curve passing through one or more lobes of each atomic orbital in the basis set, which makes a single connection between adjacent overlapping orbitals. N is the number of times the connectivity cycle passes through the node of an atomic orbital. The connectivity cycle can be broken down into subsets, termed components, subject only to the requirement that no atomic orbital can appear in more than one component. Then if N is even the result is suprafacial, if odd it is antarafacial. Note that for the butadiene–cyclobutene transformation suprafacial is equivalent to disrotatory and antarafacial to conrotatory.

Connectivity cycle
$N = 0$
suprafacial/disrot.

Connectivity cycle
$N = 1$
antarafacial/conrot.

Fig. 1-6.

REFERENCES

1. J. W. Baker, "Tautomerism," Van Nostrand–Reinhold, Princeton, New Jersey, 1934.
2. R. B. Woodward and R. Hoffmann, *J. Am. Chem. Soc.* **87**, 395 (1965).
3. R. B. Woodward and R. Hoffmann, *J. Am. Chem. Soc.* **87**, 2511 (1965).
4. R. Hoffmann and R. B. Woodward, *J. Am. Chem. Soc.* **87**, 2046 (1965).
5. R. B. Woodward and R. Hoffmann, "Conservation of Orbital Symmetry," Verlag Chemie, Weinheim, 1970.
6. See refer. 5, p. 1.
7. J. E. Baldwin, A. H. Andrist, and R. K. Pinschmidt, Jr., *Accts. Chem. Res.* **5**, 402 (1972).
8. J. E. Baldwin, *in* "Pericyclic Reactions" (A. P. Marchand and R. E. Lehr, eds.), Vol. II, pp. 273–302, Academic Press, New York, 1977.
9. R. E. Lehr and A. P. Marchand, "Pericyclic Reactions" (A. P. Marchand and R. E. Lehr, eds.), Vol. I, pp. 1–51, Academic Press, New York, 1977.
10. S. W. Benson, "Thermochemical Kinetics," p. 83, Wiley, New York, 1968.
11. W. von E. Doering, V. G. Toscano, and G. H. Beasley, *Tetrahedron* **27**, 5299 (1971).

12. See refer. 5, p. 65.
13. A. C. Day, *J. Am. Chem. Soc.* **97,** 2431 (1975).

CHAPTER 2
THEORY

I. ORIGIN OF THE THEORY

Important ideas sometimes result from asking the proper questions about relatively simple observations. Electrocyclic reactions are really very simple reactions and their study led to the conservation of orbital symmetry theory.[1] During studies on the synthesis of vitamin B_{12} Woodward examined an electrocyclic reaction and found that thermal and photo activation gave different stereochemical results.[2] This observation along with similar ones made by other investigators prompted the question—is there a good theoretical reason behind this observation? The conservation of orbital symmetry theory was the consequence. Too much has been written about the theory to permit more than a brief review here, but additional information may be obtained from the list of texts given.[3]

The theory is beguilingly simple. In essence the phase relations among p-orbital electrons in a fully conjugated system reacting via a cyclic transition state will permit an energetically favorable transition to bonding orbitals of the product only under certain rotational operations. The consequences can be traced stereochemically. Woodward and Hoffmann map these orbital conversions via symmetry operations and correlation diagrams.

II. ORBITAL SYMMETRY AND CORRELATION DIAGRAMS

Construction of correlation diagrams requires knowledge of the nature of molecular orbitals, specifically their symmetry properties and relative energies. In many cases the procedure is quite simple, but for more complex examples additional information is needed and more comprehensive works should be consulted.[3] Here the procedure will be outlined using the well-known example of butadiene–cyclobutene. Examination of the reaction indicates that two symmetry elements are germane, a plane, or a twofold axis. The elements chosen must bisect one or more of the bonds being made or broken during the reaction. Note that one must assume the reaction is initiated from *s-cis*-butadiene. The relevant molecular orbitals and their

2. Theory

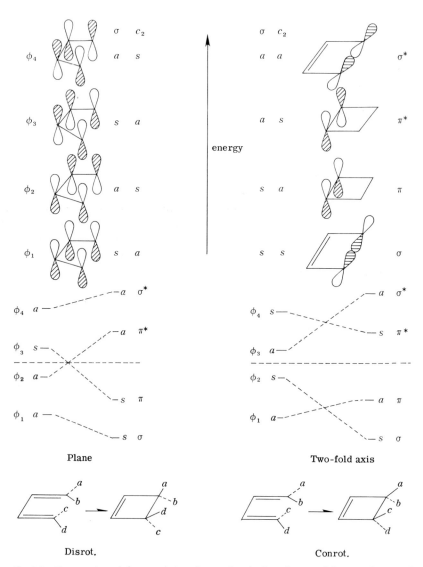

Fig. 2-1. Construction of the correlation diagram for the butadiene–cyclobutene electrocyclic reaction. (*Shaded and nonshaded lobes represent opposite phases.)

phase properties are used to ascertain their symmetry classification with respect to each symmetry element. The correlation diagram is then constructed, as shown in Fig. 2-1, by connecting the orbitals of proper symmetry. The ground state of butadiene, $\phi_1{}^2\phi_2{}^2$, correlates directly with the ground state of cyclobutene, $s^2\pi^2$, only for the conrotatory process. This route is then said to be "allowed." For the disrotatory route the ground state of butadiene correlates directly with a doubly excited state of cyclobutene, $\sigma^2\pi^{*2}$. Actually the reaction would proceed to give ground state cyclobutene, but the activation energy would be higher than for the conrotatory process. Consequently this disrotatory process is termed *forbidden*.

Photochemically the electrocyclic process might be expected to proceed via excitation of one electron from the highest occupied molecular orbital (HOMO) to the lowest unoccupied molecular orbital (LUMO). Again, if this initially formed singlet excited state can be converted to the product without loss of stereochemical control, the new correlation diagrams, Fig. 2-2, may be constructed. The electron occupation pattern shown shows that the first excited singlet state of butadiene correlates directly with the first excited singlet state of cyclobutene if the plane of symmetry is maintained, i.e., a disrotatory stereochemical result is preferred. The axially symmetric correlation relates the first excited singlet state of butadiene to the unusual excited state of cyclobutene with a sigma electron excited rather than a π electron. Clearly the conrotatory process would require a higher energy and is again termed symmetry forbidden.

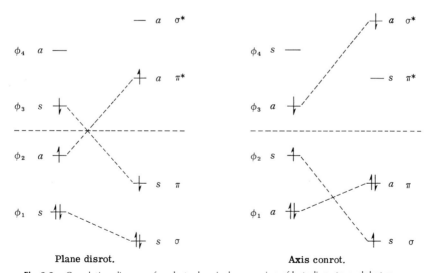

Fig. 2-2. Correlation diagrams for photochemical conversion of butadiene to cyclobutene.

Analysis via correlation diagrams can be employed with all electrocyclic reactions having requisite inherent symmetry. Naturally no trivial element of symmetry which would permit classification of all orbitals as symmetric (or alternatively as anti-symmetric) is of value. Any structural feature, such as an alkyl substituent, which would produce only a small perturbing effect on the orbital energies should be neglected. Essentially one removes such perturbing substituents, reducing the system to its highest inherent symmetry prior to setting up the diagrams. The procedure is limited to reactions having some appropriate symmetry which can be maintained throughout the reaction. For example the question of a preferred route for the reaction of Eq. (2-1), which constitutes a double cyclization of a cross-conjugated

$$\text{(2-1)}$$

system, cannot be treated by direct formation of a correlation diagram since no relevant symmetry element or elements are present.

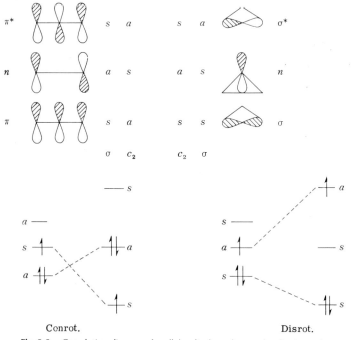

Conrot.   Disrot.

Fig. 2-3. Correlation diagrams for allyl radical–cyclopropyl radical reaction.

For electrocyclic reactions involving linear polyenes, cations or anions Woodward and Hoffmann[1,3] have proposed general rules for the preferred or allowed routes. When the total number of π electrons in the non-cyclized component n equals $4q$ where q is an integer the thermal process will be conrotatory. If $n = 4q + 2$ ($q = 0,1,2 \ldots$) the reaction will be allowed thermally in the disrotatory mode. It may be presumed from these rules that radicals are a special case. Consideration of the correlation diagram in Fig. 2-3 for the allyl radical–cyclopropyl radical reaction shows that both conrotatory and disrotatory routes are forbidden thermally.[4] Thus radicals do indeed constitute a special case and they will be treated separately in Chapter 9. Longuet-Higgins and Abrahamson[4] were the first to point out the use of correlation diagrams and also to show that radicals constitute a special case.

III. THE FRONTIER ORBITAL APPROACH

Fukui[5] has shown that the HOMO and LUMO are of prime importance in the interpretation of chemical reactions. Interactions of these so-called frontier orbitals can be employed to predict the preferred course of electrocyclic reactions. The easiest application of frontier orbitals to electrocyclic reactions involves the ring closed form. Then the interactions generated by

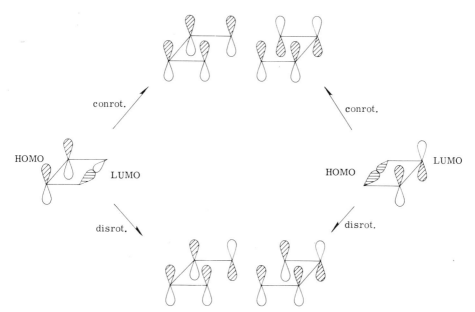

Fig. 2-4. Frontier orbital treatment of the cyclobutene–butadiene reaction.

rotations (conrotatory or disrotatory) of the HOMO (σ) LUMO (π) or vice versa will indicate the preferred path (Fig. 2-4). Both methods give the same answer, that is conrotation leads to two new bonding interactions, while disrotation gives one bond and one antibond. Clearly conrotation is preferred.

The procedure can be applied to the allyl radical–cyclopropyl radical case with interesting results (Fig. 2-5). Note that here one uses the singly occupied molecular orbital (SOMO) in both tests. The predictions now are in opposition rather than in agreement. This suggests that the course of the reaction may depend on substitution, so that if the substituents increase the HOMO-SOMO interaction the process will be disrotatory and conversely if the LUMO-SOMO interaction is more important.

One advantage to the frontier orbital method is that a lack of symmetry does not introduce problems. For example application to the reaction of Eq. (2-1) indicates that both new single bonds can be formed via disrotatory routes. Inversion of the HOMO orientation shows that both bonds can also result from conrotatory reactions.

Woodward and Hoffmann have used a simplified version of the frontier orbital procedure. In most cases correct predictions result from using solely the HOMO of the linear non-cyclic component and noting the rotation which will lead to bond formation. However when applied to the allyl radical this procedure predicts an allowed conrotatory process contrary to correlation diagrams or the full frontier orbital approach.

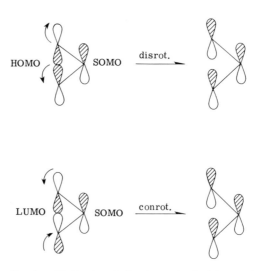

Fig. 2-5. Frontier orbital treatment of cyclopropyl radical ring opening.

IV. TRANSITION STATE AROMATICITY—HÜCKEL OR MÖBIUS

It is particularly appealing to transfer the concept of aromaticity and the special stability associated with aromatic compounds to transition states. Evans[6] suggested that the Diels–Alder reaction might have an aromatic-like transition state, and the concept has often been carried over to the Cope and Claisen rearrangements. Both Zimmerman[7] and Dewar[8] have extended the concept to electrocyclic reactions in particular and to pericyclic reactions in general. Dewar sets forward the very simple rule—"thermal electrocyclic reactions take place preferentially via aromatic transition states." It is not immediately obvious, however, how this rule can be translated into stereochemical predictions equivalent to those derived from correlation diagrams or frontier orbital relations.

Aromaticity as it applies to transition states falls in two categories. A Hückel system is a cyclic array having $4n + 2$ electrons and a basis set with zero or an even number of sign inversions. A Möbius system[9] has $4n$ elec-

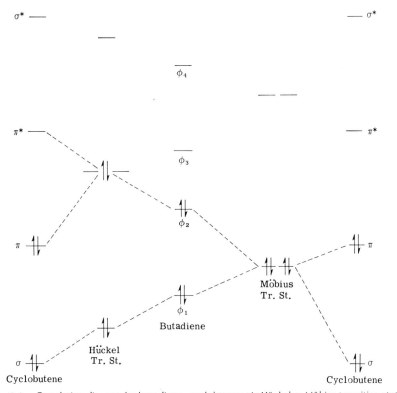

Fig. 2-6. Correlation diagram for butadiene–cyclobutene via Hückel or Möbius transition states.

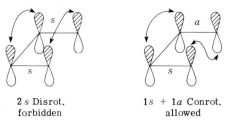

2 s Disrot.   1s + 1a Conrot.
forbidden   allowed

Fig. 2-7.   Application of the s/a procedure to the butadiene–cyclobutene reaction.

trons and a basis set with an odd number of sign inversions. Zimmerman[7] illustrates clearly how this dual concept of aromatic transition states can be related directly to the correlation diagram. Applied to the butadiene–cyclobutene case the basis set for a Hückel system leads to a disrotatory reaction, while the Möbius system gives a conrotatory closure. In this case the Hückel is anti-aromatic, while the Möbius is aromatic hence a conrotatory process is preferred. Using the inscribed polygon in a circle mnemonic (Hückel—corner down, Möbius—edge down) the relative energies of the various cyclic MO's can be ascertained. When these are put on an energy scale with the MO's for the reactant and product, a correlation diagram results (Fig. 2-6). Clearly the Möbius transition state has the lower energy and the conrotatory reaction is preferred.

Woodward and Hoffmann[3] devised a general procedure for analysis of a wide variety of systems which is a modification of the Dewar–Zimmerman approach. Two terms, suprafacial (s) and antarafacial (a), were defined according to the manner in which reactant bond orbitals were converted to product bond orbitals (Chapter 1).

The rule then states that for a total electron involvement of $4n$ electrons the number of a components, and for $4n + 2$ electrons the number of s components, must be odd. This merely constitutes a procedure for counting the number of sign inversions. For the butadiene–cyclobutene system with four electrons the number of a components must be odd, and as Fig. 2-7 shows this corresponds to the preferred conrotatory result. Kaneko[10] has devised a symbolic representation of this method using a solid curved arrow to represent an s process and a dashed arrow for an a process.

Quantitative studies using an extended valence bond approach to the determination of the aromaticity of transition states has confirmed the rules derived more qualitatively above.[11]

V.   ALTERNATIVE THEORETICAL TREATMENTS

The relatively simple theoretical approaches noted in Sections II–IV have made possible the prediction of preferred stereochemical paths for a very wide variety of concerted reactions. However the simplicity of the approaches sometimes led to a sacrifice of rigor which has resulted in a number of alternative theoretical approaches. These will be noted in this section, but for detailed information the original papers should be consulted.

Trindle[12] has pointed out that construction of correlation diagrams by symmetry analysis applies to molecules which cannot be used in experimental tests. The molecules used experimentally lack the required symmetry. As an alternative to the symmetry analysis Trindle employs a nodal analysis based on the premise that rather severe nuclear position changes will not alter the nodal properties of molecular orbitals. A mapping process follows the wave function of the reactant system as the geometry is converted from reactant to product. The overlap of the mapped wave function with the ground state of the product indicates whether the stereochemical path mapped is allowed or forbidden. Applied to the butadiene–cyclobutene reaction the state overlap for the conrotatory stereochemistry was 0.59626 whereas that for the disrotatory reaction was 0.000.

Langlet and Malrieu[13] suggested that the correlation diagram approach in general is based solely on the symmetry properties and relative energies of the initial and final states. Energies of intermediate states relevant to the allowed or forbidden transition states depend on the assumption of a regular (usually linear) variation between the extremes. Hence a non-linear variation could lead to a different prediction. In addition the use of symmetrized delocalized molecular orbitals sometimes leads to a correlation between initial and final molecular orbitals not defined on common atomic orbitals. In order to avoid such problems Langlet and Malrieu carry out a limited analysis of reasonable transition states for the two butadiene–cyclobutene processes via localized molecular orbitals. The results show that zero-order approximations differentiate little between the two routes. Second order configuration interaction with excited states produces the energy lowering which favors the conrotatory route. The results are generalized for polyenes having n double bonds. The authors point out that for large values of n the energy difference between the two stereochemical reaction modes should tend toward zero. A more sophisticated *ab initio* study of the butadiene–cyclobutene reaction[14] indicated that the stereospecificity results largely from delocalizations into excited states, but a short range electron repulsion effect on the forbidden reaction also plays a role.

Employing the generalized valence bond treatment Goddard[15] devised another method readily applicable to electrocyclic and other pericyclic reactions, which has the advantage of being applicable to non-symmetric systems. Goddard's procedure is based on the "orbital phase continuity principle." When three conditions, one, bonding pairs of orbitals remain strongly overlapped during reaction; two, bonding and non-bonding orbitals delocalized in the same region become orthogonal at the transition state; and three,

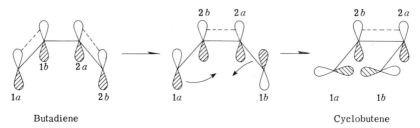

Butadiene   Cyclobutene

Fig. 2-8. Phase changes schematically illustrated in generalized valence bond method for buta-diene → cyclobutene electrocyclization.

phase relations among the orbitals change continuously, are satisfied the reaction can occur with a low activation energy and was termed *favored*. Applied to the butadiene–cyclobutene case the valence bond orbitals over-lap initially as indicated for butadiene (Fig. 2-8). During the reaction the orbitals originally bonded alter affiliations as indicated with the pair 1b and 2b which were non-bonded initially undergoing a phase alteration during the reaction before rebonding of 1a and 1b occur. This satisfies the condi-tions for a low energy process. Clearly a conrotatory reaction is predicted. For electrocyclic reactions the predictions do not differ from those of the orbital symmetry approach.

Silver and Karplus[36] have also used the valence bond approach to supply a more rigorous theoretical justification of the conservation of orbital sym-metry. Each orbital is given a symmetry label and localized pair bonds are used. One (or more) pair of bonds not individually bisected by a symmetry element must be treated as a linear combination, and the procedure does not differ greatly from that which uses delocalized molecular orbitals.

Two different indices have been proposed by Weltin to treat the electro-cyclic ring closure[16] and ring opening reactions[17] without the use of sym-metry. For ring closure a rule based on the generalized bond order, $p_{xy} = \sum_j b_j c_{jx} c_{jy}$, states that reaction will proceed in a disrotatory direction if p_{xy} is positive and conrotatory if p_{xy} is negative. In most cases the rule leads to the same prediction as the Woodward–Hoffmann rules, but several examples are cited where the predictions differ. Thus the ambiguous case [Eq. (2-2)]

$$\text{[structure]} \longrightarrow \text{[structure]} \quad \text{or} \quad \text{[structure]} \qquad (2\text{-}2)$$

has p_{xy} negative for the ground state thus suggesting a conrotatory result. The prediction will not be easily checked experimentally. A group of reac-tions having $p_{xy} \simeq 0$, hence predictably non-stereospecific, were pointed out.

For example all closures between centers separated by an odd number of atoms in neutral polyenes [Eq. (2-3)] belong to this non-stereospecific class

$$\text{[structure]} \longrightarrow \text{[structure]} \tag{2-3}$$

This prediction is not surprising either theoretically, since Eq. (2-3) really corresponds to ring closure of a pentadiene radical, or stereochemically since the configuration of only one reacting center is specified. The index T, used to predict the preferred path for ring opening, is defined in terms of the energy difference $\Delta E = E^{\text{conrot.}} - E^{\text{disrot.}}$ at small values of the rotation angle.

Using a topological approach, Day[18] has shown that the generalized Woodward–Hoffmann rules are equivalent to the Hückel–Möbius aromaticity rules. In the process of developing the proof of equivalency Day generates a very simple set of additional rules, and provides a general and precise definition of the terms suprafacial and antarafacial. Day's approach starts with a chosen basis set of AO's for the electrons involved in the pericyclic reaction, and a "connectivity cycle" is then developed by connecting one or more lobes of each AO in the pericycle in a continuous closed curve. An example (more than one connectivity cycle is possible) is shown here for *cis*-hexatriene (Fig. 2-9). If $2m$ is the number of electrons involved, Z is the number of negative (out of phase) overlaps in the connectivity cycle, and N is the number of local nodal intersections within the AO's of the cycle, then the reaction is thermally allowed if $m + N$ is odd. For the example shown $2m = 6$, $Z = 0$, $N = 2$ hence $m + N = 5$ and the reaction is allowed thermally as shown.

Mathieu and Rassat[19] have pointed up the relation between parity and stereochemistry in a broader sense than Woodward and Hoffmann. Thus any concerted reaction involving an odd number of electron pairs make and break bonds "in the same half-space" (i.e., suprafacially), and an even number of electron pairs make or break bonds in "different spaces" (i.e., antarafacially). Thus the S_N2 reaction with two electron pairs occurs antara-

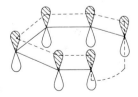

Fig. 2-9.

facially (with inversion), while the S_N2' process occurs suprafacially. Concerted addition to a double bond (2 pairs) must be preferentially antarafacial. Rassat[20] has developed a graph representation of the generalized W-H rules.

It is interesting that the most common criticism of the correlation diagram method has been that the use of symmetry is too often unjustified for most useful molecules, but Halevi[21] has argued that in some cases the symmetry analysis of Woodward and Hoffmann uses too low a symmetry to permit properly restrictive answers. Halevi has proposed a procedure for analysis in terms of the full symmetry to which the reacting system belongs, and he terms the process "orbital correspondence analysis in maximum symmetry" or OCAMS. For thermal electrocyclic reactions the method does not appear to provide any new insights, but in some cases of pericyclic reactions, namely $(\pi^2s + \pi^2a]$ cycloaddition, reactions which are allowed by Woodward–Hoffmann analysis are forbidden by OCAMS.

VI. ALLOWED, FORBIDDEN, AND NON-CONCERTED ROUTES

Orbital symmetry analysis divides concerted pericyclic reactions of systems containing $2\,m$ (where m is a positive integer) reactive electrons into two classes, allowed and forbidden. Forbidden reactions are not truly forbidden, but normally require a higher activation energy than their allowed counterparts. For neutral molecules or systems involving only neutral entities there also will exist a third category–non-concerted routes to the same pericyclic product. Thus the question of the order of activation energies for the three competitive routes arises, allowed–forbidden–non-concerted or allowed–non-concerted–forbidden. In terms of the aromaticity of transition states one is led to expect the order-allowed, non-concerted, forbidden as aromatic, non-aromatic, antiaromatic, respectively. A related question pertains to the magnitudes of the energy differences among the three routes.

For most thermal reactions the nonconcerted route is presumed to proceed via a diradical, or under appropriate conditions, the equivalent zwitterion. Berson and Salem[22] analyzed the three routes for a 1,3-sigmatropic shift and concluded that interaction between the p-orbital of the migrating group and a bonding π orbital of the allylic radical leads to a significant stabilization of the forbidden route, rendering its transition state more stable than the non-interacting diradical. This was termed the subjacent orbital effect. Whether this is applicable to pericyclic reactions in general depends on the extent of the stabilizing effect, the size of the energy difference between the diradical and the forbidden transition state in the absence of such a stabilization, whether electron repulsions will alter the situation, etc. Borden and

Salem[23] have assessed the influence of electron repulsion on the subjacent orbital effect and have concluded that in some special cases, i.e., trimethylene-methane, the electron repulsion negates the subjacent orbital stabilization. Epiotis[24] has also shown that polar substituents can negate the subjacent orbital effect. Goddard[15] and Halevi[21] have suggested that for the $\pi^2 s + \pi^2 a$ cycloaddition the diradical route is preferred over the concerted. The question has not been studied with respect to electrocyclic reactions in general.

VII.   SUBSTITUENT EFFECTS

Predictions of the orbital symmetry theory are derived from unsubstituted systems, namely, butadiene–cyclobutene, cis-hexatriene–cyclohexadiene. Perturbations which result from the introduction of substituents could perhaps alter the prediction, particularly if the activation energy difference, $\Delta E = E^{\text{forbidden}} - E^{\text{allowed}}$, happened to be relatively small. It is of interest therefore to consider the potential influence of substituents on the stereochemistry of the electrocyclic reactions. Epiotis has published a series of papers on his studies of this question on pericyclic reactions in general.[25] Of these only one[26] specifically relates to electrocyclic reactions, and that study utilizes both a perturbation treatment and a configuration interaction approach.

Essentially the two approaches lead to the same qualitative conclusions, and Epiotis' study was applied to the butadiene–cyclobutene case. As an inspection of the correlation diagrams for that reaction shows, the disrotatory ring closure would become allowed if two electrons were promoted from the HOMO to the LUMO of butadiene. Thus as the energy difference between these MO's becomes smaller mixing of the two (configuration interaction) would reduce the energy barrier for the reaction. Both electron attracting (W) and electron donating substituents (D) are predicted to reduce the energy difference between the HOMO and LUMO for butadiene, hence to reduce the barrier for the forbidden reaction. Epiotis predicts that the combined effect of both types as illustrated (Fig. 2-10) would prove particularly effective. Heteroatom substitution in the polyene skeleton

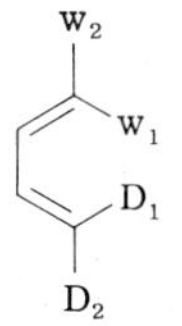

Fig. 2-10.

should also prove effective. The perturbation treatment indicates that these substitutions should also reduce the stabilization of the allowed transition state. As a result Epiotis concludes that increased substitution, particularly with the proper pattern of electron attracting and donating substituents should reduce the $\Delta E = E^{\text{forbidden}} - E^{\text{allowed}}$ hence increasing the possibility of a lowered stereoselectivity for the electrocyclization.

VIII. TRANSITION METAL COMPLEXES

Prior to development of orbital symmetry theory it had been found that some reactions which normally proceed with difficulty thermally were greatly facilitated by transition metal catalysis.[27,28] Attention was directed to such reactions in 1967 when the forbidden nature of the process in the absence of the catalyst was recognized, and the possible conversion to an allowed reaction by the metal catalyst was rationalized.[29] Further ramifications of the theoretical aspects of transition metal catalysis of pericyclic reactions have been developed by Mango[30,33-35] and Mango and Schachtschneider.[31,32] It must be stressed at this point that the catalysis being considered here does not disqualify the reaction as an electrocyclic process, because coordination with the metal is maintained during the reaction, and the reaction of the ligand and metal together is a thermal electrocyclic process. While the overall scheme of coordination, ligand alteration and removal of the ligand may be termed a catalysis which transforms a forbidden to an allowed process, the real electrocyclic reaction is of the complex, and the theory should show that that step is allowed or forbidden. Mango has carefully pointed out that the coordination must be maintained throughout reaction[35] and has also noted that symmetry restrictions still provide restraints on the reactions of complexes.[32]

Consider once again the butadiene–cyclobutene correlation diagrams. It is obvious that if two electrons were added to the system the disrotatory reaction would be allowed, though it would correlate excited states. If then the metal had two electrons in an orbital of proper symmetry to enter into combination with the LUMO of butadiene, it would create a system with the proper number of electrons, and the proper symmetry. There is however a complication which appears here since the electron pair added from the metal to the butadiene LUMO end up in the cyclobutene HOMO. Thus at the conclusion of the ligand alteration process the metal atom must be prepared to accept back two electrons from the LUMO of cyclobutene, i.e., electrons with different symmetry properties from those originally donated. Consequently the metal must have a second orbital, devoid of electrons,

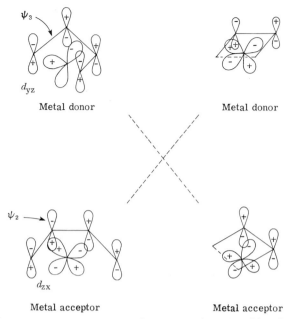

ψ_3

d_{yz}

Metal donor

Metal donor

ψ_2

d_{zx}

Metal acceptor

Metal acceptor

Fig. 2-11. Interaction between metal and HOMO's and LUMO's of butadiene and cyclobutene during electrocyclization.

with the proper symmetry to match the LUMO of cyclobutene. Essentially a sort of push–pull effect is created by the metal.

As Fig. 2-11 indicates a transition metal having an empty d_{zx} orbital can accept electrons from the ψ_2 of butadiene, and with a filled d_{yz} orbital can back bond to the ψ_3 of butadiene. In the disrotatory process the diene donor–metal acceptor combination converts to the ene acceptor–metal donor combination, while the diene acceptor–metal donor set goes to what appears to be an ene donor–metal acceptor set. However as Mango[33] has noted the metal orbital of this pair is not properly oriented for the acceptor orbital in the cyclobutene–metal coordinate bond. Interaction between the d_{yz} and cyclobutene orbital begins only as disrotation commences. Such a disadvantage does not exist in all pericyclic reactions of metal complexes. The energetics of the process would be expected to depend markedly on the metal atom and the other ligands present which can alter the relative orbital energies in the complex.

REFERENCES

1. R. B. Woodward and R. Hoffmann, *J. Am. Chem. Soc.* **87**, 395, 2046, 4388, 4389 (1965).
2. R. B. Woodward, Burgenstock Conference, June 1966.

3. R. B. Woodward and R. Hoffmann, "Conservation of Orbital Symmetry," Academic Press, New York, 1970; R. E. Lehr and A. P. Marchand, "Orbital Symmetry," Academic Press, New York, 1972; T. L. Gilchrist and R. C. Storr, "Organic Reactions and Orbital Symmetry," Cambridge Univ. Press, Cambridge, 1972; I. Fleming, "Frontier Orbitals and Organic Chemical Reactions," Wiley, New York, 1976; M. J. S. Dewar and R. C. Dougherty, "PMO Theory of Organic Chemistry," Plenum, New York, 1975; L. J. Bellamy, "Intro. to Conservation of Orbital Symmetry," Longman, London, 1974; G. B. Gill and M. R. Willis, "Pericyclic Reactions," Chapman & Hall, London, 1974.

4. H. C. Longuet-Higgins and E. W. Abrahamson, *J. Am. Chem. Soc.* **87**, 2045 (1965).

5. K. Fukui, *Tetrahedron Lett.* p. 2009 (1965).

5a. K. Fukui, *Accts. Chem. Res.* **4**, 57 (1971).

6. M. G. Evans, *Trans. Faraday Soc.* **35**, 824 (1939).

7. H. E. Zimmerman, *J. Am. Chem. Soc.* **88**, 1564, 1566 (1966); H. E. Zimmerman, *Accts. Chem. Res.* **4**, 272 (1971).

8. M. J. S. Dewar, *Tetrahedron Suppl.* **8**, 75 (1966).

9. E. Heilbronner, *Tetrahedron Lett.* p. 1923 (1964).

10. C. Kaneko, *Tetrahedron* **28**, 4915 (1972).

11. W. J. van der Hart, J. J. C. Mulder, and L. J. Oosterhoff, *J. Am. Chem. Soc.* **94**, 5724 (1972).

12. C. Trindle, *J. Am. Chem. Soc.* **92**, 3251, 3255 (1970).

13. J. Langlet and J.-P. Malrieu, *J. Am. Chem. Soc.* **94**, 7254 (1972).

14. J. P. Daudey, J. Langlet and J.-P. Malrieu, *J. Am. Chem. Soc.* **96**, 3393 (1974).

15. W. A. Goddard III, *J. Am. Chem. Soc.* **94**, 793 (1972).

16. E. E. Weltin, *J. Am. Chem. Soc.* **95**, 7650 (1973).

17. E. E. Weltin, *J. Am. Chem. Soc.* **96**, 3049 (1974).

18. A. C. Day, *J. Am. Chem. Soc.* **97**, 2431 (1975).

19. J. Mathieu and A. Rassat, *Tetrahedron* **30**, 1753 (1974).

20. A. Rassat, *Tetrahedron Lett.* p. 4081 (1975).

21. E. A. Halevi, *Helv. Chim. Acta* **58**, 2136 (1975).

22. J. A. Berson and L. Salem, *J. Am. Chem. Soc.* **94**, 8917 (1972).

23. W. T. Borden and L. Salem, *J. Am. Chem. Soc.* **95**, 932 (1973).

24. N. D. Epiotis, R. L. Yates, and F. Bernardi, *J. Am. Chem. Soc.* **97**, 4198 (1975).

25. N. D. Epiotis, *J. Am. Chem. Soc.* **94**, 1924, 1935, 1941, 1946 (1972); **95**, 1191, 1200, 1206, 1214 (1973).

26. N. D. Epiotis, *J. Am. Chem. Soc.* **95**, 1200 (1973).

27. C. W. Bird, D. L. Colinese, R. C. Cookson, J. Hudec, and R. D. Williams, *Tetrahedron Lett.* p. 373 (1961).

28. G. N. Schrauzer and S. Eichler, *Chem. Ber.* **95**, 2764 (1962).

29. F. D. Mango and J. H. Schachtschneider, *J. Am. Chem. Soc.* **89**, 2484 (1967).

30. F. D. Mango, *Tetrahedron Lett.* p. 4813 (1969).

31. F. D. Mango and J. H. Schachtschneider, *J. Am. Chem. Soc.* **91**, 1030 (1969).

32. F. D. Mango and J. H. Schachtschneider, *J. Am. Chem. Soc.* **93**, 1123 (1971).

33. F. D. Mango, *Advan. Catal.* **20**, 291 (1969).

34. F. D. Mango, *Fortschr. Chem. Forsch.* **45**, 39 (1974).

35. F. D. Mango, *Tetrahedron Lett.* p. 1509 (1973).

36. D. M. Silver and M. Karplus, *J. Am. Chem. Soc.* **97**, 2645 (1975).

CHAPTER 3
TWO ELECTRON SYSTEMS

I. CYCLOPROPYL–ALLYL CATIONS

The smallest conjugate system entering an electrocyclic process has only two electrons. If homoconjugation and other forms of cycloconjugation are barred from consideration, the sole example is the interconversion of allyl and cyclopropyl cations. Most of our knowledge of this reaction has been obtained since 1951, when Roberts and Chambers [13] measured the solvolysis rates of cyclopropyl tosylate and chloride, showing that both react exceedingly slowly and produce only allylic products. A rather impressive series of examples appeared during the decade prior to 1965, but these were generally concerned with the utility of the reaction in synthesis (see below). Orbital conservation theory predicts that the reaction should proceed in disrotatory fashion, which permits two possible reaction modes [Eq. (3-1)].

$$ \text{(3-1)} $$

When de Puy found that *trans*-2-phenylcyclopropyl tosylate reacted some fourteen times faster than the cis isomer,[1] and *exo*-7-chlorobicyclo[4.1.0]-heptane was unreactive under conditions where the endo chloride reacted readily,[25] he was prompted to speculate that only one of the two disrotatory modes was active, namely, groups trans to the leaving group rotate outward and those cis move inward. Calculations by Hoffmann supported this unanticipated extension of the theory. The rate enhancement provided by a 2-phenyl suggested that ring opening is concerted with loss of the tosylate.[1] This idea had been anticipated by Schleyer and Nicholas,[26] who based their suggestion on calculations of the Schleyer–Foote[27,27a] type.

A. Theoretical Studies

The simplicity of this electrocyclic reaction coupled with the unusual restriction to a single disrotational direction generated great interest and prompted a spate of theoretical studies.[2-8] The calculations agreed uni-

TABLE 3-1

Physical Properties of the Cyclopropyl Cation

C_1-C_2 bond length (Å)	C_2-C_1-C_3 bond angle (°)	Relative energy[a]	Reference
1.40	80	48 kcal/mol	8
1.485	61.5	39.2 kcal/mol	7

[a] Relative to the allyl cation equal to zero.

versally that ring opening via the trans-out, cis-in disrotatory mode is preferred, but the details have varied considerably. The most sophisticated *ab initio* calculations[7,8] introduce a minimum of geometrical restraints, essentially only that C_s symmetry be maintained in the disrotatory mode and C_2 in the conrotatory, and thus probably provide the most accurate description of the mechanistic details. Structures for both reactant and product, and their relative energies were established first. Both groups agree that the allyl cation is completely planar with C-C bond lengths of 1.385–1.40 Å and a C-C-C angle of 118.9–120°. Conversely they differ markedly on the properties of the cyclopropyl cation (1) (cf. Table 3-1).

Early studies[2-4] found that the cyclopropyl cation does not lie in an energy minimum. MINDO/2 calculations[5] did show a minimum for that ion, but the result is probably an artifact of the MINDO/2 tendency to over-estimate the stability of cyclic molecules. In fact those calculations made the cyclopropyl more stable than the allyl cation. The unadorned ions favor the allyl cation as being more stable, but substituents on the central carbon can reverse this order.[9] Such strong electron donors as alkoxyl, thioalkyl, or dialkylamino make the cyclopropyl cation the more stable one. Even the most recent calculations do not agree on whether the cyclopropyl cation itself represents a minimum on the energy surface, the 6-31G basis shows no minimum,[7] while an SCF method showed a shallow minimum.[8]

Extended Hückel[2] and *ab initio*[4,4a,7] calculations showed that the cyclo-propyl cation undergoes disrotatory conversion to the allyl cation without activation. The MINDO/2 study necessarily produced a transition state ($E_a = 7.4$ kcal/mol). Merlet *et al.*[8] found a low activation energy (1.6 kcal/mol). Under normal conditions then (100°–200°) the results show that ring opening is concerted with cation formation. The theoretical basis for the single direction disrotation was uncovered by Kutzelnigg,[2] who found

that as ring opening proceeds the hydrogen on C_1, initially coplanar with
the ring, moves out of the plane, and preferentially toward the hydrogens
of the methylene groups which move outward [Eq. (3-2)]. Later work has

$$(3\text{-}2)$$

generally confirmed this result[7,8] and has shown further that the rotations
are synchronized with the C_2-C_3 bond cleavage.

The orbital symmetry forbidden conrotatory process is more complex.
This arises partly because the energies of the highest occupied and lowest
unoccupied orbitals cross during the reaction. For some values of the rotational angles there are two minima on the energy surface, one resulting when
the highest occupied M.O. is bonding for the C-C bond being broken, the
other when it is antibonding.[7] Bond breaking is not concerted with methylene
rotations,[7] but this could be a result of the imposition of a C_2 symmetry
constraint. Recent MINDO/2 calculations which removed that constraint[5a]
indicate methylene rotations occur consecutively. In any event conrotation
is energetically very unfavorable with respect to the preferred disrotatory
process, $\Delta\Delta E^{\ddagger} = \Delta E^{\ddagger}_{conrot.} - \Delta E^{\ddagger}_{disrot.}$ being > 23.3 kcal/mol[7] or 85 kcal/mol.[8]
The results suggest that a conrotatory ring opening is not likely to be experi-
mentally observable.

There remain two additional disrotational processes to consider, one
where the hydrogen on the central carbon moves out-of-plane toward the
methylene hydrogens bending in (disrot. 1), or the hydrogen on the central
carbon remains in the plane of the carbons throughout (disrot. 0). Both
Clark and Armstrong[4,4a] and Merlet et al.[8] agree that disrot. 1 is less favorable than disrot. 0, by more than 10 kcal/mol according to the latter group.
The disrot. 0 route is only 3.1 kcal/mol less favorable than the preferred
disrotatory route of the Woodward–Hoffmann–de Puy (W-H-deP) rule![8]
Thus it appears to offer an experimentally viable alternative to the preferred
disrotatory path, but this may be misleading since the preferred route
receives a considerable measure of assistance from ring opening in concert
with ionization.[26] Whether this can occur with the disrot. 0 route is uncertain.
This makes the disrot. 1 route of increased importance, and if it is less
favorable than the W-H-deP route by more than 13 kcal/mol, it is an unlikely
observable.

B. Experimental Results

Experimental work is divided here into two classes, one, studies illum-
inating the mechanism, and two, those which illustrate use of the reaction
in synthesis.

1. Mechanistic Studies

Mechanistic studies of electrocyclic reactions of reactive intermediates such as cations, anions, and radicals pose special problems. Neither the desired reactant nor the direct product can be examined explicitly in most cases. Consequently most evidence is circumstantial and requires interpretation. The early work has been reviewed[10] with coverage up to 1968. Following the pioneering work of Roberts and Chambers,[13] and the publication of the W-H-de Puy rule,[1] experimenters concentrated on the extent to which the rule would apply and the degree of selectivity it imposed. Evidence came from two distinct sources, one, study of monocyclic compounds and two, study of bicyclic [n.1.0] derivatives.

The foundation on which all interpretations rest is the set of predictions derived from the W-H-deP rule. As the rule is couched in stereochemical terms, the stereochemical predictions are clear for the allylic ion, but need clarification with respect to the isolated product. Monocyclic derivatives bearing a single reference substituent are useful only if the product with the more substituted double bond is isolable and is kinetically significant [Eq.

$$(3\text{-}3)$$

(3-3)]. A similar restriction holds for 2,2,3-trisubstituted examples, while 2,3-disubstituted molecules are stereochemically always useful if no interconversion occurs. For bicyclo[n.1.0]alkanes with a cis ring juncture the endo derivatives should give products with a cis double bond, and exo derivatives those with a trans double bond. Kinetically the predictions for monocyclic derivatives are steric in origin, i.e., bulky groups moving inwards increase steric hindrance and retard the rate, moving outward they decrease steric interactions and accelerate. With bicyclic molecules if $n < 5$ the sole stable product must have a cis double bond, hence only the endo isomer should react, and if $n > 5$ both exo and endo isomers should react at relative rates which depend on n.

TABLE 3-2

Solvolysis Rates for Substituted Cyclopropyl Derivatives

Leaving group and substituents	Solvent	Temperature (°)	Rate	$\Delta H^{\ddagger}$	$\Delta S^{\ddagger}$	Reference
A. Tosylates						
None	HOAc	175	7.05×10^{-5}			22
None	HOAc	150	6.3×10^{-6}	33.8	-2.9	23
None	HOAc	150	8.17×10^{-6}	32.3	-6.1	24
None	50% aq. EtOH	150	1.03×10^{-4}	29.3	-8.3	24
1-Methyl	HOAc	100	1.30×10^{-5}	28.7	-4.6	24
cis-2-Methyl	HOAc	100	2.33×10^{-7}	30.8	-6.7	24
trans-2-Methyl	HOAc	100	5.76×10^{-6}	29.9	-2.9	24
cis-2-Ethyl	HOAc	100	3.06×10^{-7}	30.9	-6.1	30
trans-2-Ethyl	HOAc	100	7.05×10^{-6}	29.8	-2.7	30
cis-2-Isopropyl	HOAc	100	3.30×10^{-7}	31.8	-3.3	30
trans-2-Isopropyl	HOAc	100	9.63×10^{-6}	29.8	-2.0	30
cis-2-tert-Butyl	HOAc	100	2.11×10^{-7}	32.3	-2.9	30
trans-2-tert-Butyl	HOAc	100	1.38×10^{-5}	29.0	-3.6	30
2,2-Dimethyl	HOAc	100	1.79×10^{-5}	30.3	$+0.6$	24
cis-2-trans-3-Dimethyl	HOAc	100	1.93×10^{-5}	29.8	-0.7	24
cis-2-cis-3-Dimethyl	HOAc	100	9.01×10^{-8}	35.9	$+5.0$	24
trans-2-trans-3-Dimethyl	HOAc	100	1.59×10^{-3}	27.3	$+1.2$	24
2,2-cis-3-Trimethyl	HOAc	100	3.44×10^{-6}	31.0	-0.8	24
2,2-trans-3-Trimethyl	HOAc	100	1.86×10^{-3}	26.9	$+0.7$	24
2,2,3,3-Tetramethyl	HOAc	100	3.10×10^{-4}	29.8	$+4.9$	24
cis-2-Phenyl	HOAc	123	8.3×10^{-6}			20
cis-2-Phenyl	HOAc	150	1.22×10^{-4}	32.7	0.0	23
trans-2-Phenyl	HOAc	109	3.21×10^{-5}			20
trans-2-Phenyl	HOAc	150	1.75×10^{-3}	30.6	-0.9	23
trans-2-(p-Tolyl)	HOAc	109	7.47×10^{-5}			20
trans-2-(m-Tolyl)	HOAc	109	6.56×10^{-5}			20
trans-2-(m-Chlorophenyl)	HOAc	109	5.38×10^{-5}			20
1-Phenyl	HOAc	108	1.01×10^{-2}			20
1-(1-Propynyl)	50% aq. EtOH	70	1.91×10^{-5}			16
1-(2-Phenylethynyl)	50% aq. EtOH	70	8.38×10^{-5}			16
1-(2-Cyclopropylethynyl)	50% aq. EtOH	70	1.88×10^{-3}			16
B. Bromides						
None	HOAc	100	1.55×10^{-9}	33.0	-10.8	24
None	HOAc	160	6.92×10^{-7}			26
None	50% aq. EtOH	100	2.19×10^{-7}	29.0	-11.7	24
cis-2-Methyl	HOAc	100	7.17×10^{-8}	31.4	-7.7	30
trans-2-Methyl	HOAc	100	9.87×10^{-7}	30.9	-3.6	30
cis-2-tert-Butyl	HOAc	100	1.42×10^{-7}	31.7	-5.3	30
trans-2-tert-Butyl	HOAc	100	1.62×10^{-6}	30.6	-3.5	30
cis-2-trans-3-Dimethyl	HOAc	100	1.59×10^{-5}	28.5	-4.6	30
cis-2-cis-3-Dimethyl	HOAc	100	1.28×10^{-7}	33.3	-1.3	30
trans-2-trans-3-Dimethyl	HOAc	100	7.72×10^{-4}	27.1	-0.7	30

(continued)

TABLE 3-2   *(Continued)*

Leaving group and substituents	Solvent	Temperature (°)	Rate	$\Delta H^{\ddagger}$	$\Delta S^{\ddagger}$	Reference
B. Bromides *(Continued)*						
2,2-*cis*-3-Trimethyl	HOAc	100	1.05×10^{-5}	28.0	-6.8	30
2,2-*trans*-3-Trimethyl	HOAc	100	3.72×10^{-3}	23.8	-6.3	30
2,2,3,3-Tetramethyl	HOAc	100	2.04×10^{-3}	28.3	$+4.5$	30
cis-2-Phenyl	HOAc	119	1.46×10^{-5}	28.9	-9.9	26
trans-2-Phenyl	HOAc	119	6.31×10^{-5}	31.4	$+1.7$	26
trans-2-Ethyl	50% aq. EtOH	110	2.13×10^{-4}	20.6	-22.2	28
cis-2-Vinyl	50% aq. EtOH	95	2.46×10^{-4}	19.3	-27.6	28
trans-2-Vinyl	50% aq. EtOH	95	2.65×10^{-4}	19.9	-25.8	28
cis-2-Cyclopropyl	50% aq. EtOH	95	3.93×10^{-5}	25.9	-8.8	28
trans-2-Cyclopropyl	50% aq. EtOH	95	1.02×10^{-3}	23.9	-7.7	28
C. Chlorides						
None	HOAc	150	5.5×10^{-10}			23
None	50% aq. EtOH	200	1.4×10^{-5}			22
cis-2-*tert*-Butyl	HOAc	100	4.04×10^{-9}	32.6	-10.1	30
trans-2-*tert*-Butyl	HOAc	100	1.05×10^{-7}	30.7	-8.7	30
cis-2-*trans*-3-Dimethyl	HOAc	100	3.18×10^{-7}	31.5	-4.3	30
cis-2-*cis*-3-Dimethyl	HOAc	100	2.96×10^{-9}	34.6	-5.2	30
trans-2-*trans*-3-Dimethyl	HOAc	100	2.43×10^{-5}	29.1	-2.2	30
2,2-*cis*-3-Trimethyl	HOAc	100	1.18×10^{-7}	33.3	-1.5	30
2,2-*trans*-3-Trimethyl	HOAc	100	7.98×10^{-5}	28.2	-2.2	30
cis-2-Phenyl	HOAc	150	5.91×10^{-6}	30.2	-11.8	23
trans-2-Phenyl	HOAc	150	3.01×10^{-5}	31.4	-5.7	23
2,2-Diphenyl	HOAc	150	8.47×10^{-5}	34.6	$+4.0$	23
cis-2-*cis*-3-Diphenyl	HOAc	95	3.59×10^{-6}			31
trans-2-*trans*-3-Diphenyl	HOAc	95	1.15×10^{-2}			31
cis-2-*trans*-3-Diphenyl	HOAc	95	2.46×10^{-4}			31
cis-2-*cis*-3-Dipropyl	Ag^{+}/EtOH	80	$<2.2 \times 10^{-8}$			33
trans-2-*trans*-3-Dipropyl	Ag^{+}/EtOH	80	8.20×10^{-5}			33
cis-2-*trans*-3-Dipropyl	Ag^{+}/EtOH	80	1.26×10^{-6}			33

Kinetically the monocyclic derivatives conform to the predictions with monotonous regularity (Table 3-2). Rate accelerations by 1-methyl,[15] 1-phenyl,[11] 2-methyl,[15,21] 2-phenyl,[11,14] 2-vinyl,[18,19] and 2-cyclopropyl[19] prove that charge dispersal over all three carbons occurs, and thus ionization is accompanied by ring opening. In all cases with a single substituent at C_2 the trans isomer reacts faster than the cis isomer (Table 3-4). This effect is magnified by a factor of 10–15 in 2,2,3-trisubstituted compounds, and by 500- to 800-fold with 2,3-disubstituted molecules. Rate accelerations by resonance interactive substituents, vinyl, phenyl, and cyclopropyl are modest compared with methyl (Table 3-3), and the k_{trans}/k_{cis} value for phenyl is surprisingly small. The near equality of rates for cis-2-alkyl substituents from methyl to tert-butyl[15,21] implies only a moderate degree of rotation at the transition state, which is in agreement with a calculated value of 25°.[8] A major point of disagreement is the amount of ring opening at the transition state. Consideration of the relative influence of a 2-methyl vs. a γ-methyl on an allyl cation, and a 1-methyl vs. a methyl directly on a simple carbonium ion indicated a "considerable degree of ring opening."[15] Ong and Robertson[18] suggest that their results with 2-vinyl substitution indicate an early transition state. They cite the ρ^+ of -1.75 found by de Puy et al.[11] in support of that conclusion. Products with an intact cyclopropyl ring were recovered

TABLE 3-3

Relative Rates for Some Monocyclic Cyclopropyl Derivatives

Substituents	Relative rate	Reference
I. Tosylates		
A. Acetolysis at 100°		
None	1.0	24
1-Methyl	310	24
cis-2-Methyl	5.7	24
trans-2-Methyl	138	24
cis-2-Ethyl	7.4	30
trans-2-Ethyl	180	30
cis-2-Isopropyl	8.0	30
trans-2-Isopropyl	230	30
cis-2-tert-Butyl	5.0	30
trans-2-tert-Butyl	330	30
2,2-Dimethyl	430	24
cis-2-trans-3-Dimethyl	460	24
cis-2-cis-3-Dimethyl	2.2	24
trans-2-trans-3-Dimethyl	38000	24
2,2-cis-3-Trimethyl	83	24
2,2-trans-3-Trimethyl	44700	24
2,2,3,3-Tetramethyl	7450	24
B. Acetolysis at 150°		
None	1.0	23
cis-2-cis-3-Dimethyl	4.0	21
trans-2-trans-3-Dimethyl	18000	21
cis-2-trans-3-Dimethyl	260	21
cis-2-Phenyl	20	23
trans-2-Phenyl	280	23

(continued)

TABLE 3-3 (Continued)

Substituents	Relative rate	Reference
II. Bromides		
A. Acetolysis at 100°		
None	1.0	24
cis-2-Methyl	46	30
trans-2-Methyl	630	30
cis-2-tert-Butyl	92	30
trans-2-tert-Butyl	1040	30
cis-2-trans-3-Dimethyl	9700	30
cis-2-cis-3-Dimethyl	82	30
trans-2-trans-3-Dimethyl	497000	30
2,2,cis-3-Trimethyl	6800	30
2,2,trans-3-Trimethyl	2400000	30
2,2,3,3-Tetramethyl	1320000	30
B. Acetolysis at 119°		
None	1.0	26
cis-2-Phenyl	1000	26
trans-2-Phenyl	4300	26
2,2-Diphenyl	13600	26
C. 50% aq. EtOH at 95°		
None	1.0	24
cis-2-Ethyl	4.3	28
trans-2-Ethyl	91	28
cis-2-Vinyl	49	28
trans-2-Vinyl	525	28
cis-2-Cyclopropyl	78	28
III. Chlorides		
A. Acetolysis at 100°		
None	1.0[a]	30
cis-2-trans-3-Dimethyl	20000	30
cis-2-cis-3-Dimethyl	167	30
trans-2-trans-3-Dimethyl	1330000	30
2,2,cis-3-Trimethyl	6700	30
2,2,trans-3-Trimethyl	4400000	30
cis-2-tert-Butyl	254	30
trans-2-tert-Butyl	6600	30
B. Acetolysis at 95°		
None	1.0[a]	31
cis-2-Phenyl	10000	31
trans-2-Phenyl	40000	31
2,2-Diphenyl	60000	31
cis-2-cis-3-Diphenyl	1450000	31
trans-2-trans-3-Diphenyl	4.65×10^9	31
cis-2-trans-3-Diphenyl	1.0×10^8	31

[a] Values used for comparisons were 1.8×10^{-11} in refer. 30 and 2.8×10^{-12} in refer. 31.

TABLE 3-4

Values for $k(\text{trans})/k(\text{cis})$ for Substituted Cyclopropyl Derivatives

Substituents	Temperature (°)	$k(\text{trans})/k(\text{cis})$	Solvent
I. Tosylates			
2-Methyl	100	24	HOAc
2-Ethyl	100	23	HOAc
2-Isopropyl	100	29	HOAc
2-*tert*-Butyl	100	65	HOAc
2,3-Dimethyl	100	17300	HOAc
2,2,3-Trimethyl	100	539	HOAc
2-Phenyl	150	14	HOAc
II. Bromides			
2-Methyl	100	14	HOAc
2-*tert*-Butyl	100	11.4	HOAc
2-Ethyl	95	21	aq. EtOH
2-Vinyl	95	11	aq. EtOH
2-Vinyl	95	11	H_2O
2-Cyclopropyl	95	26	aq. EtOH
2-Phenyl	119	4.3	HOAc
2,3-Dimethyl	100	6030	HOAc
2,2,3-Trimethyl	100	354	HOAc
III. Chlorides			
2-*tert*-Butyl	100	26	HOAc
2-Phenyl	95	4.0	HOAc
2,3-Dimethyl	100	8210	HOAc
2,3-Dipropyl	80	>3700	Ag^+/EtOH
2,3-Diphenyl	150	3200	HOAc
2,2,3-Trimethyl	100	676	HOAc

during decomposition of *cis*- and *trans*-2-phenylcyclopropyl-*N*-nitroso-urea.[28] In methanol or formic acid small amounts of 2-phenylcyclopropyl methyl ether or formate ester were obtained. Recovered product was predominantly (85–90%) trans irrespective of the geometry of the reactant. In the presence of azide ion 80% of the product contained 2-phenylcyclopropyl azide [Eq. (3-4)].

$$\text{Ph} \quad \overset{}{\underset{NO}{N}}\!-\!CONH_2 + N_3^- \longrightarrow \text{Ph} \quad N_3 \qquad (3\text{-}4)$$

Solvolysis of 1-arylcyclopropyl 3,5-dinitrobenzoates yields increasing amounts of 1-arylcyclopropanol as the para substituent on the aryl (phenyl) ring was altered from methyl (5%) to thiomethyl (70%) to methoxy (87%).[28a]

Stereochemically the monocyclic derivatives have proven disappointingly uninformative. With *cis*-2-phenyl derivatives only traces at best of *cis*-cinnamyl products have been found.[11,14,28] Usually products interconvert under the conditions of the solvolyses. The ions can isomerize since (2) is converted to 3 at $-10°$ $(E_a = 17.5$ kcal/mol).[29] Conversion 3 to 4 requires an E_a of 24.0 kcal/mol [Eq. (3-5)].[38] It has also been calculated that *cis*-

$$(3-5)$$

$$2 \qquad\qquad 3 \qquad\qquad 4$$

methylallyl cation has $E_a = 25$ kcal/mol for double bond mutation.[30] However the stereochemistry of the ring opening has been found to conform to the W-H-deP rule since (5) is converted to 2 in strong acid medium.[29] Both 6 and 7 also followed the rule [Eq. (3-6)], and within the limits of the experiment the process appears stereospecific.

$$(3-6)$$

Not all cyclopropyl cations undergo ring opening. When treated with methoxide ion in methanol 8 gave 25% of 9 [Eq. (3-7)].[31] Replacement of

$$(3-7)$$

25%   75%

the thiophenyl by a phenoxy moiety reduced the stability of the cyclopropyl cation and no cyclopropyl product was obtained. Dialkylamino groups at C_1 stabilize the cyclopropyl cation as effectively as the phenylthio group [Eq. (3-8) and (3-9)].[32,33] The nmr spectrum of the fluorosulfonate of

$$(3\text{-}8)$$

$$(3\text{-}9)$$

1-dimethylaminocyclopropyl cation has been observed at 25°.[33] Formation of cyclopropyl rings from open chain reactants has been observed in such cases.[34,35,36] Most unexpected is formation of **11** from **10** [Eq. (3-10)].[37]

$$(3\text{-}10)$$

    Contributions to understanding the mechanism of ring opening via the study of bicyclic compounds of the [n.1.0] type have been significant, but not as extensive as from monocyclic molecules. The initial discovery[1] that 1-phenyl-exo-7-tosyloxybicyclo[4.1.0]heptane was resistant to acetolysis was part of the evidence leading to the W-H-deP rule. Further work on exo-(**13**) and endo-7-chlorobicyclo[4.1.0]heptanes (**14**) added support for the theory.[25] The endo form undergoes acetolysis at 125° with $k_1 = 1.4 \times 10^{-6}$ sec^{-1}, whereas **13** shows no reaction after 692 hrs at 125° or 2 hrs at 210° in the presence of silver acetate.[25] exo-7-Tosyloxybicyclo[4.1.0]heptane undergoes acetolysis at 150° some 5600 times slower than the endo isomer.[12] Whether the exo compound reacts via the disrotatory path with rotation in reverse of the rule is not known, but this result sets a lower limit of 6.6 kcal/mol for the difference in activation energy for the two rotational directions. Further studies with exo- and endo-tosylates of bicyclo[n.1.0]alkanes[38,39] indicated a rate differential between the isomers of bicyclo[3.1.0]-hexane to be greater than 2.5×10^6, hence raising the energy differential to at least 10.5 kcal/mol (Table 3-5). The unusually rapid reaction for the exo derivative with $n = 2$ (Table 3-5) has been attributed to a rapid interconversion of the exo to the endo form.[39] Similar results with a p-nitrobenzoate were observed and given an equivalent explanation.[40] The formation of 1,3-diacetoxy products from exo-tosylates with $n = 4,5$ is assumed to result from capture of the trans-cycloheptene (or octene) by solvent, a known reaction.[41]

TABLE 3-5

Relative Rates for Acetolysis of Isomeric Bicyclo[n.1.0]alkyl Tosylates at 100°[a]

n	Relative rate[b]	Products
I. *endo*-(n + 3)-Tosylates		
3	34000	3-Acetoxycyclohexene
4	62	3-Acetoxycycloheptene
5	3.1	3-Acetoxy-*cis*-cyclooctene
6	3.5	3-Acetoxy-*cis*-cyclononene
II. *exo*-(n + 3)-Tosylates		
2	16300	
3	<0.01	
4	1.7	(50%) *exo*-7-Acetoxybicyclo[4.1.0]heptane
		(50%) 1,3-Diacetoxycycloheptane
5	2500	(67%) 3-Acetoxy-*cis*-cyclooctene
		(33%) 1,3-Diacetoxycyclooctane
6	10000	3-Acetoxy-*cis*-cyclononene

[a] Data from refers. 38 and 39.
[b] Rates relative to cyclopropyl tosylate.

Clark and Smale[42] studied **13** and **14** having a 7-aryl substituent. The *endo*-chloride gave solely 2-aryl-1,3-cycloheptadiene as the product, while the main product from the *exo*-chloride was a 1:1 mixture of *exo*- and *endo*-7-acetoxy-7-arylbicyclo[4.1.0]heptanes. Both the product with retained structure and the rate effects of the 7-aryl substituent convinced these workers that solvolysis of the *exo*-chloride proceeded via an unopened cyclopropyl cation. The energy difference between exo and endo isomers of bicyclic systems with small n is large enough to force solvolysis of a fluoride in preference to a chloride [Eq. (3-11)].[43]

$$\text{(3-11)}$$

Fleming and Thomas[44] found that methanolysis of **15** and **16** was stereo-selective, but reaction always proceeded via stereospecific capture of the cation by the leaving chloride ion [Eq. (3-12)]. The final product was then

$$(3-12)$$

formed by solvolysis of the allylic chloride. Participation by an internal group was indicated by the accelerated reaction of **17** in base [Eq. (3-13)].

$$(3-13)$$

Creary[45] has postulated that a double bond can participate in the solvolysis of an *exo*-7-triflate [Eq. (3-14)]. However kinetic studies showed that par-

$$(3-14)$$

ticipation was providing only modest support energetically.

Acetolysis of the *exo*-tosylate of bicyclo[5.1.0]octane[38] gave a mixture (Table 3-5), possibly derived from the *trans*-cyclooctene derivative [Eq. (3-15)].[46,46a] In the presence of silver ion the trans olefin was converted to the

$$(3-15)$$

cis isomer. If the reaction is carried out at a low temperature with a large excess of silver ion, the trans olefin can be obtained.[47,47a] The massive

$$(3-16)$$

amount of catalyst accelerates ring opening more than double bond stereo-mutation because catalysis of ring opening can be both first and second order in silver ion.[48] Where silver ion catalysis can affect both endo and exo halides, the influence on the exo halide is greater for bicyclic molecules with $n = 5$ or 6.[49] Ledlie and Nelson[50] found that silver catalyzed reaction of **18** and **19** gave products with an unaltered skeleton from both isomers, but in greater amounts from the exo compound [Eq. (3-17)]. Additional results of a similar

$$(3-17)$$

sort have been reported by Groves and Ma.[51] In some cases it appears that silver ion catalysis of endo halides does not occur.[52]

Following the early discovery[53] that attempted distillation of the indene-dichlorocarbene adduct gave 2-chloronaphthalene, many examples of thermal ring opening of cyclopropyl derivatives have been found. The mechanism of this process is not clear, possibly a $[\sigma2_s + \sigma2_a]$ cycloaddi-tion[54,55,56] or an ion pair process involving the partially opened cyclopropyl cation.[44,57,58] In a few cases a radical mechanism has been suggested.[55,59] Monocyclic cases have been relatively rarely studied. Duffey et al.[60] heated cis-2,3-dimethyl-1,1-dibromocyclopropane at 160° in kerosene and obtained 3,4-dibromo-2-pentene (stereochemistry unknown). Their kinetic studies gave $k_1 = 1.8 \times 10^{-5}$ sec^{-1} (160°) and $E_a \simeq 26$ kcal/mol, $\Delta S^{\ddagger} \simeq -22$ eu. Parry and Robinson[55] pyrolyzed 1,1-dichlorocyclopropane, obtaining only 2,3-dichloropropene with $E_a = 57.8$ kcal/mol and log $A = 15.1$. Baird and

Reese[61] found that *trans*-2-*trans*-3-dimethylcyclopropyl bromide gave pure *trans*-4-bromo-2-pentene at 165°, while both cis-2-cis-3, and cis-2-trans-3 isomers gave mixtures at 200°.

Most of the investigations of the thermal ring opening have involved bicyclic or tricyclic molecules. Wynberg[62] found that **20** gave cycloheptatriene when passed through a hot tube [Eq. (3-18)]. Attempted distillation

$$(3-18)$$

of 6,6-dibromobicyclo[3.1.0]hexane at 150° gave 2,3-dibromocyclohexene.[63,64] The earliest relation between the ease of this type of rearrangement and the stereochemistry of the halogen was provided by Schweizer and Parham[65] [Eq. (3-19)]. Their stereochemical assignment was the inverse of

$$(3-19)$$

that given here, and the correction was made by de Puy,[11] and later by Ando and co-workers.[66] Vogel[64] made similar observations with **21** and **22** [Eq. (3-20)].

$$(3-20)$$

Several groups reported[67,68,69,58,69a] that dihalocarbene adducts with bicyclo[2.2.1]heptene, bicyclo[2.2.1]heptadiene, and bicyclo[2.2.2]octene will undergo ring expansion very readily, often preventing isolation of the initial

$$(3\text{-}21)$$

adduct [Eq. (3-21)]. DeSelms and Combs[58] found that the dichloro adduct of bicyclo[2.2.1]heptene reacted faster in more polar solvents and suggested an ionic mechanism. Once again further work confirmed the stereoisomer variation in activity [Eq. (3-22)].[70] It has also been established that the

$$(3\text{-}22)$$

mobile halogen ends up on the same side of the expanded ring as in the original carbene adduct, i.e., an endo adduct gives an exo allylic halide.[58,68,70,70a] Baird et al.[71] studied the influence of ring size in bicyclo[n.1.0]-alkyl halides on the ease of ring expansion and showed that the temperature required increased with increasing n. They suggested that with $n \geq 5$ the mechanism of the reaction may be altered. For thermal reactions which occur below 200° the ion pair mechanism and the W-H-deP rule correlate all the evidence and should provide correct predictions for additional examples, though the ease of conversion of cis and trans allyl cations at elevated temperatures must be considered.

Aside from the purely thermal examples, ring expansion of cyclopropyl halides occurs in the presence of tertiary bases such as quinoline, lepidine, pyridine, etc. This procedure gives rise to dienes, rather than allylic derivatives. Mechanistic details are rare, but Schweizer and Parham,[65] de Puy et al.,[11] Ando et al.,[66] Baird and Reese,[61] Baird and Reese,[72] and Pandit and de Graaf[73] have confirmed the relation between ease of reaction and the stereochemistry of the leaving group. It seems safe to conclude that these reactions proceed via the cation, which undergoes elimination under the conditions of the reaction.

All of the work considered up to this point has restricted the routes for ring opening sterically via bulky substituents or geometrically by incorporating the 2,3-bond in a ring. Additional restrictions can be developed by adding a second ring as in 23. Formation of an allylic cation then runs into the restrictions of Bredt's rule[74,75] as applied to 24 [Eq. (3.23)]. It seems

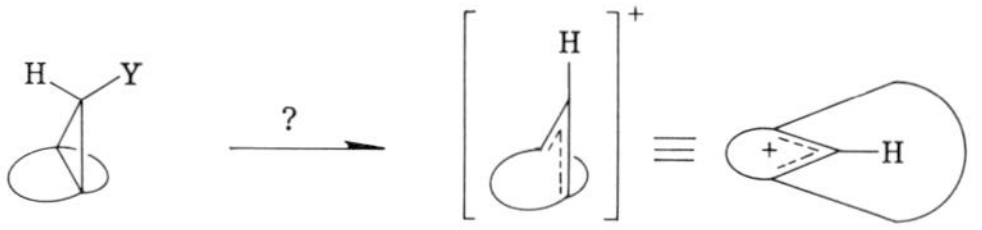

$$(3\text{-}23)$$

23   24

reasonable to presume that sufficient restriction can be applied in this way to prevent ring opening. Reese and Stebles[76] treated 25 with silver ion in aqueous acetone and obtained 26 [Eq. (3-24)]. It appeared that the positive

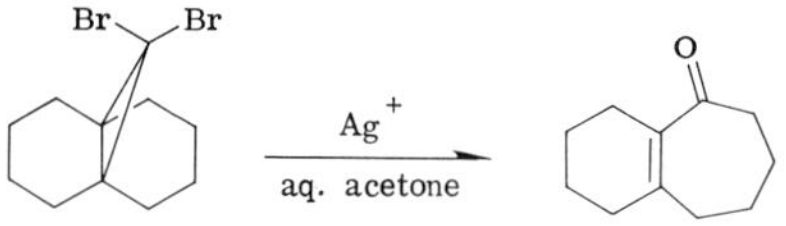

$$(3\text{-}24)$$

25   26

charge was largely restricted to C_{11} and the ketone was formed via the route of Scheme 1. Note that the one carbon bridge ends up as the carbonyl carbon

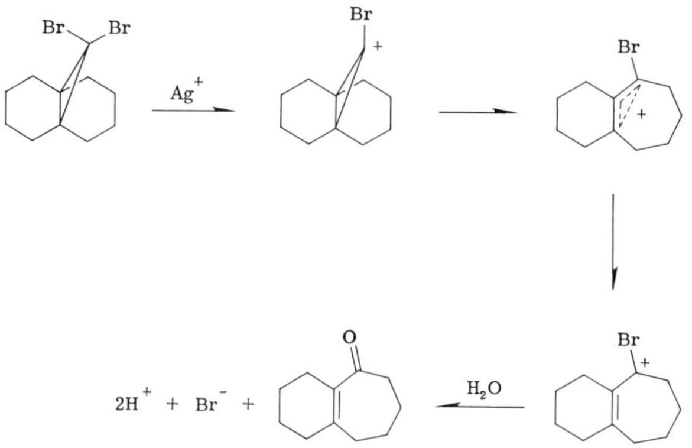

Scheme 1.

(compare Scheme 3). Further study by Reese and Stebles[77] almost immediately cast some doubt on this idea, since a more highly strained tricyclic molecule gave products indicative of a direct ring opening of the cyclopropyl ring [Eq. (3-25) and Scheme 2]. Additional products formed under slightly

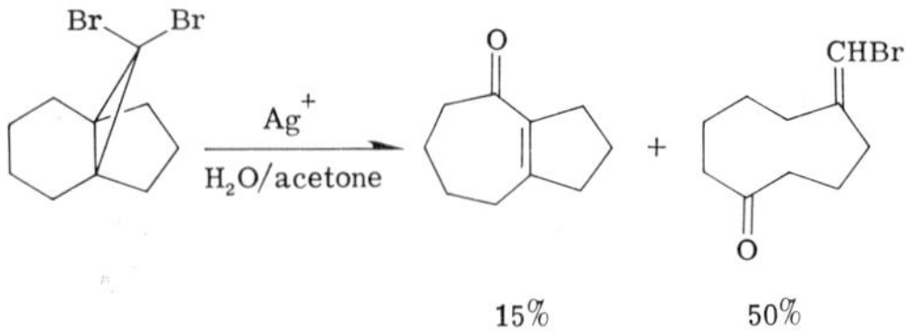

$$(3-25)$$

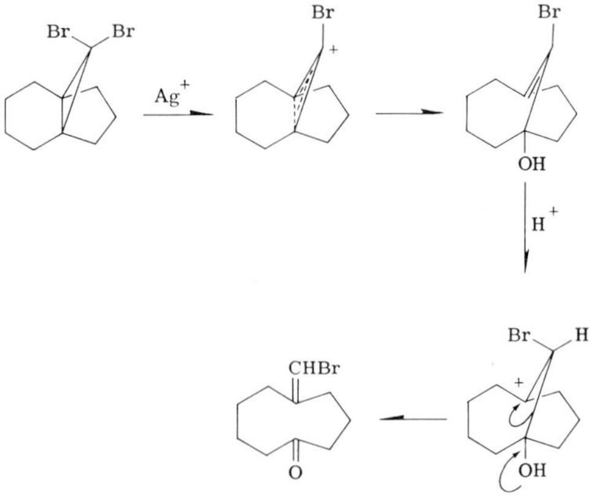

15%   50%

Scheme 2.   **27**

altered conditions were rapidly uncovered. Products containing the intact carbon skeleton were obtained[78,79] [Eqs. (3-26) and (3-27)] and also products

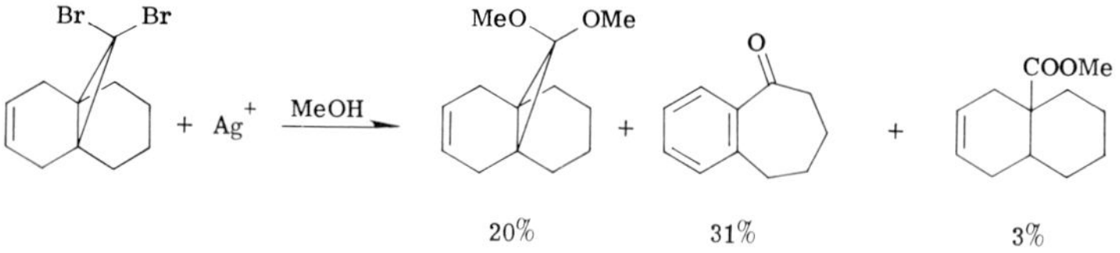

20%   31%   3%

$$(3-26)$$

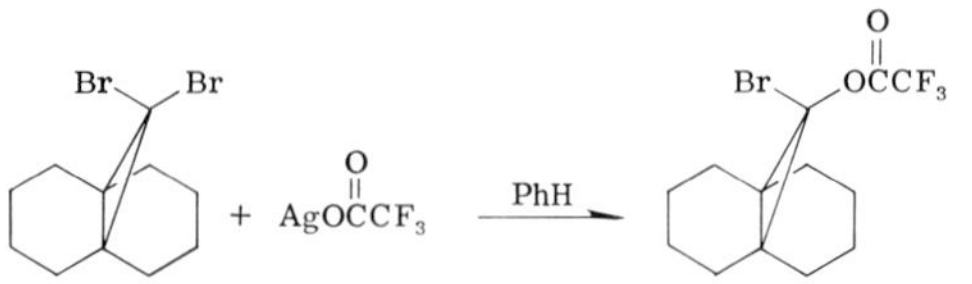

$$(3-27)$$

clearly derived from trapping the bridgehead double bond [Eq. (3-28)][80]

This helped to confirm the process of Scheme 2, and the ratio of capture of the ion **27** by water vs. cleavage was changed notably by addition of a double

28

bond as in **28**.[80] The bridgehead double bond can be captured by cycloaddition [Eq. (3-29)] or dimer formation.[81] The use of silver ion is not needed

to bring about reaction [Eq. (3-30)].[82] Warner and Lu[83] have used a ^{13}C

enriched reactant to eliminate the mechanism of Scheme 1, and show that the process of Scheme 3 fits the data. Comparative rate studies have also

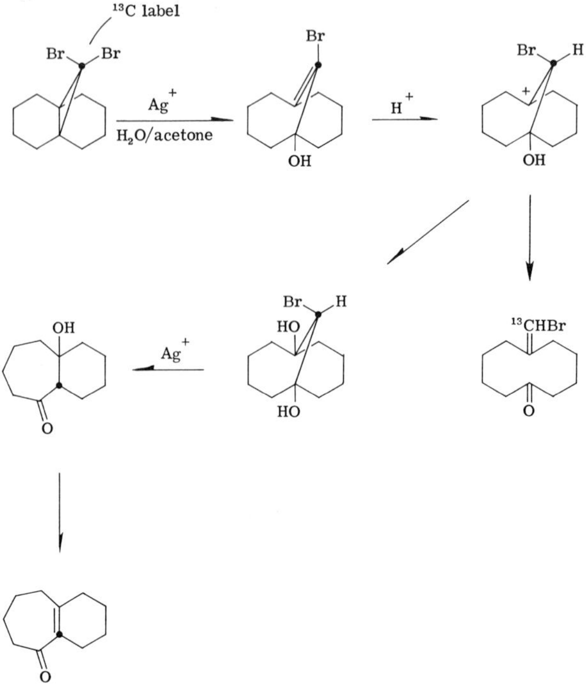

Scheme 3.

been used.[165] Tricyclic compounds with smaller rings lead to products derived from cleavage of a lateral bond of the cyclopropyl ring [Eq. (3-31)].[84]

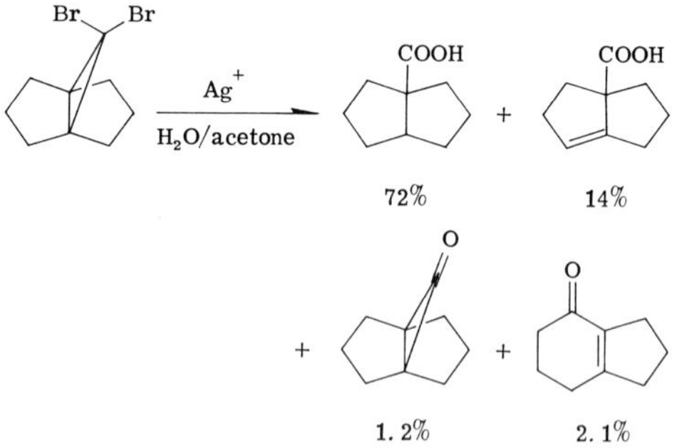

(3-31)

When a single bromine is present on the cyclopropyl ring, the products with an intact structure have fully retained stereochemistry at the substituted carbon [Eq. (3-32)].[85,86] Additional examples have been reported by Ledlie

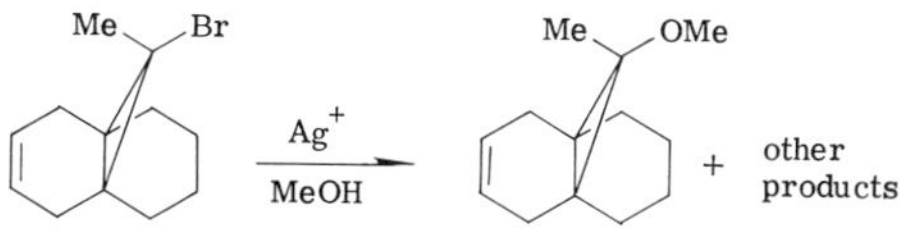

$$(3-32)$$

and co-workers[88,89,90] including an interesting formation of Vogel's aromatic 10 π compound.[90]

Olah et al.[87] have generated a cation from **29** in solution and have presented evidence that the ion has a bent cyclopropyl cation structure **30**.

$$(3-33)$$

29   **30**

2. Uses in Synthesis

The pioneering study of Parham and Rieff[53] was followed by a series of papers by Parham and his students[91–94] devoted to exploring the value of the ring expansion process with aromatic compounds [Eq. (3-34)]. When

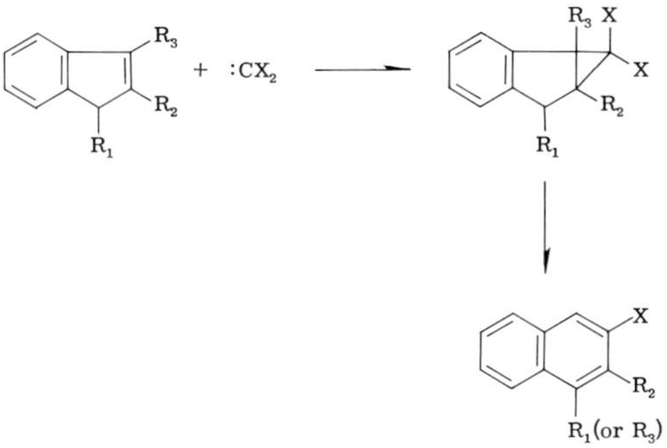

$$(3-34)$$

$R_1 = $ Me or $R_3 = $ Me, rapid interconversion of 1- and 3-methylindene, led to isolation of only 1-methyl-3-halonaphthalenes.[91] If $R_2 = $ Me,OEt reaction proceeded in modest yield, but for $R_2 = $ Cl,COOEt no ring expanded product was obtained.[93] The bonds in 1- or 2-methoxynapthalene or 9-methoxyphenanthrene were reactive enough to permit ring expanded products to be obtained, albeit in modest yield [Eqs. (3-35) and (3-36)].[94]

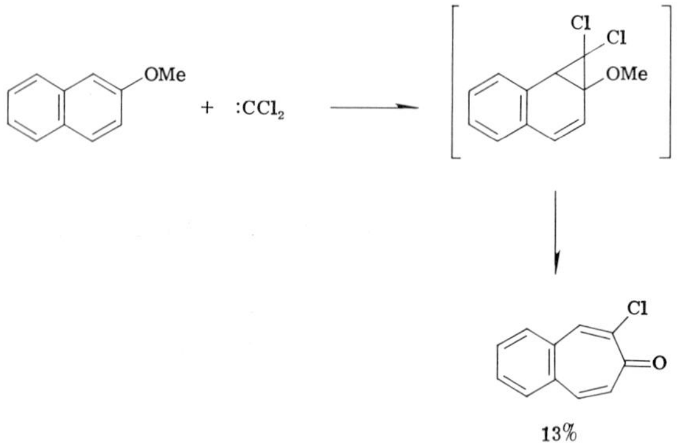

$$(3\text{-}35)$$

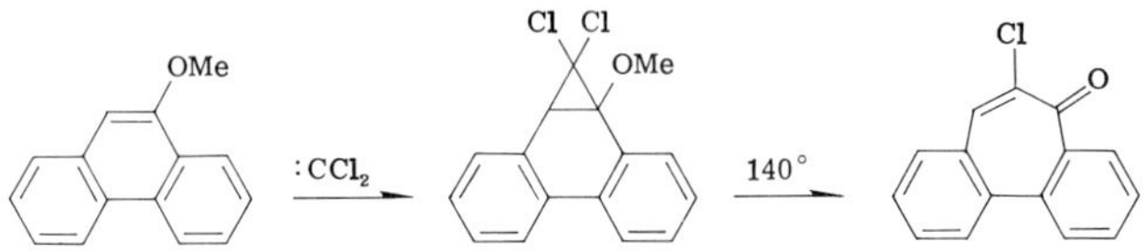

13%

$$(3\text{-}36)$$

The ring expansion route has been used for synthesis of some *m*-cyclo-phanes,[170,171] but the route is not always successful.[172] Schweitzer and Parham[65] found that dihydrooxepins can be prepared conveniently from dihydropyran [Eq. (3-37)]. Robinson[95] improved Wynberg's route to

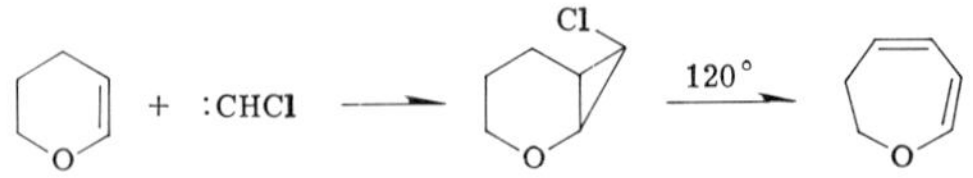

$$(3\text{-}37)$$

cycloheptatriene,[62] and in conjunction with the further ring expansion of cycloheptatriene by TerBorg and Bickel[96] this provides a clever route from cyclohexene to benzocyclobutenes [Eq. (3-38)]. Examples of this route to

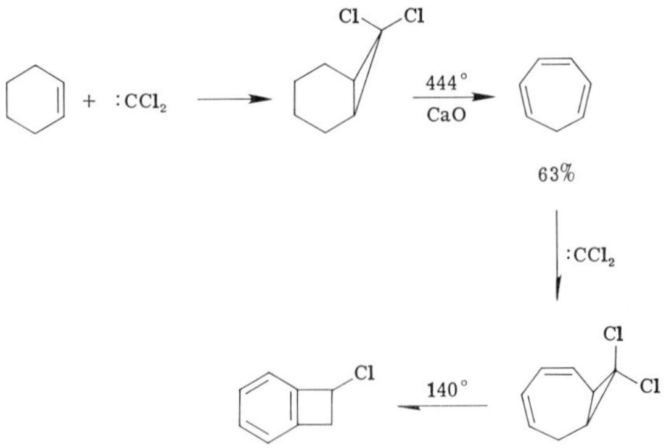

63%

$$(3\text{-}38)$$

benzocyclobutenes from 1,4-cyclohexadienes using silver ion catalysis are known,[97,98,99] but the reaction can also lead to substituted styrenes [Eq.

$$\text{(3-39)}$$

(39)].[100] An interesting synthesis of 3,5-cycloheptadienone has been developed by this ring expansion route [Eq. (3.40)].[101]

$$\text{(3-40)}$$

The formation of a single product in this case is surprising unless it is a stable thermodynamic product. That is probably the case since White and Wade[102] found that a single product is obtained in the double ring expansion of Eq. (3.41). In this case a hydrogen migration is necessitated.

$$\text{(3-41)}$$

Skell and Sandler[103] pointed out that the addition of a halocarbene followed by ring opening constitutes a route to insertion of a carbon between doubly bonded atoms [Eq. (3-42)]. Using silver ion in acetic acid Sandler[104]

$$\text{(3-42)}$$

investigated a series of dihalocarbene adducts. Tetrasubstituted alkenes give diene products [Eq. (3-43)], and unsymmetrical cases lead to mixtures

$$
\underset{Me_2}{\overset{Me_2}{\triangleright}}\!\!-Br_2 + Ag^+ \xrightarrow{\ HOAc\ } \underset{Me}{\overset{Me}{>}}C\!=\!\underset{Br}{\overset{Me}{C}}{-}C\!=\!CH_2
\tag{3-43}
$$

$$
\underset{Me}{\overset{Me}{\triangleright}}\!\!-Br_2 + Ag^+ \xrightarrow{\ HOAc\ } \underset{Me}{\overset{Me}{>}}C\!=\!\underset{Br}{\overset{CH_2OAc}{C}}
$$
$$
+
$$
$$
\underset{Me}{\overset{Me}{>}}\underset{OAc}{\overset{|}{C}}{-}C\!\underset{Br}{\overset{CH_2}{<}}
\tag{3-44}
$$

[Eq. (3-44)], or if a group is present which conjugates effectively with a double bond a single product is obtained [Eq. (3-34)]. The adducts from enol ethers

$$
\underset{Ph}{\triangleright}\!\!-Br_2 + Ag^+ \xrightarrow{\ HOAc\ } PhCH\!=\!CH\!-\!CH_2OAc
$$
$$
\text{trans}
\tag{3-45}
$$

provide a route to unsaturated acetals,[104,105] and with ketene acetals reaction gives unsaturated esters [Eq. (3-45)][106] Schlosser and Chan[107,108] have

$$
\underset{Me}{\overset{Me}{>}}C\!=\!\underset{Ph}{\overset{C}{\underset{Cl}{|}}}\!C\!\underset{OMe}{\overset{OMe}{<}} \xrightarrow{\ \Delta\ } \underset{Me}{\overset{Me}{>}}C\!=\!C\!\underset{Ph}{\overset{COOMe}{<}}
\tag{3-46}
$$

used the route to prepare a series of fluorinated allyl acetates, and have shown that the allyl cations can be used to alkylate aromatic rings [Eq. (3-45)].

$$
\underset{Et}{\overset{Et}{\triangle}}\!\!\underset{Cl}{\overset{F}{}} \xrightarrow[PhMe]{AgBF_4} \underset{Et}{\overset{Et}{>}}C\!=\!C\!\underset{CH_2}{\overset{F}{<}}{-}\!\!\langle\!\!\rangle\!{-}Me
\tag{3-47}
$$

$$
o, p\ 1/1
$$

Regioselectivity in capture of the allylic ion has also been studied.[108a] The reaction has been used to prepare a variety of α-halo-α-,β-unsaturated ketones and aldehydes.[166,167] Stable cyclopropyl cations have been used in a Mannich reaction to prepare cyclopropyl substituted ketones [Eq.

$$
\underset{NMe_2}{\overset{OH}{\triangleright\!\!<}} + CH_3COCH_3 \xrightarrow{\ HOAc\ } \underset{NMe_2}{\triangleright\!\!<}\!CH_2\!-\!\overset{\overset{\displaystyle O}{\|}}{C}\!-\!CH_3
\tag{3-48}
$$

(3-48)].[109] Alkylidenecyclopropanes with a halogen on the ring react to form mainly allenic products [Eq. (3-49)],[110,111,111a,112] but the use of silver

$$\text{(3-49)}$$

ion alters not only the ratio but also the nature of the products. Formation of the unopened cyclopropyl cation is observed when substitution destabilizes the ring opened cation [Eq. (3-50)].[113] A most unusual ring opening was

$$\text{(3-50)}$$

observed by Babb and Gardner,[114] with a result which is up to the present unique [Eq. (3-51)].

$$\text{(3-51)}$$

Numerous examples of the formation of bicyclo[3.2.1]octenes from di-halocarbene adducts with bicyclo[2.2.1]heptenes,[58,67–70a] and bicyclo[2.2.1]-heptadienes[67] have been reported. Thermal ring expansion normally occurs rapidly enough so that the adducts are not isolated. The reaction is stereospecific. The rearrangement products can be converted to 3-bicyclo[3.2.1]-octanones [Eq. (3-52)].[67] This ring expansion has been applied to 7-oxabi-

$$(3-52)$$

cyclo[2.2.1]heptene[115] and to bicyclo[2.2.2]octene.[69,69a] As expected **31** leads to mixtures [Eq. (3-53)],[69,69a,116] and **32** gives a diene [Eq. (3-54)].[69,69a,116]

$$(3-53)$$

This ring expansion has also been used to prepare an unsaturated ketone in the bicyclo[4.2.1]nonane system.[116a]

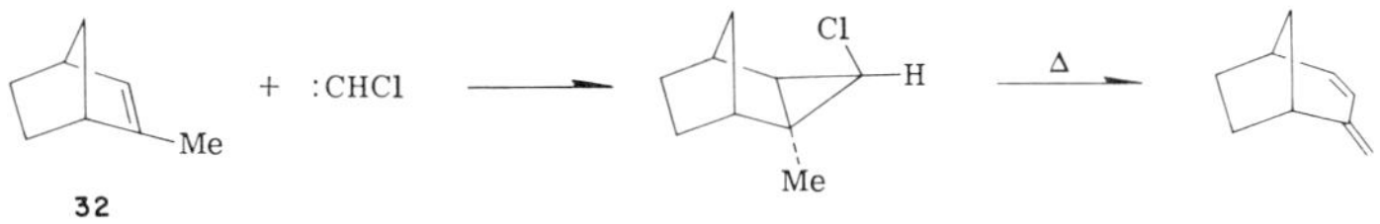

$$(3\text{-}54)$$

32

Perhaps the greatest value of this reaction in synthesis is in the stereoselective formation of a variety of medium ring compounds having either cis
or trans double bonds. The considerable problems with synthesis of medium
rings, the ready availability of reactants such as *cis*-cyclooctene, and the
control of stereochemistry has made this a very valuable route. Furthermore
the product contains an additional and useful function in the allylic halide
or alcohol. The first attempt to use this method for synthesis of medium ring
alkenes was made by Whitham and Wright[46,46a] who showed that **33** produced the expected *trans*-2-cycloocten-1-ol [Eq. (3-55)] when solvolysis was

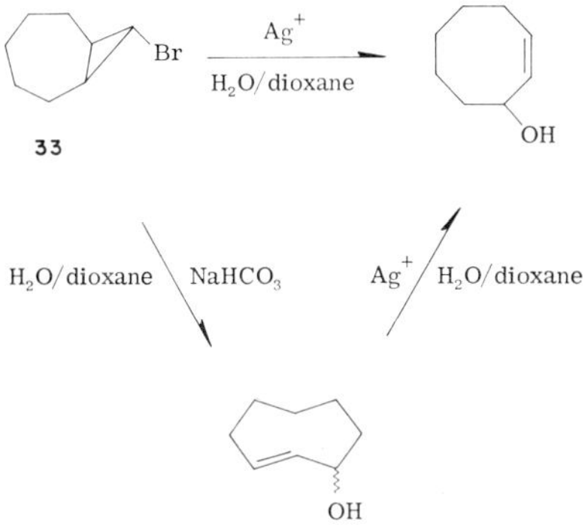

33

$$(3\text{-}55)$$

carried out in aqueous dioxane and sodium bicarbonate. Use of silver ion
gave only *cis*-2-cyclooctenol. Acetolysis of *exo*-8-tosyloxybicyclo[5.1.0]-
octane gave only the acetate of the *cis*-cyclooctenol,[38] indicating the ease
with which the geometric isomers interconvert at elevated temperatures.
In accord with this expectation, thermal reaction of *exo*-9- and 9,9-dibromo-
bicyclo[6.1.0]nonanes gives only 3-bromo-*cis*-cyclononenes.[47,47a,117] However the rather less strained *trans*-2,3-dibromocyclodecene can be obtained
via the thermal route [Eq. (3-56)].[117] Use of the eliminative expansion with

$$(3\text{-}56)$$

tertiary amine solvents leads to *cis,cis*-1,3-cycloalkadienes [Eq. (3-57)].[118]

$$(3\text{-}57)$$

Presence of an additional double bond did not alter the stereochemistry, but elimination of the allyl halide did not occur [Eq. (3-58)].[119]

$$(3\text{-}58)$$

The solvolysis rates for *exo*- and *endo*-tosylates of bicyclo[*n*.1.0]alkanes (Table 3-5) lead to the expectation that with dibromo or dichloro derivatives it will be the *exo*-halogen which reacts preferentially when $n \geq 5$. This is indeed realized [Eq. (3-59)] and may be considered a general rule.[47,47a,49,52]

$$(3\text{-}59)$$

The less reactive *endo*-bromides generally lead to *cis*-cycloalkenes in only moderate yield.[120] When the barrier to rotation of a trans double bond through the circle of the ring becomes high enough, the stereoselectivity of capture of the allylic cation becomes apparent [Eqs. (3-60) and (3-61)].[47,47a, 49,52,120] Formation of the *trans*-cyclooctene in the presence of silver ion [Eqs. (3-59) and (3-60)][49] requires a large excess of silver ion. Use of a two-

$$(3\text{-}60)$$

$$(3\text{-}61)$$

fold excess with an initial concentration of 2 M has been recommended.[49] The basis for the important concentration effect probably rests in the second order term in silver ion observed by Blackburn and Ward.[48] In some cases the ratio of trans to cis product depends on the nucleophile.[120a]

Though the above ring expansion procedure has made *trans*-cycloalkenes rather readily available, very little has been done with carbene adducts to the *trans*-olefins. Parham and Sperly[121] reported that **34** (Y = H) did not react with silver ion in ethanol under conditions where the cis isomer reacted readily. However **34** (Y = OEt) did undergo ring expansion thermally [Eq. (3-62)]. Successful thermal expansion of *trans*-13,13-dibromobicyclo-

$$(3\text{-}62)$$

34

[10.1.0]tridecane has been reported, though the cis isomer reacted more readily.[122] Optically active **35** gives active *cis*-2-bromo-3-methoxycyclo-nonene [Eq. (3-63)].[123] The authors suggested solvent capture of the partly

(-) 1*R*, 8*R*   (+) *R*

35

$$(3\text{-}63)$$

opened cation, but models indicate that the planar allylic cation could not rotate through the ring and would lead to the same active isomer.

The very important questions for synthesis (i.e., how substituents will influence the course of the reaction) has received only modest attention. The availability of 1,3- and 1,4-cycloheptadienes led to the investigation of monoadducts of these. Silver ion catalyzed reaction of **36** gave equal parts of **37** and **38** [Eq. (3-64)].[124] However **39** gave exclusively **40** [Eq. (3-65)].[120]

36

MeO   OMe

50%   50%

37   **38**

$$(3\text{-}64)$$

Steric factors appear to be the dominant cause for these contrasting results.

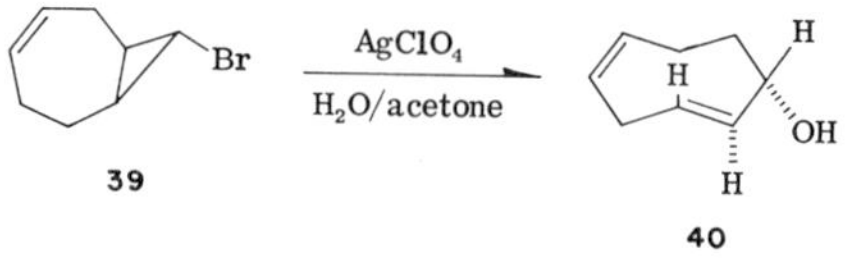

$$(3\text{-}65)$$

In the bicyclo[6.1.0]nonane system a 4-ene expedites the reaction and, while it does not participate under normal conditions,[49,52] it can lead to trans-annular ring closure [Eq. (3-66)].[125] A 4,5-oxide does not participate in the

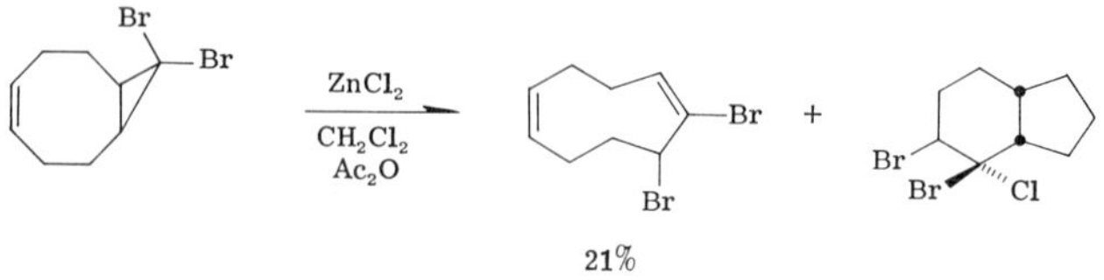

$$(3\text{-}66)$$

21%

reaction when syn to the cyclopropyl ring [Eq. (3-67)].[126] A methyl group

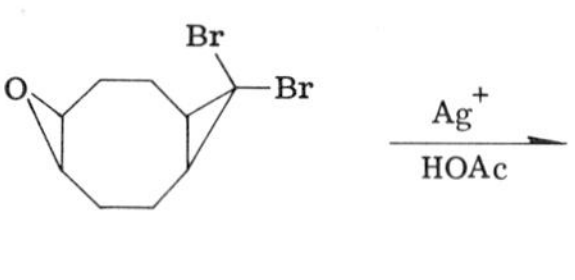

$$(3\text{-}67)$$

at C-1 does not permit exclusive capture by solvent at one end of the unsym-metrically substituted allylic cation [Eq. (3-68)].[127] Double expansion of **41**

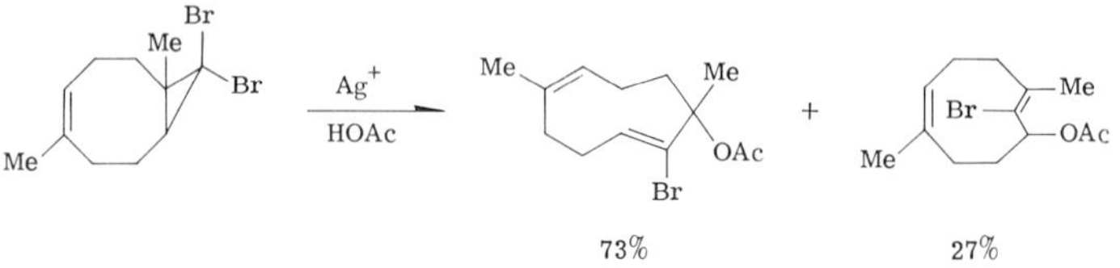

$$(3\text{-}68)$$

73%   27%

leads to three products with surprisingly little transannular ring closure and exclusive formation of the 1,6-cyclodecadiene skeleton [Eq. (3-69)].[47,47a]

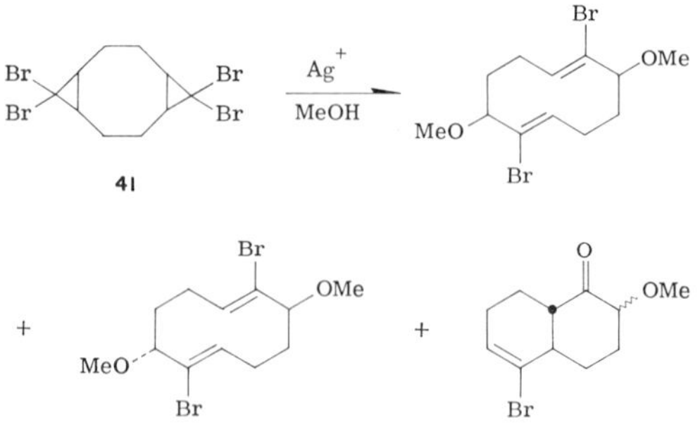

$$(3\text{-}69)$$

Considerable effort has been expended to develop this ring expansion process to a useful synthesis of enlarged ring ketones. *trans*-2-Cyclodecenone was prepared conveniently by aqueous solvolysis of *exo*-10-bromobicyclo-[7.1.0]decane followed by oxidation of the allylic alcohol.[128] Casanova and Waegell prepared cyclotridecanone from the dibromocarbene adduct of cyclododecene by thermal ring expansion, reductive removal of the allylic bromide, and conversion of the vinyl bromide to the ketone with concentrated sulfuric acid.[122] Cyclononanone was prepared from cyclooctene by the procedure shown in Eq. (3-70).[129] Seebach *et al.*[130] have also developed a

novel route to ring expansion leading directly to a ketone from a bisthiomethyl derivative **42** [Eq. (3-71)]. The method is applicable to preparation

of acyclic ketones as well.

II. CYCLOPROPANONE–OXYALLYL AND RELATED SYSTEMS

Cyclopropanone would be expected to be a highly strained molecule, hence prone to ring scission. By analogy with the well-known degenerate methylenecyclopropane rearrangement cyclopropanone might be expected to interconvert with methyleneoxirane [Eq. (3-72)]. Conceivably the inter-

$$(3\text{-}72)$$

conversion could be concerted, but oxyallyl might play an intermediary role. Oxyallyl could be written as a biradical or a zwitterion, but the zwitterion should be the strongest contributor to the hybrid. Since odd electron species are not strongly predisposed toward either con or disrotatory reactions, the contribution of the zwitterion form should turn the tide toward a two electron, i.e., disrotatory electrocyclic process.

Unknown for many years cyclopropanones were considered as possible intermediates in the Favorskii rearrangement,[131] or as intermediates in the formation of cyclobutanones from ketenes and diazomethane.[132] About 1965 the isolation of tetramethylcyclopropanone,[133-136] 2,2-dimethylcyclopropanone,[137] and finally cyclopropanone itself,[138] opened the door to investigations of these interesting molecules.[139,139a]

Theory has not provided very helpful information about the cyclopropanone–oxyallyl system. An extended Hückel treatment indicated that oxyallyl would undergo electrocyclic closure to cyclopropanone in disrotatory fashion, and would also undergo a concerted 1,4-cycloaddition.[140] However oxyallyl was found in that calculation to be more stable than either cyclopropanone or methyleneoxirane. A second study produced results at the opposite extreme, i.e., oxyallyl was more than 200 kcal/mol less stable than cyclopropanone, and the C-C-C angle for oxyallyl was 90°.[141] An *ab initio* SCF calculation set oxyallyl 83 kcal/mol above cyclopropanone, and again the C-C-C angle for oxyallyl was in the range of 90° to 105°.[142] This study again showed that the disrotatory ring opening would be preferred to a conrotatory one. MINDO calculation placed oxyallyl 36 kcal/mol above cyclopropanone, yet gave the C-C-C angle for oxyallyl as 70°.[143] MINDO/2 made oxyallyl 78 kcal/mol less stable than cyclopropanone.[144] The most recent study, via CNDO/2, suggested that oxyallyl could not be an intermediate in the methyleneoxirane → cyclopropanone reaction, and the authors calculated that the preferred path would be an extreme bending–rotation–realignment process [Eq. (3.73)].[145] The general lack of internal

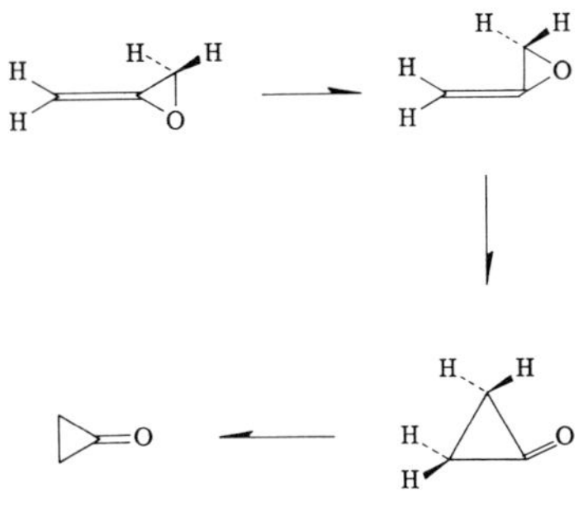

(3-73)

consistency does not engender any great confidence in the theoretical results.

That ring opening of cyclopropanones with cleavage of the C_2-C_3 bond can occur is shown by the cycloaddition reactions of Eq. (3-74)].[145a] Theory

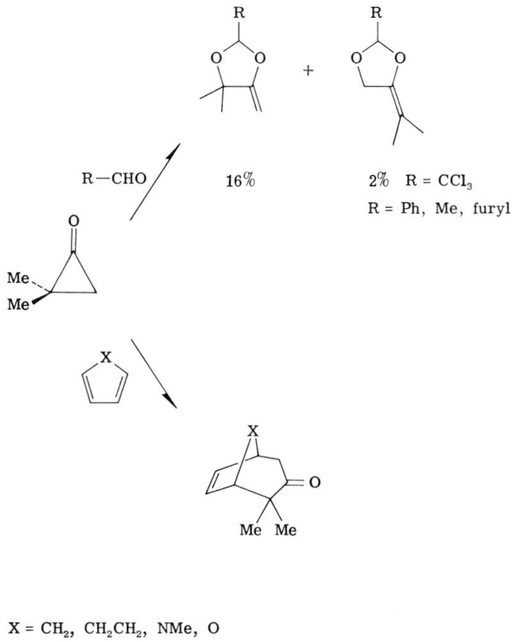

(3-74)

X = CH$_2$, CH$_2$CH$_2$, NMe, O

suggests that oxyallyl should undergo 1,2-cycloaddition via the C-C-O group as the results above indicate.[139,139a] The 1,4-cycloaddition process is not very selective [Eq. (3-75)]. While a case could be made that ring opening is

$$(3\text{-}75)$$

induced by attack of the aldehyde or diene directly on the cyclopropanone, the rates of reaction of a series of substituted cyclopropanones with furan (Table 3-6) are in much better accord with a prior disrotatory ring opening to oxyallyl followed by a 1,4-cycloaddition [Eq. (3-76)]. The reaction rates,

$$(3\text{-}76)$$

however, were not independent of the furan concentration, as the first step was not fully rate determining.

The very interesting question of whether oxyallyl can reclose to cyclopropanone has received a partial answer via investigation of an optically

TABLE 3-6

Rates of Reaction of Substituted Cyclopropanones with Furan

R_1	R_2	R_3	R_4	Relative rate
H	H	H	H	$<1 \times 10^{-4}$ (not observed)
Me	H	H	H	~ 1
Me	Me	H	H	1
Me	Me	Me	H	$>10^2$
Me	Me	Me	Me	$\sim 1\text{-}2$

active cyclopropanone. *trans*-1,2-di-*tert*-Butylcyclopropanone is a rather stable molecule,[146] and it has been possible to prepare it in optically active form.[147] The racemic molecule appears to be stable to 150°, at which temperature it decomposes to carbon monoxide and *trans*-1,2-di-*tert*-butylethylene. The active material, on the other hand, racemizes at 80° [Eq. (3-77)].

$$
\underset{t\text{-Bu}\quad\quad t\text{-Bu}}{\overset{O}{\triangle}} \quad \xrightarrow{\;80°\;} \quad \underset{t\text{-Bu}}{\overset{O^-}{\diagup}} \underset{+}{\diagdown} \overset{t\text{-Bu}}{} \tag{3-77}
$$

If the oxyallyl isomer is an intermediate in this reaction, it appears to be too hindered to permit cycloadditive trapping. The racemization process cannot be attributed to enolization, since no deuterium is incorporated if the reaction is run in *tert*-butanol–OD. The rate of the racemization is notably solvent dependent, being about 13 times faster in acetonitrile than in isooctane.

Crandall and Machleder[148] made the first attempt to prepare a methyleneoxirane (allene oxide). Though the desired allene oxides were not isolated, the products could be rationalized as being derived from such an intermediate [Eq. (3-78)]. Further work with 1,1-dimethylallene and 1,2-cyclo-

$$
Me_2C{=}C{=}CMe_2 \quad \xrightarrow[\substack{CO_3^{2-}}]{\overset{\overset{O}{\parallel}}{MeCOOH}} \quad [\,Me_2C\overset{O}{\diagup}\!\diagdown C{=}CMe_2\,] \tag{3-78}
$$

$$
Me_2C\overset{\overset{O}{\parallel}}{\underset{O}{\diagup\!\diagdown}}CMe_2 \qquad\qquad Me_2CH{-}\overset{\overset{O}{\parallel}}{C}{-}\underset{OAc}{CMe_2}
$$

nonadiene gave similar results.[149] More interesting results were obtained when the bulky *tert*-butyl substituent was used. Thus *tert*-butyltrimethylallene with peracetic acid gave an unrearranged diepoxide.[150] 1,1-di-*tert*-butylallene gives the relatively stable **43** [Eq. (3-79)],[151] but the presumed

$$
CH_2{=}C{=}C\underset{t\text{-Bu}}{\overset{t\text{-Bu}}{\diagdown}} \quad \xrightarrow[\text{buffer}]{\overset{\overset{O}{\parallel}}{MeCOOH}} \quad \left[\, CH_2{=}C\overset{O}{\diagup}\!\diagdown C\underset{t\text{-Bu}}{\overset{t\text{-Bu}}{\diagdown}} \,\right] \tag{3-79}
$$

$$
\underset{t\text{-Bu}\quad t\text{-Bu}}{\overset{O}{\triangle}}
$$

43

intermediate allene oxide was not isolated. However when 1,3-di-*tert*-butylallene was treated with *m*-chloroperbenzoic acid the allene oxide **44** was isolated [Eq. (3-80)].[147,152] At 100° the allene oxide rearranges to **45**.

(3-80)

Racemization of active **45** is faster than the above rearrangement. Finally, 1-*tert*-butylallene oxide has been isolated, but rearrangement to the cyclo-propanone does not occur.[168]

Crandall and Conover[153] have studied the low temperature reaction between some allenes and ozone. Both allene oxides and cyclopropanones were obtained in some cases. All of the products can be conveniently rationalized by Eq. (3-81). A clever approach to the synthesis of an allene oxide[154] shows

$$R_1 = R_2 = t\text{-Bu} \quad R_3 = H \qquad \begin{array}{c} O \\ \triangle \\ R_1 \quad R_2 \end{array} \qquad 27\%$$

$$R_1 = R_2 = R_3 = t\text{-Bu} \quad \begin{array}{l} 1 \text{ equiv. } O_3 \\ \text{allene oxide } 5\% \\ \text{diepoxide } 41\% \end{array}$$

(3-82)

that these compounds generally decompose to form oxyallyls very readily even at room temperature [Eq. (3-82)]. Directed synthesis of oxyallyls by

$$\underset{H}{\overset{Ph}{>}}C=C\underset{CH_2Cl}{\overset{SiPh_3}{<}} \quad \xrightarrow{MCPA} \quad PhHC\underset{}{\overset{O}{-}}C\underset{CH_2Cl}{\overset{SiPh_3}{<}}$$

$$N_2, \text{ dark} \Big| F^-, 25° \quad + \quad MeCN \quad + \quad \text{(cyclopentene with X)}$$

(3-83)

$$\text{(bicyclic structure with X, Ph, O)}$$

$$X = CH_2, \ O, \ NCOOMe$$

debromination of 2,2'-dibromoketones leads to the conclusion that when the oxygen atom of an oxyallyl is coordinated to a metal atom the ring closure to a cyclopropanone is prevented, though 1,4-cycloaddition does occur [Eqs. (3-83) and (3-84)]. Surprisingly 1,2-cycloaddition has also been ob-

$$\underset{Br \quad Br}{PhHC\overset{O}{\underset{|}{-}}C\overset{||}{\underset{|}{-}}CH-Ph} \quad \xrightarrow[\text{or NaI}]{Zn/Cu} \quad \text{(bicyclic structure with X, Ph, O, Ph)}$$

$$\text{(cyclopentene with X)}$$

refer. 155

$$X = O, \ CH_2$$

$$\underset{Br \quad Br}{Me_2C\overset{O}{\underset{|}{-}}C\overset{||}{\underset{|}{-}}CMe_2} \quad \xrightarrow{Fe_2(CO)_9} \quad \text{(seven-membered ring with Me groups and O)}$$

(3-84)

refer. 156

served with retention of the olefin stereochemistry.[157] Fort[158] has shown that an intermediate trappable by cycloaddition can be generated from an α-haloketone [Eq. (3-85)].

$$(3\text{-}85)$$

Some evidence for the interconversion of cyclopropanethione and methylenethiirane via the thioallyl moiety has been obtained.[169] A variety of nitrogen containing analogs of the cyclopropanone–allene oxide–oxyallyl system have been investigated directly or proposed as intermediates in other reactions. The simplest of these, where the oxygen is replaced by a nitrogen, appears to have played a role only as a set of unobserved intermediates [Eq. (3-86)].[159] On the other hand aziridinones (or α-lactams) are isolable entities.

$$(3\text{-}86)$$

The aziridinones decompose thermally to give ketones and isocyanides [Eq. (3-87)].[160,161,161a] The relative stabilities of a series of substituted aziridi-

$$(3\text{-}87)$$

nones are in good accord with the mechanism of Eq. (3-88). Thus, for ex-

$$(3-88)$$

ample, when $R_1 = R_2 = Ph$ the aziridinone can be isolated only by operating at low temperature ($< 20°$), when $R_1 = R_2 = Me$ the aziridinone is stable at $25°$, and when $R_1 = tert$-butyl, $R_2 = H$ the aziridinone is stable to $100°$.

Diaziridinones with *tert*-alkyl substituents have been obtained from the reaction between isocyanides and nitroso compounds.[162,163] The mechanism proposed for this reaction [Eq. (3-89)] is the reverse of that suggested for the

$$(3-89)$$

lactam decomposition. Evidence which supports the scheme was derived from capture of the first intermediate by phenyl isocyanate [Eq. (3-90)]. If

$$(3-90)$$

the diaziridinones form diazaoxyallyls these are not captured by TCNE, maleic anhydride, or cyclopentadiene.

Even the trinitrogen equivalents, the diaziridimines, have been isolated and shown to undergo the equivalent of the allene oxide–cyclopropanone rearrangement. In the present case this rearrangement becomes degenerate and can become recognizable only when properly labeled [Eq. (3-91)].[164]

$$
\text{(3-91)}
$$

REFERENCES

1. C. H. de Puy, L. G. Schnack, J. W. Hausser, and W. Wiedemann, *J. Am. Chem. Soc.* **87,** 4006 (1967).
2. W. Kutzelnigg, *Tetrahedron Lett.* p. 4965 (1967).
3. D. T. Clark and G. Smale, *Tetrahedron* **25,** 13 (1969).
4. D. T. Clark and D. R. Armstrong, *Theor. Chim. Acta* **13,** 365 (1969).
4a. D. T. Clark and D. R. Armstrong, *Theor. Chim. Acta* **14,** 370 (1969).
5. M. J. S. Dewar and S. Kirschner, *J. Am. Chem. Soc.* **93,** 4290, 4291 (1971).
5a. M. J. S. Dewar and S. Kirschner, *J. Am. Chem. Soc.* **96,** 5244 (1974).
6. L. Farnell and W. G. Richards, *Chem. Commun.* p. 334 (1973).
7. L. Radom, P. C. Hariharan, J. A. Pople, and P. v. R. Schleyer, *J. Am. Chem. Soc.* **95,** 6531 (1973).
8. P. Merlet, S. D. Peyerimhoff, R. J. Buenker, and S. Shih, *J. Am. Chem. Soc.* **96,** 959 (1974).
9. L. Radom, J. A. Pople, and P. v. R. Schleyer, *J. Am. Chem. Soc.* **95,** 8193 (1973).
10. D. Wendisch, *in* Houben-Weyl "Methoden der Organischer Chemie", Vol. 4, Part 3, pp. 575–664, Thieme, Stuttgart, 1971.
11. C. H. de Puy, L. G. Schnack, and J. W. Hausser, *J. Am. Chem. Soc.* **88,** 3343 (1966).
12. P. v. R. Schleyer, G. W. Van Dine, U. Schöllkopf, and J. Paust, *J. Am. Chem. Soc.* **88,** 2868 (1966).
13. J. D. Roberts and V. C. Chambers, *J. Am. Chem. Soc.* **73,** 5034 (1951).
14. J. W. Hausser and N. J. Pinkowski, *J. Am. Chem. Soc.* **89,** 6981 (1967).
15. P. v. R. Schleyer, W. F. Sliwinski, G. W. Van Dine, U. Schöllkopf, J. Paust, and K. Fellenberger, *J. Am. Chem. Soc.* **94,** 125 (1972).
16. J. Salaün, *J. Org. Chem.* **41,** 1237 (1976).
17. J. W. Hausser and M. J. Grubber, *J. Org. Chem.* **37,** 2648 (1972).
18. J. H. Ong and R. E. Robertson, *Can. J. Chem.* **52,** 2660 (1974).
19. J. A. Landgrebe and J. W. Becker, *J. Org. Chem.* **33,** 1173 (1968).
20. D. C. Duffey, J. P. Minyard, and R. H. Lane, *J. Org. Chem.* **31,** 3865 (1966).
21. W. F. Sliwinski, T. M. Su, and P. v. R. Schleyer, *J. Am. Chem. Soc.* **94,** 133 (1972).
22. J. W. Hausser and J. T. Uchic, *J. Org. Chem.* **37,** 4087 (1972).
23. W. E. Parham and K. S. Yong, *J. Org. Chem.* **33,** 3947 (1968).
24. W. E. Parham and K. S. Yong, *J. Org. Chem.* **35,** 683 (1970).
25. S. F. Cristol, R. M. Seguira, and C. H. de Puy, *J. Am. Chem. Soc.* **87,** 4007 (1965).
26. P. v. R. Schleyer and R. D. Nicholas, *J. Am. Chem. Soc.* **83,** 182 (1961).
27. C. S. Foote, *J. Am. Chem. Soc.* **86,** 1853 (1964).
27a. P. v. R. Schleyer, *J. Am. Chem. Soc.* **86,** 1854, 1856 (1964).
28. W. Kirmse and H. Schätte, *J. Am. Chem. Soc.* **89,** 1284 (1967).
28a. H. C. Brown, C. G. Rao, and M. Ravindranathan, *J. Am. Chem. Soc.* **99,** 7663 (1977).

29. P. v. R. Schleyer, T. M. Su, M. Saunders, and J. C. Rosenfeld, *J. Am. Chem. Soc.* **91**, 5174 (1969).
30. V. Buss, R. Gleiter, and P. v. R. Schleyer, *J. Am. Chem. Soc.* **93**, 3927 (1971).
31. U. Schöllkopf, E. Ruban, P. Tonne, and K. Riedel, *Tetrahedron Lett.* p. 5077 (1970).
32. H. H. Wasserman and M. S. Baird, *Tetrahedron Lett.* p. 1729 (1970).
33. E. Jongejan, W. J. M. van Tilborg, C. H. V. Dusseau, H. Steinberg, and Th. de Boer, *Tetrahedron Lett.* p. 2359 (1972).
34. J. Szmuskovicz, D. J. Duchamp, E. Cerda, and C. G. Chidester, *Tetrahedron Lett.* p. 1309 (1969).
35. J. C. Blazejewski, D. Cantacuzene, and C. Wakselman, *Tetrahedron* **29**, 4233 (1973).
36. D. Cantacuzene and M. Tordeux, *Tetrahedron Lett.* p. 4807 (1971).
37. P. Weyerstahl, G. Blume, and C. Müller, *Tetrahedron Lett.* p. 3869 (1971).
38. U. Schöllkopf, K. Fellenberger, M. Patsch, P. v. R. Schleyer, T. Su, and G. W. van Dine, *Tetrahedron Lett.* p. 3639 (1967).
39. K. Fellenberger, U. Schöllkopf, C. A. Bahn, and P. v. R. Schleyer, *Tetrahedron Lett.* p. 359 (1972).
40. J. J. Tufariello, A. C. Bayer, and J. J. Spadaro, Jr., *Tetrahedron Lett.* p. 363 (1972).
41. E. J. Corey, F. A. Carey, and R. A. Winter, *J. Am. Chem. Soc.* **87**, 935 (1965).
42. D. T. Clark and G. Smale, *Chem. Commun.* p. 868 (1969).
43. C. W. Jefford, A. N. Kabengele, and U. Burger, *Tetrahedron Lett.* p. 4799 (1972).
44. I. Fleming and E. J. Thomas, *Tetrahedron* **28**, 4989 (1972).
45. X. Creary, *J. Am. Chem. Soc.* **98**, 6608 (1976).
46. G. H. Whitham and M. Wright, *Chem. Commun.* p. 294 (1967).
46a. G. H. Whitman and M. Wright, *Chem. Commun.* p. 883 (1971).
47. C. B. Reese and A. Shaw, *Chem. Commun.* p. 1365 (1970).
47a. C. B. Reese and A. Shaw, *J. Chem. Soc. Perkin Trans. 1* p. 2422 (1975).
48. G. M. Blackburn and C. R. M. Ward, *Chem. Commun.* p. 74 (1976).
49. C. B. Reese and A. Shaw, *J. Am. Chem. Soc.* **92**, 2566 (1970).
50. D. Ledli and E. A. Nelson, *Tetrahedron Lett.* p. 1175 (1969).
51. J. T. Groves and K. W. Ma, *Tetrahedron Lett.* p. 909 (1974).
52. H. J. J. Loozen, W. M. M. Robben, T. L. Richter, and H. H. Buck, *J. Org. Chem.* **41**, 384 (1976).
53. W. E. Parham and H. E. Reiff, *J. Am. Chem. Soc.* **77**, 1177 (1955).
54. C. W. Jefford, *Chimia* **24**, 357 (1970).
55. K. A. W. Parry and P. J. Robinson, *J. Chem. Soc. B* p. 49 (1969).
56. R. Fields, R. N. Hazeldine, and D. Peter, *J. Chem. Soc. C* p. 165 (1969).
57. K. A. Holbrook and K. A. W. Parry, *J. Chem. Soc. B* p. 1019 (1970).
58. R. C. DeSelms and C. M. Combs, *J. Org. Chem.* **28**, 2206 (1963).
59. N. P. Neureiter, *J. Org. Chem.* **24**, 2044 (1959).
60. D. C. Duffey, J. P. Minyard, and R. H. Lane, *J. Org. Chem.* **31**, 3865 (1966).
61. M. S. Baird and C. B. Reese, *Tetrahedron Lett.* p. 2117 (1969).
62. H. Wynberg, *J. Org. Chem.* **24**, 264 (1959).
63. J. Sonnenberg and S. Winstein, *J. Org. Chem.* **27**, 748 (1962).
64. E. Vogel, *Angew. Chem. Internl. Edit.* **2**, 1 (1963).
65. E. E. Schweizer and W. E. Parham, *J. Am. Chem. Soc.* **82**, 4085 (1960).
66. T. Ando, H. Yamanaka, and W. Funasaka, *Tetrahedron Lett.* p. 2587 (1967).
67. W. R. Moore, W. R. Moser, and J. E. LaPrade, *J. Org. Chem.* **28**, 2200 (1963).
68. E. Bergmann, *J. Org. Chem.* **28**, 2210 (1963).
69. C. W. Jefford, *Proc. Chem. Soc.* 64 (1963).
69a. C. W. Jefford, S. Mahajan, J. Waslyn, and B. Waegell, *J. Am. Chem. Soc.* **87**, 2183 (1965).

70. C. W. Jefford and R. Medary, *Tetrahedron Lett.* p. 2069 (1966).

70a. C. W. Jefford, E. Huang Yen, and R. Medary, *Tetrahedron Lett.* p. 6317 (1966).

71. M. S. Baird, D. G. Lindsay, and C. B. Reese, *J. Chem. Soc. C* p. 1173 (1969).

72. M. S. Baird and C. B. Reese, *Tetrahedron Lett.* p. 1379 (1967).

73. U. K. Pandit and S. A. G. de Graaf, *Chem. Commun.* p. 659 (1972).

74. F. S. Fawcett, *Chem. Rev.* **47**, 219 (1950).

75. C. B. Quinn and J. R. Wiseman, *J. Am. Chem. Soc.* **95**, 6120 (1973).

76. C. B. Reese and M. R. D. Stebles, *Tetrahedron Lett.* p. 4427 (1972).

77. C. B. Reese and M. R. D. Stebles, *Chem. Commun.* p. 1231 (1972).

78. D. B. Ledlie and J. Knetzer, *Tetrahedron Lett.* p. 5021 (1973).

79. J. T. Groves and K. W. Ma, *Tetrahedron Lett.* p. 909 (1974).

80. P. Warner, J. Fayos, and J. Clardy, *Tetrahedron Lett.* p. 4473 (1973).

81. P. M. Warner, R. C. LaRose, R. F. Palmer, C. Lee, D. O. Ross, and J. C. Clardy, *J. Am. Chem. Soc.* **97**, 5507 (1975).

82. P. Warner, S. Lu, E. Myers, P. de Haven, and R. A. Jacobsen, *Tetrahedron Lett.* p. 4449 (1975).

83. P. Warner and S. Lu, *J. Am. Chem. Soc.* **97**, 2536 (1975).

84. P. Warner and S. Lu, *J. Am. Chem. Soc.* **98**, 6752 (1976).

85. D. B. Ledlie, T. Swan, J. Pite, and L. Bowers, *J. Org. Chem.* **41**, 419 (1976).

86. D. B. Ledlie, W. Barker, and F. Switzer, *Tetrahedron Lett.* p. 607 (1977).

87. G. Olah, G. Liang, D. B. Ledli, and M. G. Costopoulos, *J. Am. Chem. Soc.* **99**, 4196 (1977).

88. D. Ledlie, *J. Org. Chem.* **37**, 1439 (1972).

89. D. B. Ledlie, J. Knetzer, and A. Gitterman, *J. Org. Chem.* **39**, 708 (1974).

90. D. B. Ledlie, and L. Bowers, *J. Org. Chem.* **40**, 792 (1975).

91. W. E. Parham, H. E. Reiff, and P. Schwartzentruber, *J. Am. Chem. Soc.* **78**, 1437 (1956).

92. W. E. Parham and R. R. Twelves, *J. Org. Chem.* **22**, 730 (1957).

93. W. E. Parham and C. D. Wright, *J. Org. Chem.* **22**, 1473 (1957).

94. W. E. Parham, D. A. Bolon, and E. E. Schweitzer, *J. Am. Chem. Soc.* **83**, 603 (1961).

95. G. C. Robinson, *J. Org. Chem.* **29**, 3433 (1964).

96. A. P. TerBorg and A. F. Bickel, *Proc. Chem. Soc.* p. 283 (1958).

97. A. J. Birch, J. M. Brown, and F. Stansfield, *J. Chem. Soc.* p. 5343 (1964).

98. G. M. Iskander and F. Stansfield, *J. Chem. Soc.* p. 1390 (1965).

99. A. J. Birch, G. M. Iskander, B. I. Magboul, and F. Stansfield, *J. Chem. Soc. C.* p. 358 (1967).

100. G. M. Iskander, *J. Chem. Soc. Perkin Trans. 1* p. 2202 (1973).

101. W. E. Parham, R. W. Soeder, and R. M. Dodson, *J. Am. Chem. Soc.* **84**, 1755 (1962).

102. J. D. White and L. G. Wade, Jr., *J. Org. Chem.* **40**, 118 (1975).

103. P. S. Skell and S. R. Sandler, *J. Am. Chem. Soc.* **80**, 2024 (1958).

104. S. R. Sandler, *J. Org. Chem.* **32**, 3876 (1967).

105. L. Skattebøl, *J. Org. Chem.* **31**, 1554 (1966).

106. S. M. McElvain and P. L. Weyna, *J. Am. Chem. Soc.* **81**, 2579 (1959).

107. Le Van Chan and M. Schlosser, *Synthesis* p. 115 (1974).

108. M. Schlosser and L. V. Chan, *Helv. Chim. Acta* **58**, 2595 (1975).

108a. M. Schlosser and Y. Bessiere, *Helv. Chim. Acta* **60**, 590 (1977).

109. W. J. M. van Tilberg, G. Dooyewaard, H. Steinberg, and Th. J. de Boer, *Tetrahedron Lett.* p. 1677 (1972).

110. J. Meinwald, J. W. Wheeler, A. A. Nimetz, and J. S. Liu, *J. Org. Chem.* **30**, 1038 (1965).

111. H. Monti, G. Leandri, and M. Bertrand, *C. R. Acad. Sci. Paris, Ser. C* 274 (1972).

111a. G. Leandri, H. Monti, and M. Bertrand, *Tetrahedron* **30**, 289 (1974).

112. G. Leandri, H. Monti, and M. Bertrand, *Bull. Soc. Chim. Fr.* p. 1919 (1974).

113. M. Vidal, A. Dussauge, and M. Vincens, *Tetrahedron Lett.* p. 313 (1977).

114. R. M. Babb and P. D. Gardner, *Tetrahedron Lett.* p. 6197 (1968).

115. L. Ghosez, G. Slinckx, M. Glineur, P. Hoet, and P. Laroche, *Tetrahedron Lett.* p. 2773 (1967).

116. C. W. Jefford and W. Wojnarowski, *Tetrahedron Lett.* p. 119 (1968).

116a. G. Klein and W. Kraus, *Tetrahedron* **33**, 3121 (1977).

117. C. B. Reese and A. Shaw, *Chem. Commun.* p. 271 (1972).

118. M. S. Baird and C. B. Reese, *J. Chem. Soc. C* p. 1803 (1969).

119. M. S. Baird and C. B. Reese, *J. Chem. Soc. C* p. 1808 (1969).

120. M. S. Baird and C. B. Reese, *Chem. Commun.* p. 1644 (1970).

120a. H. J. J. Loozen, J. W. de Haan, and M. M. Buck, *J. Org. Chem.* **42**, 418 (1977).

121. W. E. Parham and R. J. Sperley, *J. Org. Chem.* **32**, 924 (1967).

122. J. Casanova and B. Waegell, *Bull. Soc. Chim. Fr.* p. 2669 (1972).

123. R. D. Bach, U. Mazur, R. N. Brummel, and L.-H. Liu, *J. Am. Chem. Soc.* **93**, 7120 (1971).

124. M. S. Baird and C. B. Reese, *Tetrahedron Lett.* p. 4637 (1971).

125. L. W. Boyle and J. K. Sutherland, *Tetrahedron Lett.* p. 839 (1973).

126. D. Duffin and J. K. Sutherland, *Chem. Commun.* p. 626 (1970).

127. W. Heggie and J. K. Sutherland, *Chem. Commun.* p. 957 (1972).

128. G. H. Whitham and M. Zaidlewicz, *J. Chem. Soc. Perkin Trans. 1* p. 1509 (1972).

129. D. Seebach and H. Newmann, *Chem. Ber.* **107**, 847 (1974).

130. D. Seebach, M. Braun, and N. Du Preez, *Tetrahedron Lett.* p. 3509 (1973).

131. A. S. Kende, *Org. Reactions* **11**, 261 (1960).

132. D. A. Semenow, E. F. Cox, and J. D. Roberts, *J. Am. Chem. Soc.* **78**, 3221 (1956).

133. N. J. Turro, W. B. Hammond, and P. A. Leermakers, *J. Am. Chem. Soc.* **87**, 2774 (1965).

134. H. C. Richey, J. M. Richey, and D. C. Clagett, *J. Am. Chem. Soc.* **86**, 3906 (1964).

135. I. Haller and R. Srinivasan, *J. Am. Chem. Soc.* **87**, 1144 (1965).

136. R. C. Cookson, M. J. Nye, and G. Subrahmayan, *Proc. Chem. Soc.* p. 144 (1964).

137. W. B. Hammond and N. J. Turro, *J. Am. Chem. Soc.* **88**, 2880 (1966).

138. N. J. Turro and W. B. Hammond, *J. Am. Chem. Soc.* **88**, 3672 (1966); S. E. Schaafoma, H. Steinberg, and T. J. De Boer, *Rec. Trav. Chim.* **85**, 1170 (1966).

139. N. J. Turro, *Accts. Chem. Res.* **2**, 25 (1969).

139a. N. J. Turro, *Topics Curr. Chem.* p. 47 (1974).

140. R. Hoffmann, *J. Am. Chem. Soc.* **90**, 1475 (1968).

141. J. F. Olsen, S. Kang, and L. Burnello, *J. Mol. Struct.* **9**, 305 (1971).

142. A. Liberles, A. Greenberg, and A. Lesk, *J. Am. Chem. Soc.* **94**, 8685 (1972).

143. A. Liberles, S. Kang, and A. Greenberg, *J. Org. Chem.* **38**, 1922 (1973).

144. N. Bodor, M. J. S. Dewar, A. Harget, and E. Haselbach, *J. Am. Chem. Soc.* **92**, 3854 (1970).

145. M. E. Zandler, C. E. Choc, and C. K. Johnson, *J. Am. Chem. Soc.* **96**, 2317 (1974).

145a. N. J. Turro, S. S. Edelson, J. R. Williams, T. R. Darling, and W. B. Hammond, *J. Am. Am. Chem. Soc.* **91**, 2283 (1969).

146. J. F. Pazos and F. O. Greene, *J. Am. Chem. Soc.* **89**, 1030 (1967).

147. D. B. Sclove, J. F. Pazos, R. L. Camp, and F. D. Greene, *J. Am. Chem. Soc.* **92**, 7488 (1970).

148. J. K. Crandall and W. H. Machleder, *Tetrahedron Lett.* p. 6037 (1966).

149. J. K. Crandall, W. H. Machleder, and S. A. Sojka, *J. Org. Chem.* **38**, 1149 (1973).

150. J. K. Crandall, W. H. Machleder, and M. J. Thomas, *J. Am. Chem. Soc.* **90**, 7346 (1968).

151. J. K. Crandall and W. H. Machleder, *J. Am. Chem. Soc.* **90**, 7347 (1968).

152. R. L. Camp and F. D. Greene, *J. Am. Chem. Soc.* **90**, 7349 (1968).

153. J. K. Crandall and W. W. Conover, *J. Chem. Soc. Chem. Commun.* p. 340 (1973).
154. T. H. Chan, M. P. Li, W. Mychajlowkÿ, and D. N. Harpp, *Tetrahedron Lett.* p. 3511 (1974).
155. R. C. Cookson and M. J. Nye, *Proc. Chem. Soc.* p. 129 (1963).
155a. R. C. Cookson, M. J. Nye, and G. Subrahmanyan, *J. Chem. Soc. C.* p. 473 (1967).
156. R. Noyori, S. Makino, and H. Takaya, *J. Am. Chem. Soc.* **93,** 1272 (1971).
157. R. Noyori, K. Yokoyama, and Y. Hayakama, *J. Am. Chem. Soc.* **95,** 2722 (1973).
158. L. W. Fort, *J. Am. Chem. Soc.* **84,** 4979 (1962).
159. J. A. Deyrup and R. B. Greenwald, *Tetrahedron Lett.* p. 5091 (1966).
160. J. C. Sheehan and J. H. Beeson, *J. Am. Chem. Soc.* **89,** 362, 366 (1967).
161. I. Lengyel and J. C. Sheehan, *Angew Chem. Internl. Edit.* **80,** 449 (1969).
161a. I. Lengyel and J. C. Sheehan, *Chem. Ber.* p. 103, 539 (1970).
162. F. D. Greene, J. C. Stowell, and W. R. Bergmark, *J. Org. Chem.* **34,** 2254 (1969).
163. F. D. Greene and J. F. Pazos, *J. Org. Chem.* **34,** 2269 (1969).
164. H. Quast and E. Schmitt, *Angew. Chem. Int. Ed. Engl.* **8,** 449 (1969).
165. P. Warner, S.-L. Lu, and S. C. Chen, *Tetrahedron Lett.* p. 1947 (1978).
166. Y. Bessiere, D. N.-H. Savary, and M. Schlosser, *Helv. Chim. Acta* **60,** 1739 (1977).
167. W. Kraus, H. Patzelt, and G. Sawitzki, *Tetrahedron Lett.* p. 445 (1978).
168. T. H. Chan and B. S. Ong, *J. Org. Chem.* **43,** 2994 (1978).
169. E. Block, R. E. Penn, M. D. Ennis, T. A. Owens, and S.-L. Yu, *J. Am. Chem. Soc.* **100,** 7436 (1978).
170. W. E. Parham and J. K. Rinehart, *J. Am. Chem. Soc.* **89,** 5668 (1967).
171. W. E. Parham, D. R. Johnson, C. T. Hughes, M. Meilahn, and J. K. Rinehart, *J. Org. Chem.* **35,** 1048 (1970).
172. P. Grice and C. B. Reese, *Tetrahedron Lett.* p. 2563 (1979).

CHAPTER 4
FOUR ELECTRON–THREE ATOM
SYSTEMS

Classical four π electron electrocyclic processes occurring within a three atom framework, namely, interconversion of cyclopropyl and allyl anions, raise no questions about the number of electrons expected to participate in the reaction. However, ring opening of cyclopropylidene for example might proceed to give an allyl cation with a negatively charged central carbon with only two electrons participating in the electrocyclic reaction, or on the other hand reaction might lead directly to an allene with four electrons involved. For classification purposes only, the number of electrons is considered to be equal to those used in the overall reaction. Thus, for the cyclopropylidene (presumed reactant) to allene conversion four electrons are utilized; hence the reaction is classed as a four electron reaction.

I. THE ALLENE–CYCLOPROPYL CARBENE ELECTROCYCLIZATION

Energy considerations immediately dictate that this electrocyclic process will be observable only in the retro direction. Since its discovery by Doering and Laflamme[1] in 1958, the reaction has become a valuable synthesis of allenes, and is also of mechanistic interest. The main problem which exists in the experimental investigation is the nature of the cyclopropyl entity generated. Whether a carbenoid, a metal ion complexed entity, or a free carbene is the actual reactant has not always been completely answered.

Carbenes can exist in both singlet and triplet states,[2] and while the singlet is the state obtained directly from a ground state precursor by non-photochemical means, the triplet is the lower energy state. Borden[3] has calculated that the triplet cyclopropyl carbene can go only to the planar form of the triplet allene, and this reaction is non-allowed in any concerted process whether conrotatory, disrotatory or, nonrotatory. The singlet carbene presents a more interesting case. A MINDO/2 calculation[4] gave 50 kcal/mol as the activation energy for the ring opening to give the allene. The authors note that this could be overestimated by ca. 12 kcal/mol. Even the 38 kcal/mol is considerably too high if the entity which reacts in the experiments noted below is a singlet carbene.

The question which orbital symmetry usually poses, i.e., conrotation vs. disrotation really becomes irrelevant in this case. Assume, for example, that reaction occurs via ring opening to a linear system followed there by rotation [Eq. (4-1)]. A 45° conrotatory motion of each terminal carbon leads to an allene as shown. Conversely a disrotatory motion of any number of degrees, provided only that both terminal carbons move equally, is equivalent to a rotation of the entire molecule, since the center carbon provides no frame to which rotations can be referred. As a result, it becomes most convenient to consider the stereochemical results as a single methylene rotation

$$(4\text{-}1)$$

which can be either clockwise (CW) or counter clockwise (CCW). Hoffman has noted that orbital symmetry can provide no prediction for the direction which would be preferred.[5] It has been reported[6] that Borden has concluded that the preferred process is a single methylene rotation with that group which rotates being the one best fitted to furnish the pair of electrons to the carbene center.

Jones and Brinker[212] have reviewed the researches on cyclopropylidenes and they define conrotatory and disrotatory terms with respect to a planar allene. A related theoretical study[213] proposes a mixed process for ring opening, i.e., disrotatory initially to give a planar allene or if steric restraints pertain, the rotation of that group ceases and the other group rotates in a reverse direction to give the normal allene.

For a single methylene rotation, the relevant processes are illustrated in Scheme 1. It is irrelevant which group rotates and, in the absence of optically

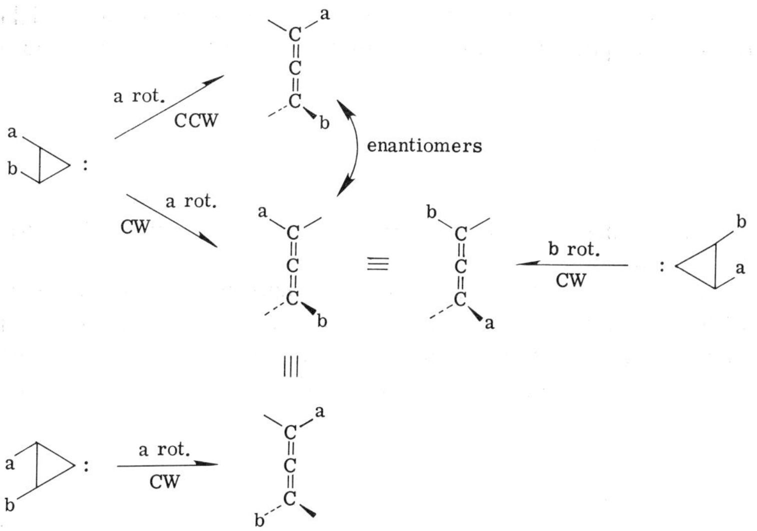

Scheme 4-1.

active materials, it is impossible to ascertain the direction of rotation. The lack of symmetry control of the rotational process implies that other more subtle features must supply any stereoselectivity realized during ring opening.

The prediction that a triplet carbene has no allowed route to ring opening has received experimental verification from the work of Skell on reactions of carbon atoms with olefins. For example, reaction of $C(^3P)$ with cis-2-butene leads to formation of spiropentanes, but no allenes [Eq. (4-2)].[7] On

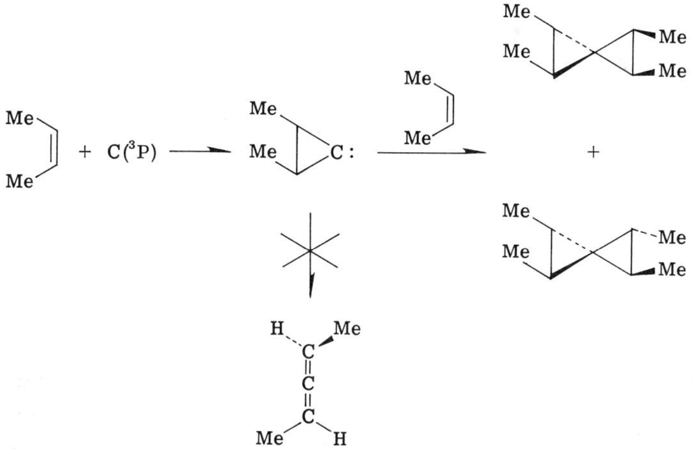

$$\text{(4-2)}$$

the other hand, reaction with $C(^1S)$ permits formation of dimethylallene [Eq. (4-3)].[8] This latter reaction shows that a free carbene can undergo con-

$$\tag{4-3}$$

version to an allene, but the excess thermal energy which this carbene possesses is uncertain.

Elegant work by W. M. Jones and his students has shown that for trans disubstituted cyclopropyl carbenes formation of the allene is controlled by steric factors. Initially Jones[9] found that attempted formation of a diazo-cyclopropane [Eq. (4-4)] gave an allene and that the intervening carbene

$$\tag{4-4}$$

(or carbenoid) could be trapped with added olefin. The trapping was later shown to be stereoselective, leading to the presumption that the carbene was a singlet.[10,11] The intermediacy of the diphenyldiazocyclopropane was also confirmed since in the presence of diethyl fumarate a pyrazoline was formed.[11] Thus armed with evidence that a carbene-like intermediate was present, Jones et al.[12] prepared and rearranged an optically active compound [Eq. (4-5)], and the allene retained a considerable measure of optical activity.

$$\tag{4-5}$$

The absolute configuration of **46** was shown[13] to be 2S,3S, and the di-phenylallene has been assigned the S configuration.[14] Thus, the dominant rotation here is clockwise, or the phenyl rotates outward and away from the cleaving bond. This mode of reaction has also been shown to hold for a *trans*-dimethyl-cyclopropylidene.[5,15] The stereoselectivity in this case was high.[16]

Another example of the same behavior was uncovered by Kleveland and Skattebøl,[17,17a] but using a different route to the cyclopropylidene moiety [Eq. (4-6)]. In both cases these investigators found the larger group moved

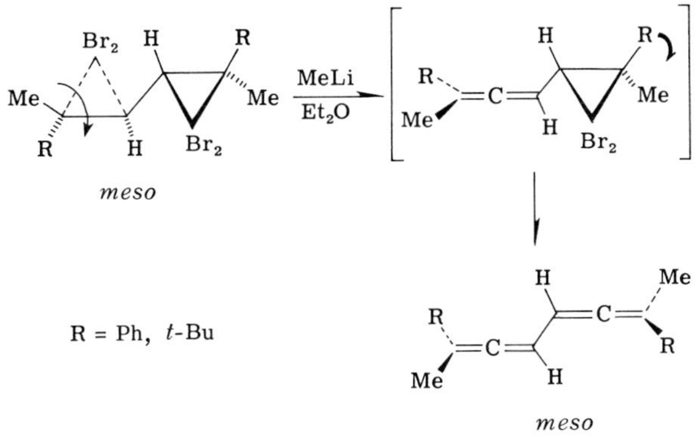

$$R = Ph, \ t\text{-Bu}$$

(4-6)

meso

outward, and within the limits of experimental error the reaction was stereospecific. Moore and his collaborators have found a case where rotation in the opposite sense was enforced with a high degree of selectivity. Addition of dibromocarbene to $(+)S$-*trans*-cyclooctene[18] gave an adduct which was treated with methyllithium to give an optically active 1,2-cyclononadiene [Eq. (4-7)],[19] which has the $(-)S$ configuration.[20] The optical purity of the

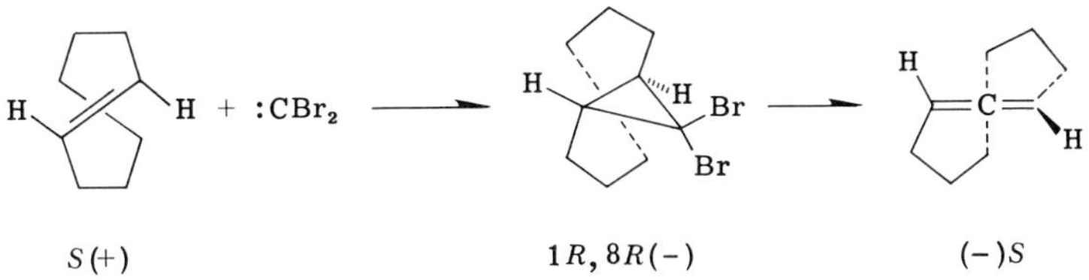

(4-7)

product formed at low temperature was 95–98% from an optically pure cyclooctene.[16] In this example the strain in the bicyclic molecule would be increased by an outward rotation of the methylene group, hence it moves inward to relieve the strain. All of the above examples show that rotation accompanies bond cleavage, as would be expected of a single group rotation, since this permits partial formation of the second double bond during the rotation process.

The two routes to formation of an allene may or may not proceed via a free singlet carbene, but both lead to the same optically active carbene.[16] Starting with optically active **47**, Moore and Bach converted it directly to the allene with methyllithium, and indirectly via the nitrosourea as well. Both gave the same active 1,2-cyclononadiene with only a slight difference in stereoselectivity. If some moiety other than a singlet carbene is acting as an intermediate, it behaves very differently from the ion produced in the N-chloroaziridine solvolyses (see Section II). Observation of a CIDNP effect during the generation of dimethylallene from 2,2-dimethyl-1,1-dichloro-cyclopropane,[23] places an upper limit on the lifetime of 1-lithio-1-chloro-2,2-dimethylcyclopropane of a few seconds at 40°. This does not preclude the action of this entity as an intermediate. Some sort of metal coordinated intermediate is suggested by the asymmetric induction produced when reaction is carried out in the presence of (−)sparteine.[24] Early reports[11,21] that two competitive routes to the allene from a diazocyclopropane existed were retracted[22] when the basis for the report was shown to be a solvent effect.

Unlike the ring opening of an optically active trans disubstituted cyclo-propylidene, any stereoselectivity produced during opening of a cis disubstituted cyclopropylidene must be controlled by more subtle features than pure steric bulk. Since outward rotation of either group is expected, and this must lead to enantiomeric allenes depending on which group rotates [Eq. (4-8)], stereoselectivity depends entirely on which group rotates. Con-

$$(4-8)$$

trol could be exerted by ponderal effects, or (because the rotating group supplies electrons)[6] by electronic effects. Jones and Krause[6] have investigated such a case and have suggested that electronic effects are more important (Scheme 2). In general, however, the selectivity appears to be quite low compared to the trans disubstituted examples. The results in Scheme 2 show that the same group, (tolyl in both cases, or phenyl in one and bromophenyl in the other) rotates preferentially in both examples. Ponderal effects would predict that the phenyl would be more likely to rotate than the tolyl in one case, but the tolyl over the bromophenyl in the other. Thus, ponderal effects are subordinated to electronic.

Trapping of a cyclopropylidene by olefin insertion[9,10,11] gave evidence

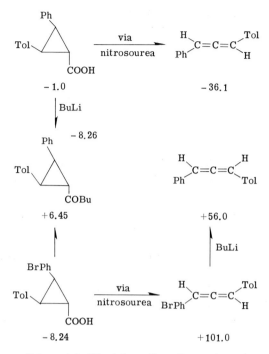

Scheme 4.2 (Absolute configurations unknown).

for the intermediacy of a non-ring opened intermediate in the allene synthesis. Competition between allene formation and routes leading to other products is subject to structural features which may slow or inhibit the allene route. Tetrasubstituted cyclopropylidenes, where no single group rotation free from steric restraints exists, generally fail to give allenes, carbon–hydrogen insertion leading to a bicyclobutane being preferred [Eq. (4-9)].[25,26,27] A reported synthesis of tetramethylallene[29] proved erron-

(4-9)

eous,[30] although traces of that allene are apparently formed.[26] It is interesting that with two aryl substituents at C_2 some allene can be formed [Eq. (4-10)].[28] This was tentatively attributed to stabilization of the electron

$$
\underset{\substack{\text{Br}_2}}{\overset{\substack{\text{Ph}\\\text{Ph}}}{\bigvee}}\overset{\substack{\text{Me}\\\text{Me}}}{}\quad\xrightarrow[\substack{\text{Et}_2\text{O}\\0°}]{\text{MeLi}}\quad\overset{\substack{\text{Me}}}{\underset{\substack{\text{Ph}}}{\triangle}}\!\!\text{Ph}\quad+\quad\underset{\text{Ph}}{\overset{\text{Ph}}{>}}\text{C}=\text{C}=\text{C}\underset{\text{Me}}{\overset{\text{Me}}{<}}\quad(4\text{-}10)
$$

deficiency which develops at C_2 during slow rotation of the methyl-bearing carbon. In at least one case a less substituted cyclopropylidene was trapped by carbon–hydrogen insertion [Eq. (4-11)], and the competition was shown

$$
\underset{\text{Me}}{\overset{\text{EtO}-\text{H}_2\text{C}}{>}}\!\!\underset{\text{Br}}{\overset{\text{Br}}{\triangle}}\quad\xrightarrow{\text{MeLi}}\quad\text{(bicyclic structure with O)}\quad+\quad\underset{\text{Me}}{\overset{\text{EtOH}_2\text{C}}{>}}\text{C}=\text{C}=\text{CH}_2\quad(4\text{-}11)
$$

to be temperature dependent with more allene being formed at higher temperatures.[31]

Properly situated double bonds may act as internal traps for the cyclopropylidene. For example 2-vinyl-1,1-dibromocyclopropane reacts with methyllithium to give cyclopentadiene as the main product [Eq. (4-12)].[32]

$$
\underset{\substack{\text{Br}\quad\text{Br}}}{\text{(vinyl dibromocyclopropane)}}\quad\xrightarrow{\text{MeLi}}\quad\underset{86\%}{\text{(cyclopentadiene)}}\quad+\quad\underset{14\%}{\text{(allene)}}\quad(4\text{-}12)
$$

The product composition is subject to substituent effects, a particularly striking example being illustrated by Eq. (13).[32] A reasonable mechanism

$$
\underset{\substack{\text{Br}_2}}{\overset{\substack{\text{Me}\\\text{Me}}}{>}}\!\!\overset{\substack{\text{Me}\\\text{Me}}}{\text{(structure)}}\!\!\underset{\text{Br}_2}{}\quad\xrightarrow{\text{MeLi}}\quad\underset{\text{CH}=\text{C}=\text{CMe}_2}{\text{Me}_2\text{C}=\text{C}=\text{CH}}\tag{}
$$

$$(4\text{-}13)$$

$$
\underset{\substack{\text{Br}_2}}{\overset{\substack{\text{Me}}}{\text{(structure)}}}\!\!\overset{\text{Me}}{}\!\!\underset{\text{Br}_2}{}\quad\xrightarrow{\text{MeLi}}\quad\underset{\text{Me}}{\overset{\text{Me}}{\text{(cyclopentadiene)}}}
$$

for the cyclopentadiene formation is shown in Eq. (4-14). The position of

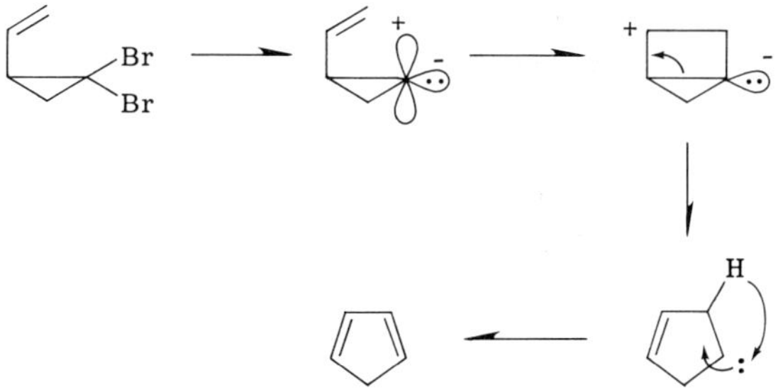

(4-14)

the double bond with respect to the cyclopropylidene carbon exerts a strong influence on product composition.[33] Results of varying the chain length between these is illustrated in Table 4-1.[33] A similar type of trapping can occur with a double bond in a bicyclic system [Eq. (4-15) and (4-16)].[32,34]

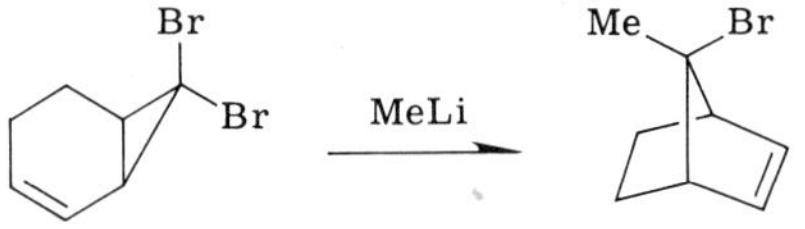

(4-15)

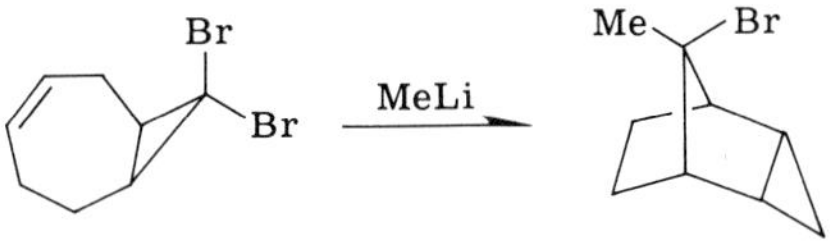

(4-16)

TABLE 4-1

Variation of Trapping of Cyclopropylidenes by Internal Double Bonds as a Function of Chain Length between the Groups

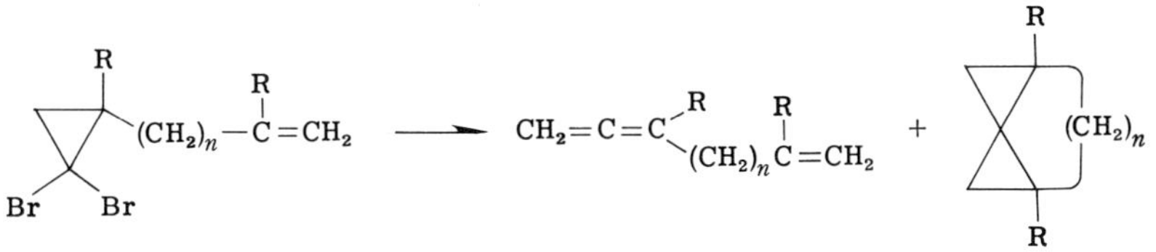

$$CH_2=C=C\underset{(CH_2)_n}{\overset{R}{C}}=CH_2 \quad + $$

R	n	Temperature (°)		
H	1	0	100	0
H	2	−78	52	48
		0	84	16
Me	2	0	38	62
H	3	0	90	10
H	4	0	100	0

The basic mechanism of Eq. (14) has been advanced to explain these products,[34] but the stereospecificity of the insertion of the carbene into the
carbon-bromine bond of methyl bromide rests on tenuous ground.

Dibromides of bicyclo[5.1.0]octane and bicyclo[4.1.0]heptane, which
would give rise to badly strained allenes, undergo carbon–hydrogen inser-
tions or other reactions. With **48** both insertion products and a dimer have
been obtained [Eq. (4-17)],[35,36,37] The intermediate cyclopropylidene can

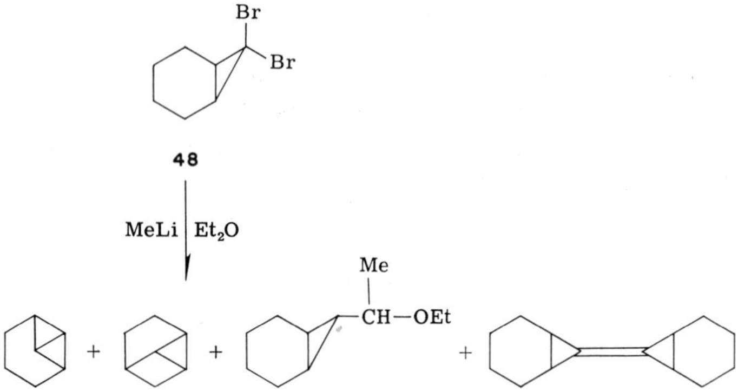

(4-17)

be trapped with olefins,[35,36] but apparently the presence of lithium iodide
is important for the formation of the carbene.[37] If the methyllithium used is
formed from methyl chloride rather than from methyl iodide, a different reaction ensues [Eq. (4-18)]. No products related to the allene are found. With

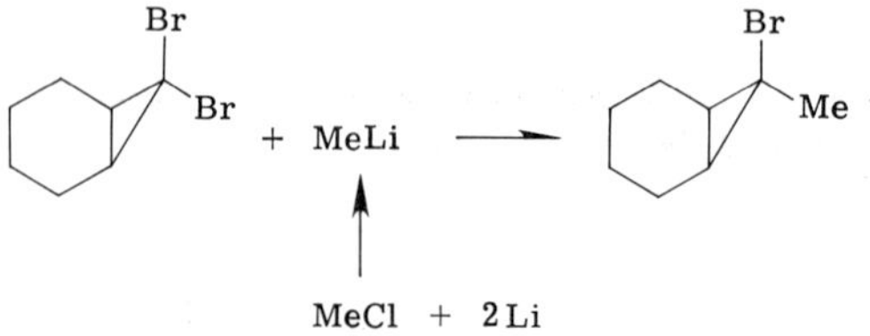

(4-18)

49 competition between allene formation and insertion occurs [Eq. (4-19)].[38]

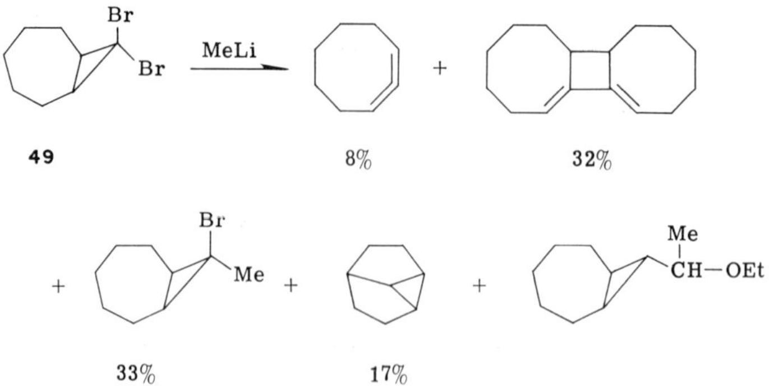

(4-19)

Again the cyclopropylidene can be trapped by added olefins. Presence of a

2-ene in the bicyclo[6.1.0]nonane system causes diversion from normal allene formation to internal carbon–hydrogen insertion [Eqs. (4-20) and (4-21)].[39,]

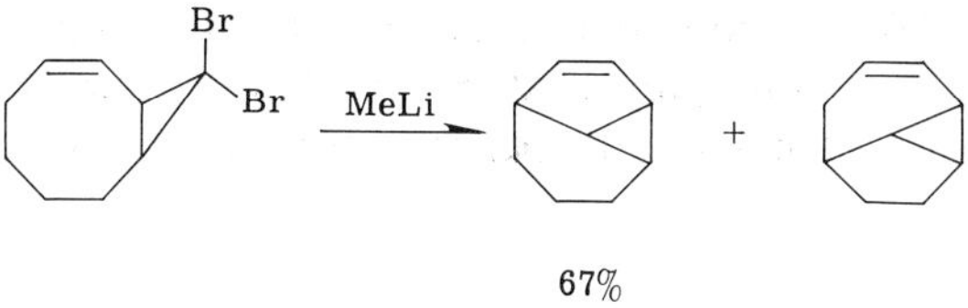

$$\text{(4-20)}$$

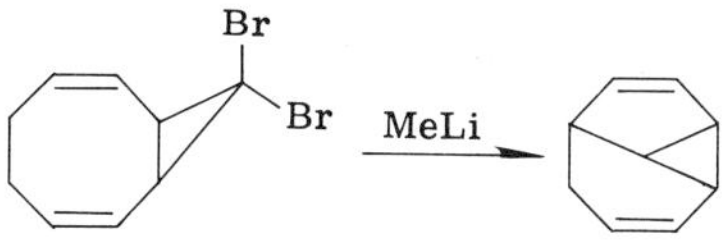

67%

$$\text{(4-21)}$$

[40,40a] With this experience in mind, one finds the behavior of **50** most surprising. Products which appear to have been derived from an unstable 1,2-cyclohexadiene are formed [Eqs. (4-22) and (4-23)].[41,42] In view of the ease

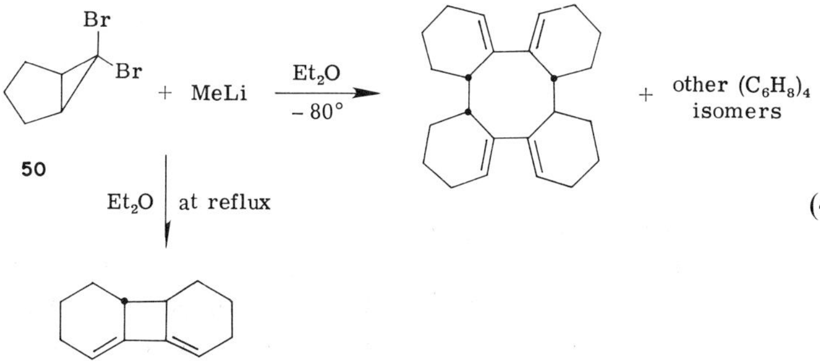

$$\text{(4-22)}$$

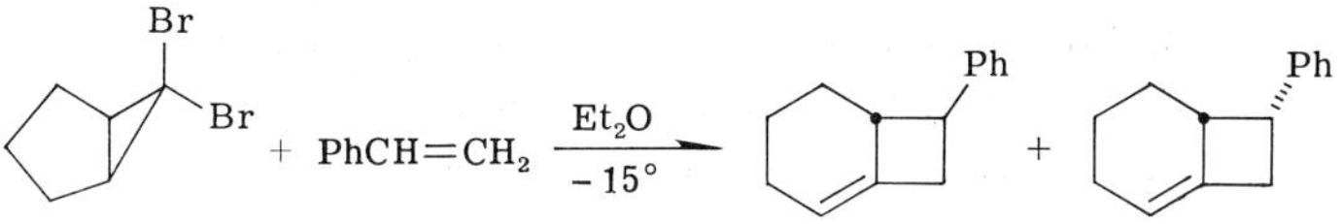

$$\text{(4-23)}$$

with which this dibromide undergoes thermal ring opening via a two-electron process[43] and the general behavior of N-chloroaziridenes (see Section II), it appears likely that this proceeds via the route of Eq. (4-24).[41] It would seem to be of interest to try to generate the same intermediate via the nitrosourea.

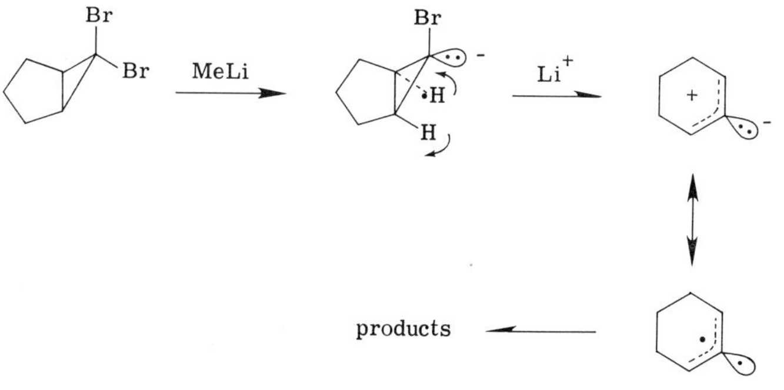

$$\text{(4-24)}$$

products

That 9,9-dibromobicyclo[6.1.0]nonane is methylated by treatment with butyllithium at $-95°$ followed by methyl iodide, while similar treatment of **51** gave only 1,2,6-cyclononatriene prompted the interesting suggestion that a homoconjugate anion is involved [Eq. (4-25)].[44] Since six electrons are involved, the ring-opening should occur in disrotatory fashion.

$$(4\text{-}25)$$

This ring opening of dihalocyclopropanes constitutes a very effective synthesis for a wide variety of allenes under conditions which do not give rise to acetylenic by-products. Some of the allenes which have been prepared in this way are listed in Table 4-2. Perhaps the most convenient route to pure *cis*-cyclononene proceeds from cyclooctene via 1,2-cyclononadiene and its reduction by sodium and ammonia.[48] A route to converting aldehydes or ketones to α,β-unsaturated carbonyl compounds with two added carbons has been developed [Eq. (4-26)].[52] Spectral evidence for the intervention of

$$(4\text{-}26)$$

two unstable intermediates was reported, and the procedure was applied to synthesis of *dl*-muscone.

A valuable extention has been developed for the preparation of cumulenes. Skattebøl[53] has used the procedure of Eq. (4-27) to prepare a triene, a tetraene and a pentaene (too unstable to be isolated). Several trisubstituted 1,2,3-propatrienes have been prepared similarly.[54] Moore and Ozretich[55] applied this reaction successfully to the synthesis of 1,2,3-cyclodecatriene. A

TABLE 4-2

Allenes Prepared by Ring Opening of Dihalocyclopropanes

Allene	% yield	Reference
Phenylallene		30
1,2-Tridecadiene		45
2,3-Heptadiene	88	46
1,2,6-Heptatriene		30
2,3-Undecadiene	81	46
2,3-Tridecadiene	68	46
2-Methyl-2,3-pentadiene		30
1,1-Diphenyl-1,2-butadiene		28
3-Methyl-1,2-pentadiene-4-ol	38	47
4-Methyl-2,3-pentadien-1-ol	43	47
3,4-Dimethyl-1,2-pentadien-4-ol	30	47
4-Methyl-2,3-hexadien-1-ol	34	47
5-Methyl-3,4-hexadien-2-ol	44	47
2,5-Dimethyl-3,4-hexadien-2-ol	65	47
2,5-Dimethyl-2,3-heptadien-5-ol	33	47
1,2-Cyclononadiene	81	46
1,2,5-Cyclononatriene		40
1,2,6-Cyclononatriene	70	49
1,2,5,7-Cyclononatetraene	(unstable above 0°)	40
1,2-Cyclononadien-5-ol	74	50
1,2-Cyclodecadiene	78	46
1,2-Cyclodecadien-4-ol		51
1,2,5,6-Cyclodecatetraene		40
1,2,6,7-Cyclodecatetraene		30, 40
1,2-Cycloundecadiene	89	46
1,2-Cycloundecadien-4-ol	58	47
1,2,9,10-Cyclohexadecatetraene		30
Bicyclo[7.1.0]deca-4,5-diene		44
1,2-Bisallenylidenecyclobutene		214
2-Allenylidene-3-methyldibenzo-bicyclo[2.2.2]octanol-3		215

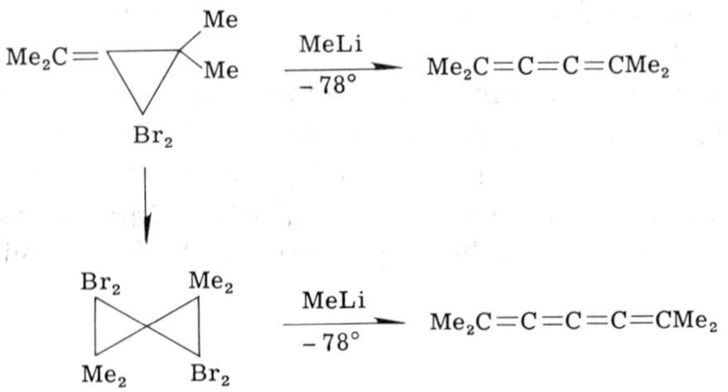

(4-27)

convenient preparation of 2,5-dimethyl-2,3,4-hexatriene from dimethyl-ketene dimer [Eq. (4-28)] was discovered independently by two groups.[56,57]

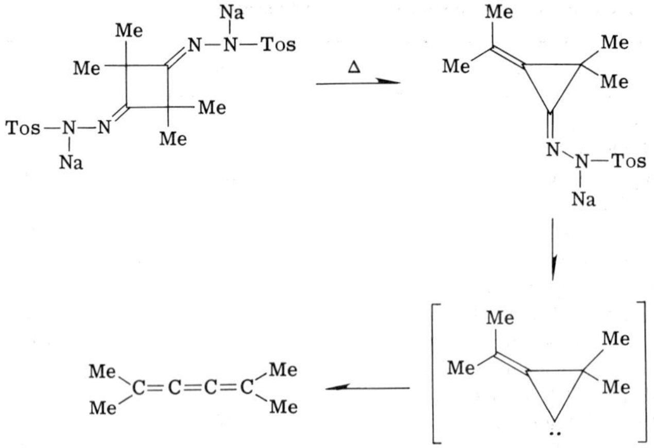

(4-28)

As part of some studies aimed at the preparation of 1,2,4,6,8,10-cycloun-decahexaene several 1,2-cycloundecadienes were prepared.[58] An unexpect-edly facile formation of an allenic ansa compound was uncovered [Eq. (4-29)].

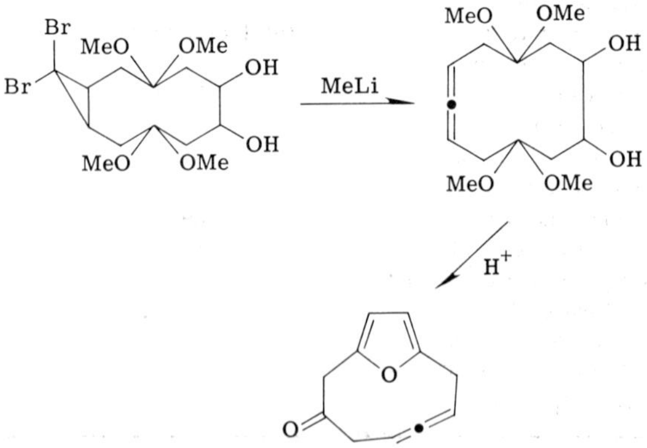

(4-29)

Formation of a symmetrical bis-allene [Eq. (4-30)] led to both meso and *dl*

$$(4\text{-}30)$$

meso   dl
67%   33%

forms. The ratio of the isomers produced was independent of whether the reactant was the cis or the trans isomer. Identity of the two stereoisomeric forms was established by carrying out an asymmetric synthesis in the presence of ($-$)sparteine and separating the active material from the inactive meso form.[59,59a]

II. THE AZIRIDINYL CATION

The aziridinyl cation is isoelectronic with the cyclopropylidene moiety, and as such it might be expected to undergo a 4π electron single methylene rotation to give an azaallenium ion [Eq. (4-31)]. The reaction does appear

$$(4\text{-}31)$$

to proceed in that fashion (see below), but the product ion is unstable and in the presence of water is immediately hydrolyzed to ketones and ammonium ion. Gassman and his students[60,61 61a] have provided the only extensive study of the reaction, but their results show that the reaction is more complex than the simple considerations above would lead one to expect.

A theoretical study of the ring opening of the aziridinyl cation by CNDO/2 has been made by Weiss.[62] The calculations covered the opening to a planar bent ion via a disrotatory process involving 2π electrons [Eq. (4-32)] and via a conrotatory process involving 4π electrons [Eq. (4-33)]. Since the sp^2

$$(4\text{-}32)$$

$$(4\text{-}33)$$

TABLE 4-3

Relative Rates of Solvolysis of N-Chloroaziridines

Compound	Relative rate	Compound	Relative rate
(aziridine) N—Cl	1.0	Me / H \ / Me N—Cl	1490
Me aziridine Cl N	15.	Me / Me aziridine Cl N	1860
H / Me aziridine Cl N	210.	Me / Me aziridine Cl N	155000

orbital of nitrogen is of lower energy than the p orbital of nitrogen the ground state for the aziridinyl cation places the unshared pair in the sp^2 orbital. Consequently the 4π electron conrotatory process corresponds to an excited state of the azaallenium ion. The 2π disrotatory reaction was found to be an allowed reaction whereas the conrotatory 4π process showed all of the characteristic behavior of a forbidden reaction. The calculations showed that the nonplanar linear azaallenium ion is lower in energy than the planar bent cation. However no calculation of the direct conversion of the aziridinyl cation to the azaallenium ion via a single rotation appears to have been made.

Solvolysis of N-chloroaziridines occurs at a reasonable rate in methanol or water at 60°. Under these conditions some of the N-chloroaziridines are configurationally stable enough to permit determination of their solvolysis rates.[60,61,61a] The relative rates for a series of substituted examples are listed in Table 4-3. The results clearly are not compatible with a single methylene rotation process, since for such a process both monomethyl compounds would react at essentially equivalent rates. Obviously a conrotatory process is eliminated by the relative rates noted for the cis- and trans-2,3-dimethyl derivatives. The clear relation between the rates observed for these N-chloro-aziridines and the cyclopropyl halides and tosylates (see Table 3-2) gives strong support to the route shown in Eq. (4-34). A rational explanation of

$$\text{(Y, X) aziridine Cl N} \longrightarrow \text{y} \diagup \atop{x} \diagdown +\!:\!N \longrightarrow \text{y} \diagdown \atop{} C\!=\!\overset{+}{N}\!=\!C \diagup \atop{} \text{x} \qquad (4\text{-}34)$$

this behavior can be based on the known pattern for the solvolysis of cyclo-propyl halides. Thus as the chloride group leaves bond cleavage and rotation occur simultaneously. In essence no low energy route to the free aziridinyl cation exists. Presumably if that free ion were formed it would behave in the same way the cyclopropylidene group does.

The only example of an aziridinyl cation in a bicyclic system is that shown in Eq. (4-35).[62] Presumably the chloride must be endo though the stereo-

$$(4\text{-}35)$$

chemistry was not specified. It is, of course, possible that the nitrogen inver-sion proceeds fast enough in this case that reaction can go via the endo chloride whatever the original position of the chloride.

III. THE CYCLOPROPENE–VINYLCARBENE REACTION

Despite the far reaching consequences of the conservation of orbital sym-metry theory, there exists quite a group of electrocyclic reactions such as the cyclopropylideneallene process where the theory provides very little informa-tion. Another case of this type is the very interesting cyclopropene–vinyl-carbene interconversion [Eq. (4-36)]. According to the Walsh[64] rules

$$(4\text{-}36)$$

vinylcarbene should be a planar moiety, hence its geometry should bear a close resemblance to that of allyl carbonium ion. Ring opening thus becomes a single methylene rotation, and the terms conrotatory and disrotatory lose significance. If there exists a preferred direction for the rotation, it would appear that steric effects might provide the controlling feature.

It is rather surprising that the first theoretical study of the cyclopropene–vinylcarbene reaction appeared fifteen years after the discovery of the reac-tion.[203,216] Singlet vinylcarbene can exist in two planar forms with the two electrons occupying the same hybrid orbital ($^1A''$) or one in the hybrid orbital and one in the non-bonding π orbital ($^1A'$). The $^1A''$ state is 2.3 kcal/mol

$^1A'$ $^1A''$

lower in energy than the $^1A'$ state. In both states the double bond becomes semi-localized, but more so in the $^1A'$ state. The localization reduces the barrier to rotation about the bond of reduced π character to ~2.0 kcal/mol in $^1A'$ and 18 kcal/mol in $^1A''$. Calculations show that cyclopropene correlates directly to the $^1A'$ state which decays to the lower $^1A''$ state, presumably without any energy barrier.

The early work on the formation of cyclopropenes via vinylcarbenes or vinylcarbenoids has been reviewed.[65] Recently a more comprehensive review has appeared.[217] Discovery of the ring closure of vinylcarbenes (or carbenoids) to form cyclopropenes was made by Closs [Eqs. (4-37) and (4-38)].[66,66a,]

$$\underset{\text{Me}}{Me_2C=\overset{\overset{\displaystyle Me}{|}}{C}-Li} + CH_2Cl_2 \longrightarrow Me_2C=\overset{\overset{\displaystyle Me}{|}}{C}-CH_2Cl$$

$$\Big| RLi \qquad\qquad (4\text{-}37)$$

$\longleftarrow Me_2C=\overset{\overset{\displaystyle Me}{|}}{\underset{\underset{\displaystyle Li}{|}}{C}}-CHCl$

$$(4\text{-}38)$$

[67,67a] Apparently both singlet and triplet vinylcarbenes will undergo the ring closure since diazopropenes will form cyclopropenes photochemically.[68] The thermal decomposition of α,β-unsaturated tosylhydrazone salts gives reasonable yields only where R_1 and R_2 [Eq. (4-38)] are both alkyl groups. Presence of a β-hydrogen markedly lowers the yield and an aryl group prevents the reaction. Though a pyrazolenine could be an intermediate, this was eliminated since the pyrazolenine was stable under the conditions of the reaction.[67,67a]

These routes generally are of preparative value, even when the yields are modest, because of the simplicity of the method and the ready availability of the starting materials. Cyclopropene can be generated, albeit in low yield, by treating allyl chloride with sodamide.[69] A similar reaction with methallyl chloride gives 1-methylcyclopropene in 50% yield [Eq. (4-39)].[70] Cyclo-

$$
\underset{\substack{|\\ \text{Me}}}{CH_2{=}C{-}CH_2Cl} \; + \; NaNH_2 \quad \longrightarrow \quad \triangle_{Me} \tag{4-39}
$$

propenes having electron attracting substituents on the olefinic carbons can be obtained via photochemical elimination of nitrogen from pyrazolenines.[71,72] The reaction proceeds via formation of an intermediate diazopropene [Eq. (4-40)], which apparently is converted to the vinylcarbene

$$\tag{4-40}$$

photochemically and not thermally.[71] Electron attracting substituents such as CN, COOR, COR, etc., at $C_4(R_1)$ or $C_5(R_2)$ permit formation of the cyclopropene except when R_2 is COR.[72] Reaction then leads to other products. Cyclopropene-1-carboxylic acids can be obtained indirectly via the carbenoid process of Closs and Closs [Eq. (4-41)].[66] The cyclopropenyllithium reagent can also be alkylated.

$$\tag{4-41}$$

The very high strain energy of the cyclopropene ring (ca. 52 kcal/mol)[73] makes possible ring opening to give the vinylcarbene under relatively mild conditions. Cyclopropene can be pyrolyzed in the gas phase to give methylacetylene.[74] Tetramethylcyclopropene also undergoes ring cleavage in the gas phase, but yields no acetylene, only 2,3-dimethyl-1,3-pentadiene was obtained [Eq. (4-42)].[75] Frey[74] has suggested the product may be the trans

$$ (4\text{-}42) $$

isomer (see below). Battiste *et al.*[76] has studied the thermolysis of tetraphenylcyclopropene [Eq. (4-43)]. The reaction requires an activation energy

$$ (4\text{-}43) $$

of 40 kcal/mol. Srinivasan[77] found that alkyl substituted cyclopropenes give mixtures of alkynes and dienes (Table 4-4). Increasing alkyl substitution increases the stability of the cyclopropene and decreases the rate of decomposition. The fifth example in the table shows a case in which two different cleavages are possible [Eq. (4-44)]. Both cleavages occur but the formation of a more highly substituted carbene is preferred.

$$ (4\text{-}44) $$

York *et al.*[79,79a] used an optically active cyclopropene in a very pretty experiment to show that the ring opening is a reversible reaction at 160°–190°. Over the range 160°–190° ring closure of the vinylcarbene occurred nine times

TABLE 4-4

Thermolysis Reactions of Alkylcyclopropenes[77,78]

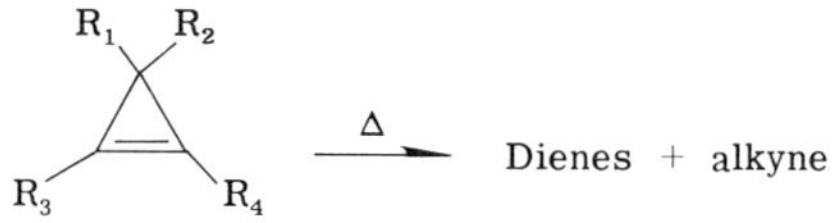

R_1	R_2	R_3	R_4	Products	E_a	log A
H	H	H	H	MeC≡CH	35.2	12.1
H	H	Me	H	 (6%) + MeC≡CMe (90%)	34.7	11.4
Me	H	H	H	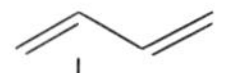 (10%) + EtC≡CH (90%)	37.6	13.5
Me	Me	H	H	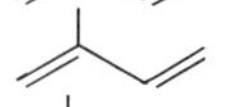 (10%) + i-PrC≡CH (90%)	36.6	13.0
Me	Me	Me	H	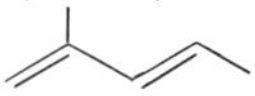 Me (71%) + 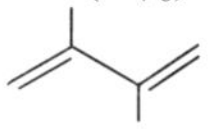 (21%)	39.0	13.4
Me	Me	Me	Me	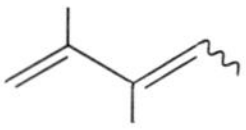 Me 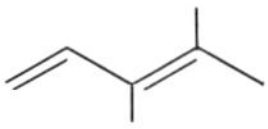 + (10%) 85% trans 5% cis	40.0	12.5

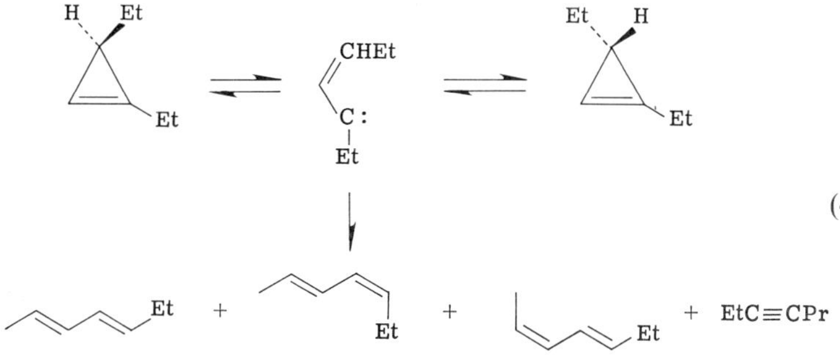

$$(4\text{-}45)$$

34%   29%   4%

faster than isomerization to products. In this case no products attributable
to formation of the less substituted carbene were obtained. Thus once again
the preferred cleavage is of the more substituted bond. With respect to the
intriguing question of the direction of rotation, this type of experiment is
precisely the type which could give useful information [Eq. (4-46)] if the

$$
\begin{array}{c}
\text{H} \quad \text{Et} \\
\triangle \quad \text{Et} \\
\xrightarrow[\text{toward H}]{\text{Et rotat.}} \quad
\begin{array}{c}
\text{Et} \\
\text{H} \\
\text{H} \\
\ddot{\text{C}}\text{—Et}
\end{array}
\end{array}
$$

Et rotat. toward Et

1,2-H shift

$$
\begin{array}{c}
\text{H} \\
\text{Et} \\
\text{H} \\
\ddot{\text{C}}\text{Et}
\end{array}
\xrightarrow[\text{shift}]{1,4\text{-H}}
\begin{array}{c}
\text{Et} \\
\text{H} \\
\text{H} \\
\text{CH}=\text{CHMe}
\end{array}
$$

(4-46)

1,2-H shift

$$
\begin{array}{c}
\text{H} \\
\text{Et} \\
\text{H} \\
\text{CH}=\text{CHMe}
\end{array}
$$

(4-47)

vinylcarbene were stereochemically stable under the reaction conditions. Since the study of tetramethylcylopropene[77] shows that the 1,4-H shift (an allowed suprafacial shift) gives rise to at least 95% of a trans double bond, we may assume that a similar result pertains in Bergman's example. Then the sole source of the *cis*-4-ene obtained is the rotation of ethyl toward ethyl [Eqs. (4-46) and (4-47)] during ring opening. The large amount of the *cis*-4-ene compared to the trans isomer suggests that if the carbene were undergoing stereomutations, equilibrium was not attained. Thus one may conclude that some rotation in this direction does occur. While it seems quite reasonable to suspect that rotation of ethyl toward hydrogen is also taking place, no conclusions can be drawn from the data.

Some results from the ring opening of a benzcyclopropene support the conclusion that rotations in both directions can occur [Eq. (4-48)].[80] The

$$
\xrightarrow{25°}
$$

(4-48)

assumption about the configurational stability of the vinylcarbene seems much safer in this case. Theory[203] did not clarify the barrier to methylene rotation in the vinylcarbene, but to the extent that double bond localization does occur the barrier between the two carbenes could be reduced below the value of 20 kcal/mol for the allyl radical.[81] Evidence in support of the vinylcarbene structure of the intermediate was noted by Steeper and Gardner[82] who found a C-H insertion product from the pyrolysis of 52 [Eq. (4-49)]. The formation of 53, result of a formal methyl migration, is unusual.

$$(4-49)$$

Further evidence favoring the general conclusion that ring closure of vinylcarbenes is faster than other product forming steps comes from the behavior of dl and meso forms of 54[Eq. (4-50)].[83,83a] In this case the meso

$$(4-50)$$

and *dl* forms interconvert faster than either one is converted to **55** or to aromatic products.

Interconversion of benzyl carbenes can occur via the intermediacy of cycloheptatrienyl carbenes with bicyclo[5.1.0]hepta-2,4,6-trienes. The pertinent material has been reviewed thoroughly by Jones and Brinker.[232] Dewar and Landman have studied the reaction theoretically by MINDO/3, and have found that the bicyclic carbene is ca. 26 kcal/mol more stable than benzyl carbene.[233]

IV. AZIRINE–VINYLNITRENE–IMINOCARBENE CONVERSIONS

The introduction of a heteroatom (here nitrogen) into the cyclopropene ring in place of a CH group adds a new dimension to the electrocyclic reaction [Eq. (4-51)]. In addition, the symmetrical azirine, which could be in

$$\overset{CH_2}{\underset{N-CH}{\diagup}} \quad \underset{b}{\overset{a}{\rightleftharpoons}} \quad \overset{CH_2}{\underset{N=CH}{\diagup}} \quad \underset{d}{\overset{c}{\rightleftharpoons}} \quad \overset{H_2C}{\underset{:N-CH}{\diagdown}} \tag{4-51}$$

equilibrium with the C-iminocarbene (**56**) could bring two more reactions into consideration. Experimental work so far has established that reactions

$$\overset{\overset{\displaystyle H}{\underset{N}{|}}}{\underset{HC=CH}{\diagup\!\!\!\diagdown}} \quad \underset{f}{\overset{e}{\rightleftharpoons}} \quad \overset{HN}{\underset{HC-CH}{\diagdown}}$$

$$\textbf{56}$$

b, c, d, and f can occur, and that the symmetrical azirine rapidly converts to the unsymmetrical form with the shift of a group from nitrogen to carbon.

Smolinsky and co-workers[84,84a,85] demonstrated in the early sixties that vinyl azides where $R_3 \neq H$ readily gave azirines on thermolysis. Presumably

$$\overset{R_1}{\underset{R_2}{\diagdown}}C=C\overset{R_3}{\underset{N_3}{\diagup}} \quad \overset{\Delta}{\longrightarrow} \quad \left[\overset{R_1}{\underset{R_2}{\diagdown}}\overset{R_3}{\underset{N:}{\diagup}} \right] \quad \longrightarrow \quad \overset{R_1 \quad R_2}{\underset{N}{\diagup\!\!\diagdown}}_{R_3} \tag{4-52}$$

the vinylnitrene is an intermediate. Groups R_1 and R_2 can be hydrogen, alkyl, or aryl and R_3 can be alkyl or aryl. Isomura[86,87] showed that azirines do form when R_3=H, but the azirines are extremely unstable in the presence of oxygen. Finally, Ford[218] obtained the parent 1-azirine. Successful reactions were carried out at 80°–125° under nitrogen. The temperature required varied with the substitution, the lowest being for R_1=R_2=Ph. An interesting route to 3-carbalkoxy-azirines from 5-alkoxyisoxazoles [Eq. (4-53)] was

$$(4-53)$$

developed by Nishiwaki.[88] The intermediacy of a vinylnitrene has been specifically confirmed. Padwa *et al.*[204] has reported trapping a vinylnitrene with a phosphine, based on a similar reaction of Nishiwaki and Saito.[205]

The presence of a symmetrical intermediate in the pyrolysis of 1,2,3-triazoles has been very neatly demonstrated.[89] Thus the two triazoles of Eq. (4-54) gave identical mixtures of azirines and their decomposition products. The symmetric azirine is certainly the most plausible intermediate, as the authors noted. The presence of the C-iminocarbene as a prior intermediate in the formation of the symmetrical azirine must be inferred.

$$(4-54)$$

Cleavage of the C-N single bond during pyrolysis of an azirine is illustrated by the formation of an indole [Eq. (4-55)].[86,90,91] The route via a

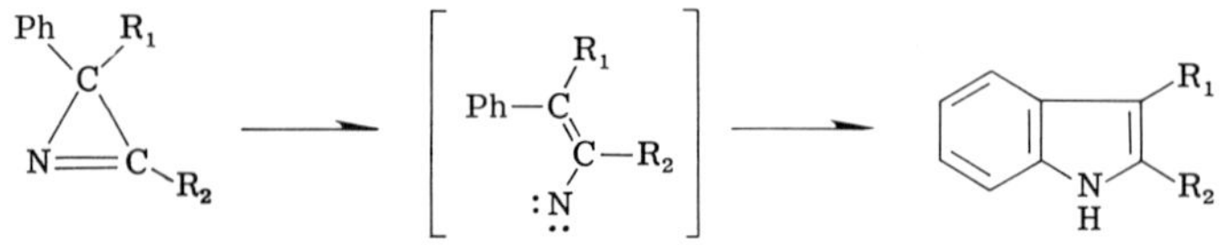

R₁ = H, R₂ = Me   refer. 86
R₁ = Pth, R₂ = Ph   refer. 90
R₁ = R₂ = Me   refer. 91

(4-55)

vinylnitrene has good precedents in cyclopropene chemistry. It has also been clearly demonstrated that cleavage of the carbon–carbon single bond of the azirine can also occur [Eq. (4.56)].[91] The route to styrene and a nitrile

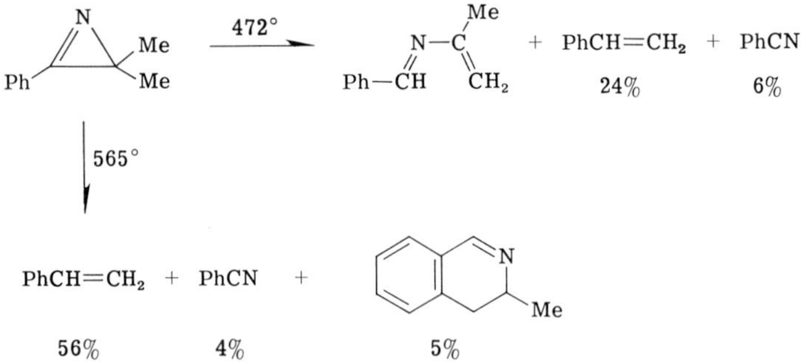

(4-56)

was postulated to involve ring closure of the vinylimine and subsequent ring cleavage [Eq. (4-57)]. A synthesis of 1,4-dihydropyridines from amino-

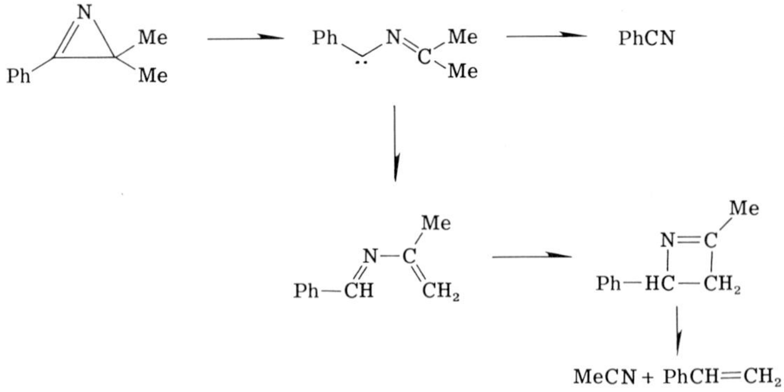

(4-57)

azirines has been developed [Eq. (4-58)].[92] Dimerization of the initial cleavage product with formation of a pyrazine has also been observed.[93]

$$(4\text{-}58)$$

V. THE CYCLOPROPYL ANION–ALLYL ANION PROCESS

The first example here of a four electron electrocyclic reaction in which two methylene rotations must occur is the cyclopropyl anion–allyl anion conversion. Simple orbital conservation predicts that the preferred stereochemistry will be conrotatory. This prediction has been in accord with all calculations of a more sophisticated type which have been made. Clark and Smale[94] found that a conrotatory reaction was favored over the disrotatory (groups cis to leaving group rotate outward, called dis 1 in refer. 96) by 15 kcal/mol. The unusual character of the forbidden process was noted,[95] with an energy increase continuing until rotation was virtually completed. In this case the $E_{\text{conrot.}} - E_{\text{disrot.}}$ was given as 35 kcal/mol.[95] *Ab initio* calculations on the cyclopropyl-allyl cation were extended to include the anion by application of Koopman's theorem.[96] These calculations gave E_a for the preferred conrotatory process as 43 kcal/mol, and $E_{\text{conrot.}} - E_{\text{disrot. 1}} \simeq$ 54 kcal/mol. The cyclopropyl anion with the electron pair in a hybrid orbital was found to be 12 kcal/mol more stable than that with the electrons in a *p* orbital. Ring opening occurs via ring bond extention, movement of the electron pair to a *p*-orbital, and when the ring C-C-C bond angle reaches ca. 85°, sudden conrotation of the methylene groups. That is, there is no concert between the bond cleavage and methylene rotation processes.

The picture which theory conveys is that the cyclopropyl anion lies in a deep energy well, and thus should be an isolable entity. Ring opening to form the more stable allyl anion (31 kcal/mol more stable[96]) should be an activated process with a relatively high activation energy. At the transition state the electron pair is in a *p*-orbital and presumably methylene rotation is in progress, hence appropriately situated stabilizing substituents should reduce the energy barrier. Though the conrotatory process should perhaps be observable, the forbidden disrotatory route ought to remain invisible.

That the cyclopropyl anion is indeed a stable entity which forms prior to ring opening, if indeed the latter reaction occurs at all has been clearly demonstrated.[97] Even in cases where ring opening is favored by substitution the intermediacy of the cyclopropyl anion can be demonstrated by deuterium

$$\text{(4-59)}$$

exchange.[98,99,100,100a] Experimental verification of the preference for con-totatory ring opening has been harder to achieve. Product structure does not appear to be particularly helpful. While **57** [Eq. (4-60)] gives products in

$$\text{(4-60)}$$

accord with a conrotatory process,[101] both **58** and **59** give the same mixture of cis and trans isomers [Eq. (4-61)].[102] Reaction rates have been more

$$\text{(4-61)}$$

helpful. Thus, for example, **58** reacts 47 times faster than **59**.[103] Both isomers undergo deuteration without ring opening at $-78°$, but at $-28.5°$, the rate of formation of the colored allylic anion can be followed conveniently.

Ring cleavage does occur in bicyclic molecules where disrotatory opening is enforced geometrically [Eqs. (4-62) and (4-63)].[98,103] The nitrile of **60** reacts some 2.7×10^{-4} times as fast as **59**. All of the information obtained

$$ \text{(4-62)} $$

$$ \text{(4-63)} $$

60

supports the preference for conrotatory reaction, but quite obviously the disrotatory ring opening can occur. Whether this latter reaction is exemplary of the concerted orbital symmetry forbidden process is a matter of conjecture.

A key to the problem of why product structures do not reflect the conrotatory process even from trans disubstituted cyclopropyl anions has been uncovered recently.[104] The unsymmetrical allylic anion [Eq. (4-64)] has two

$$ \text{(4-64)} $$

nonequivalent protons whose nmr resonances coalesce at 65°. The process by which the protons interconvert has $\Delta H^{\ddagger} + 12$ kcal/mol, and presence of the symmetrical allyl anion was established by cycloaddition capture using acenapthene.

VI. HETEROCYCLES ISOELECTRONIC WITH THE CYCLOPROPYL ANION

Three-membered rings having at least one heteroatom bearing an unshared pair of electrons are isoelectronic with the cyclopropyl anion. The group includes as heteroatoms nitrogen, oxygen, and sulfur, and along with aziridines, epoxides, and episulfides includes oxaziridines and thioxiranes. The essential foundation which permitted examination of these molecules was laid by the elegant work of Huisgen *et al.*[105] and his students in the late fifties and early sixties which established both the broad outlines and many of the fine details of dipolar cycloadditions. In fact the beautiful studies of Huisgen and co-workers provide most of the experimental verification of the predictions of the orbital symmetry theory as it applies in this section. Before proceeding to individual molecules we will find it useful to consider the general theoretical question which will be applicable here—how large a perturbation of the symmetric system can be made before the orbital symmetry analysis is rendered invalid?

This problem has been treated theoretically by Shilling and Snyder[106] for the series of molecules which could undergo the reaction of Eq. (4-65).

$$\left[\begin{array}{c} X \\ Y \quad Z \end{array} \right]^{0 \text{ or } -} \longrightarrow \left[\begin{array}{c} Y \\ Y \text{---} Z \end{array} \right]^{0 \text{ or } -} \tag{4-65}$$

Essentially the fully symmetric prototype here is the cyclopropyl anion, and it is perturbed by the introduction of heteroatoms. A second case is the aziridine molecule also perturbed by added heteroatoms. The results are shown in Table 4-5. Note that in most of the cases no experimental study

TABLE 4-5

Perturbation Effects on Orbital Symmetry Control

Carbanions X=CH		Prediction	Aziridines X=NH	
Y	Z		Y	Z
CH_2	CH_2	Conrot.	CH_2	CH_2
CH_2	NH	Conrot.		
		Both allowed }	CH_2	NH
			CH_2	0
CH_2	0	{ Both		
NH	0	{ forbidden		

could be made since the rotational groups bear unshared pairs only. The effect of strong perturbation is notable. The authors also report that reactions of the symmetrical compounds where X = CH and Y = Z = NH or O are forbidden thermally in both conrotatory and disrotatory forms. Carlsen and Snyder[219] have added sulfene ⇄ oc-sultine and oxathiirane to the list, finding both allowed.

A. Aziridines

Reports on the thermal instability of the aziridine ring appeared in the early part of this century,[107 206-209] but the main work developed after Huisgen pointed up the value of dipolar cycloadditions for the synthesis of heterocycles. Reviews which cover some of the work in this area have been published.[108,210,211] Following a series of three communications which indicated the opening of an aziridine to an azomethine ylide and subsequent capture of the ylide by a dipolarophile,[109,110,111] a complete examination of the mechanism was made by Huisgen.

$$\text{(4-66)}$$

Earlier work had shown conclusively that dipolar cycloadditions proceed in a stereospecific suprafacial–suprafacial manner as expected from orbital symmetry considerations. Thus, the geometry of the reacting dipolar group will be maintained in the cycloaddition. If then the capture of the dipolar group is faster than its isomerization, the stereochemistry of the ring opening process will be reflected by that of the product heterocycle [Eq. (4-66)]. The stereospecific conrotatory ring opening was demonstrated initially using N-aryl-2,3-aziridinecarboxylates and trapping with dimethyl acetylenedicarboxylate (DMAD).[113] In the absence of DMAD the cis-dimethyl ester isomerizes to the trans ester at 100° with $K = 4$ [Eq. (4-67)]. This isomerization was shown to proceed via ring opening, isomerization of the resultant azomethine ylide and reclosure because the presence of DMAD resulted in formation of a pyrrolidine and prevented the isomerization. At 100° $k_{obs} = 3.07 \times 10^{-5}$ sec^{-1}, $k_{-obs} = 7.69 \times 10^{-6}$ and $\Delta H^{\ddagger}$ is 26 ± 3 kcal/mol. As is illustrated in Eq. (4-67), when the cis-aziridine was heated at 100° in the

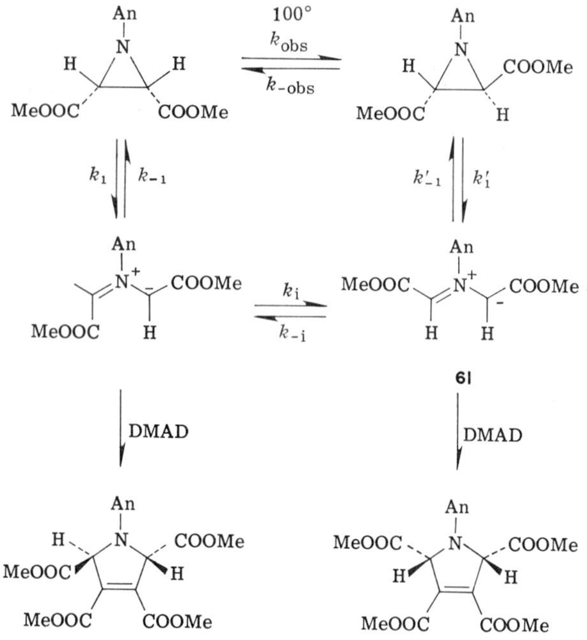

(4-67)

An = *p*-anisyl

presence of DMAD, only the *trans*-dimethyl 2,5-pyrolidinedicarboxylate was formed. Conversely, the *trans*-aziridine gave solely the *cis*-pyrrolidine. Clearly ring opening proceeds stereospecifically in a conrotatory fashion exactly as predicted by the conservation of orbital symmetry. This result quite obviously bolsters the case for conrotatory ring opening of the cyclopropyl anion.

A full kinetic analysis of the reaction system of Eq. (4-67) has been carried out.[114,115] Values for k_1 and k'_1 were obtained by trapping the azomethine ylides with tetracyanoethylene (TCNE) under conditions where the rate of adduct formation is independent of the concentration of TCNE. The ratios for k_{-1}/k_i and k'_{-1}/k_{-i} were obtained from k_1, k'_1, k_{obs} and k_{-obs} by kinetic analysis of the isomerization scheme. Finally, k_{-1} and k'_{-1} were determined by preparing the *cis*- and *trans*-azomethine ylides by flash photolysis and following their decay rates to the aziridines.[115] Measurements were made at several temperatures and the activation parameters were determined. The results of all these studies are listed in Table 4-6. This analysis shows that the interconversion of the *trans*- and *cis*-azomethine ylides is a very rapid reaction, and as a result only the strongest dipolarophiles can trap both azomethine ylides before isomerization occurs.[116] Results with a series of di-

TABLE 4-6

Rates and Activation Parameters for the Aziridine
Cleavage Processes of Eq. (4-67)

Processes	Rates (sec^{-1})	Temperature (°)	$\Delta H^{\ddagger}$	$\Delta S^{\ddagger}$
k_1	6.18×10^{-5}	100	28.6	-1.7
k'_1	5.14×10^{-5}	100	29.5	$+0.5$
k_{obs}	1.73×10^{-5}	100	28.1	-5.6
k_{-obs}	0.71×10^{-5}	100	28.6	-6.1
k_{-1}	0.136	24.6	9.1	-32
k'_{-1}	0.106	27.1	12.7	-21
k_i	0.068	24.6	($k_{-1}/k_i = 2$)	
k_{-i}	0.023	27.1	($k'_{-1}/k_{-i} = 4.5$)	

polarophiles are listed in Table 4-7. That the structure of the **61** has both
carbomethoxy groups cis to the anisyl rather than trans has been neatly
demonstrated [Eq. (4-68)].[117,117a]

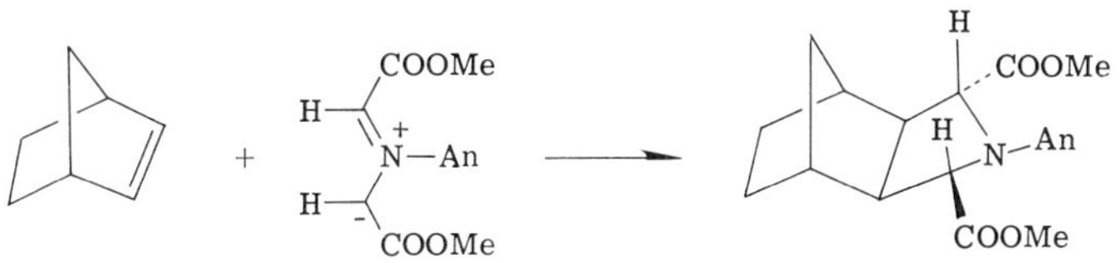

(4-68)

61

TABLE 4-7

Stereochemistry of the Capture of Azomethine Ylides by Dipolarophiles

Dipolarophile	Pyrrolidine cis	trans	cis	trans
TCNE	0	100	100	0
DMAD	0	100	100	0
Dimethyl azodicarboxylate	0	100	94	6
Dimethyl fumarate	0	100	91	9
Tetramethyl ethylene-tetracarboxylate	0	100	75	25
Norbornene	0	100	55	45
Cyclohexene	0	100	0	100

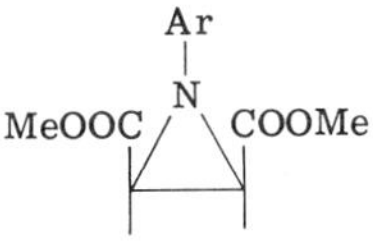

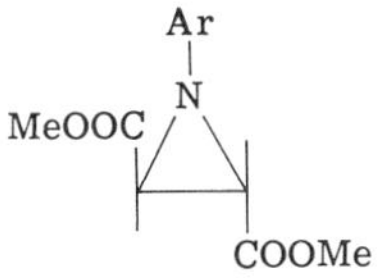

Disrotatory ring openings of aziridines which are fused to rings too small to accommodate trans double bonds do occur.[120-123] No kinetic study of the process has been made as yet, nor has the aziridine been attached to rings other than five membered ones. Reversible thermochromic behavior is noted with the ring opening of certain fused aziridines [Eq. (4-69)].[120,122] The

$$(4\text{-}69)$$

colored product has been trapped with dipolarophiles such as diethyl fumarate or maleate.[120,121] Other variants of this same theme are indicated in Eq. (4-70). It is unfortunate that no activation parameters have been determined for these reactions for comparison with the conrotatory results.

The general conrotatory ring opening, azomethine ylide isomerization and dipolarophile trapping has been observed with a number of substituted aziridines. However, the substituents cause a wide variation in the rates of these processes. An early study[109,124] of the reactions of 1,2,3-triphenyl-aziridine with several dipolarophiles was examined in greater detail by Hall and Huisgen.[125,125a] This aziridine is not as loaded in favor of ring

$$(4\text{-}70)$$

R = H   R′ = PhCH$_2$ R = Ph   R′ = Et
refer. 123   refer. 122

opening as is *N-p*-anisyl-2,3-dicarbomethoxyaziridine which should lead to a slower ring opening for the triphenyl compound. Conversely the triphenyl-azomethine ylide should be less stable, hence subject to more rapid capture. The results are in reasonable accord with expectations, cycloaddition being particularly favored. Even such modest dipolarophiles as dimethyl maleate or fumarate can capture the ylide before isomerization can occur. Not all

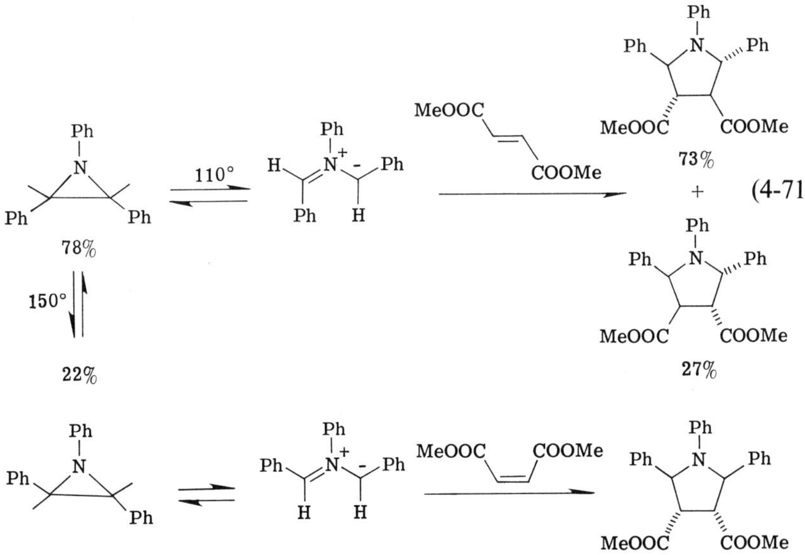

$$(4\text{-}71)$$

cases are quite as nicely predictable. For example, the azomethine ylides which are intermediates in the reactions of Eq. (4-72) are completely equi-

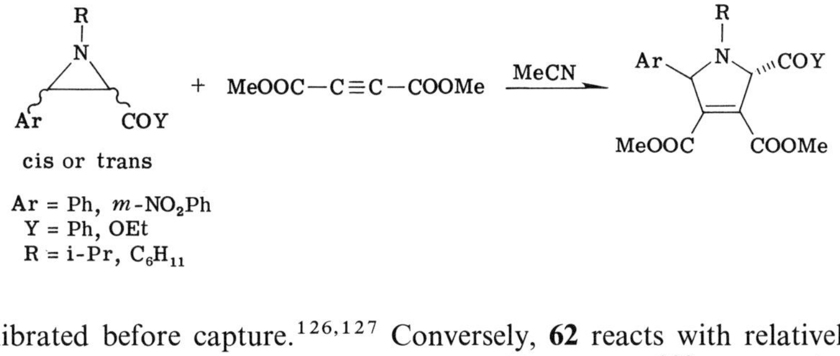

$$(4\text{-}72)$$

Ar = Ph, *m*-NO$_2$Ph
Y = Ph, OEt
R = i-Pr, C$_6$H$_{11}$

librated before capture.[126,127] Conversely, **62** reacts with relatively poor dipolarophiles in a completely stereospecific manner.[128] The results can be

$$(4\text{-}73)$$

partially rationalized via the assumption that reducing the stability of the azomethine ylide enhances the capture rate and also reduces the ylide isomerization rate.

When the aziridine nitrogen bears a hydrogen an ene reaction with the potential dipolarophile can intervene prior to ring opening [Eq. (4-73)].[129]

$$(4\text{-}74)$$

However, the presence of substituents which stabilize the ylide enables ring opening to compete [Eq. (4-75)].[129] This hydrogen atom makes possible

$$(4\text{-}75)$$

study of the behavior of the aziridinyl anion.[130–138] Ring opening of the anion does occur and the conrotatory stereochemistry still pertains [Eq. (4-76)].[135] If **63** is cooled to 0° after its formation and *trans*-stilbene is added the sole product is **64**.

$$(4\text{-}76)$$

64

Cycloaddition reactions of the azomethine ylides have proven very useful in synthesis of a variety of heterocycles. Pyrrolidines, pyrrolines, and pyrroles have been obtained depending on the conditions used and the dipolarophile [Eq. (4-77)].[109,110,118,124,129,139,140,141] Where the trapping agent is an

$$(4\text{-}77)$$

aldehyde or a ketone oxazolidines are formed [Eq. (4-78)].[119,142–145] A
ketone which acts as a stabilizing group for the azomethine ylide can act
as an internal dipolarophile [Eq. (4-79)].[146,147] Note that the oxygen of the

$$(4-78)$$

$$(4-79)$$

external dipolarophile generally goes to that carbon of the azomethine ylide
which bears less of the negative charge. Imidazolidines are obtained where
imines are the cycloadditive agent [Eq. (4-80)],[126] and thiocyanates give

$$(4-80)$$

thiazolidines [Eq. (4-81)].[148] A rather unusual route to an oxazole has been

$$(4-81)$$

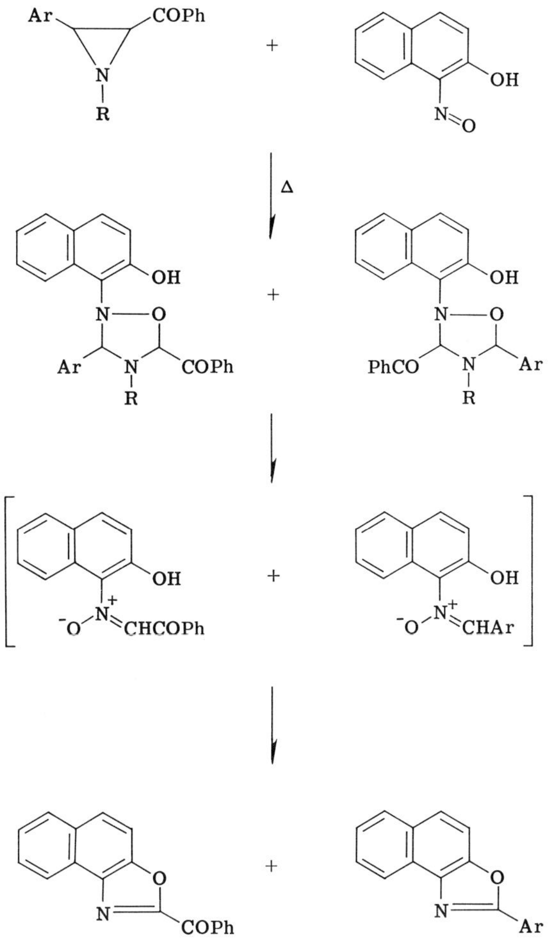

(4-82)

found [Eq. (4-82)].[149] In one case the azomethine ylide was stabilized without cycloaddition by an internal $N \rightarrow O$ acyl shift [Eq. (4-83)].[150] The presumed

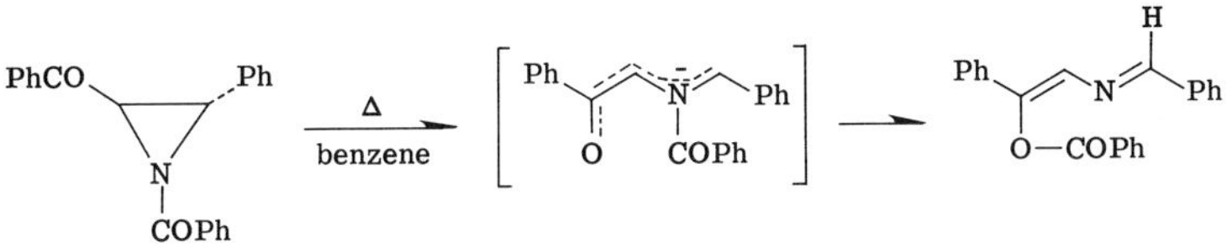

(4-83)

intermediate could not be captured by an external agent.

B. Epoxides

Scission of the C-O bond in epoxides is commonly observed, but C-C bond breaking is more unusual. Cleavage of the C-C bond does occur but it is limited to those epoxides whose substituents will stabilize the carbonyl ylide. The ylide is subject to trapping by dipolarophiles, thus opening the way to experimental investigation of the ring opening process. The observations which prompted subsequent studies were made with both a monocyclic and a bicyclic epoxide [Eqs. (4-84) and (4-85)].[151,152] Compound 65 exhibits

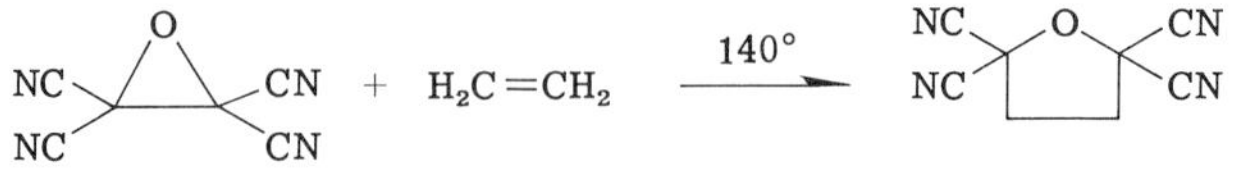

$$\text{(4-84)}$$

thermochromism and the colored carbonyl ylide form has been snared with DMAD.[153]

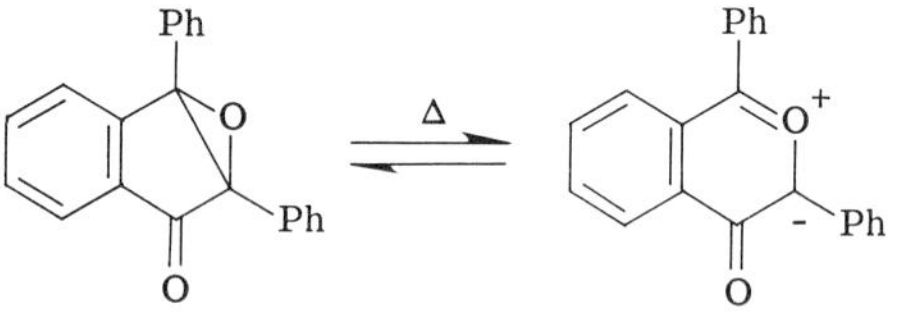

$$\text{(4-85)}$$

65

While the reactions of properly substituted epoxides are expected to follow the pattern set by the aziridines, the stereoselectivity in ring opening was predicted to be lower.[154] Increase in the percent radical character going from azomethine ylides (30%) to carbonyl ylides (38%) and the knowledge that odd electron species are less selective in concerted reactions[155] was the basis of that prediction. In light of this prediction the experimental results are most interesting. The cis and trans isomers of 66 were found to equilibrate at 100° with the trans isomer being more stable.[156] Both isomers gave the same adducts with either DMAD or diethyl maleate [Eq. (4-86)]. These

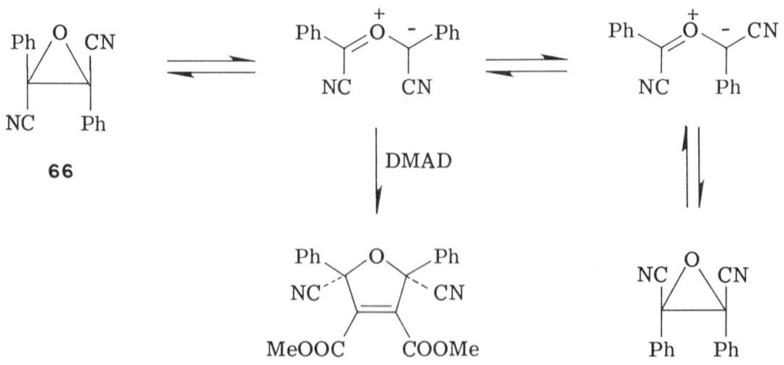

66

$$\text{(4-86)}$$

results can be interpreted as indicating stereospecific ring opening followed by rapid equilibration of the ylides and slower capture by cycloaddition. Equally suitable would be a partially stereoselective ring opening followed by rapid ylide interconversion and slow cycloaddition. That the second interpretation is not correct was shown by a study of *cis-* and *trans-***67** [Eq.

(4-87)

(4-87)].[157] In this case both isomers do not give the same products. Rate studies[220] with dimethyl fumarate showed the rate of ring opening of the cis isomer (7.5×10^{-7} sec^{-1} at 127°) is slower than the trans isomer (1.1×10^{-4} sec^{-1}). Kinetic dissection of the trapping of the ylide from the cis oxirane leads to the interesting conclusion that 36% of the ring opening process is not giving the conrotatory product. Arguments that direct forma-

TABLE 4-8

Rates of Loss of Optical Activity and Isomerization of 2-Phenyl-3-p-Tolyloxiranes

Reaction	Rate	Temperature ($°$)	E_a	Log A
k_α (trans)	1.09×10^{-5}	179.8	35.9	12.34
k_α (cis)	0.86×10^{-5}	225.3	41.1	12.96
k_i (trans $\rightarrow$ cis)	0.301×10^{-5}	229.1	39.6	11.72
k_i (cis $\rightarrow$ trans)	2.86×10^{-5}	229.1	39.6	12.68

tion of the disrotatory product is occurring have been summarized by
Markowski and Huisgen[221] and Huisgen.[222] A value of 5.6 kcal/mol for the
conrotatory–disrotatory energy differential was obtained in this system.

Evidence which supports the conrotatory reaction for most oxiranes has
been adduced from kinetic studies.[158] Optically active 2-phenyl-3-p-tolyl-
oxiranes were found to undergo loss of optical activity and isomer interconversion at rates shown in Table 4-8. Loss of optical activity is much faster
for the trans isomer than the cis, as expected for a conrotatory process. If
both isomers were to undergo an indiscriminate ring opening, the less stable
cis isomer would react faster than the trans. At first glance it appears surprising that k_i (cis $\rightarrow$ trans) is larger than k_α (cis), a result which makes it
evident that an isomerization route not leading to racemization must be
active. Presumably this results from C-O bond cleavage with a strong preference for the bond to the carbon bearing the p-tolyl group.

The disrotatory path, concerted or not, is available as shown by the ring
opening of oxiranes fused to four- or five-membered rings.[152,153,159–163]
Dunston and Yates[160] found that **68** undergoes a reversible ring cleavage

$$(4\text{-}88)$$

68

of the oxirane ring, but the 2,3,5-triphenyl derivative does not revert to the
epoxide form.[159] Thermochromism has been observed with appropriately
substituted cyclobutene oxides [Eq. (4-89)].[161,162] As would be expected,

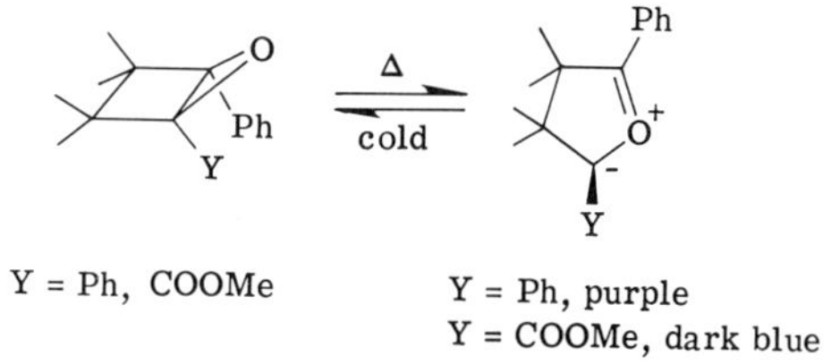

$$
\text{(4-89)}
$$

Y = Ph, COOMe

Y = Ph, purple
Y = COOMe, dark blue

the carbonyl ylide (Y = COOMe) is more stable than that with Y = Ph ($t_{\frac{1}{2}}$ = 21 and 8 min, respectively, at 22° in benzene). Cycloaddition to the carbonyl ylide form occurs readily, but the initial adduct may lose tetramethylethylene [Eq. (4-90)]. With cyclobutene oxides special substitution is

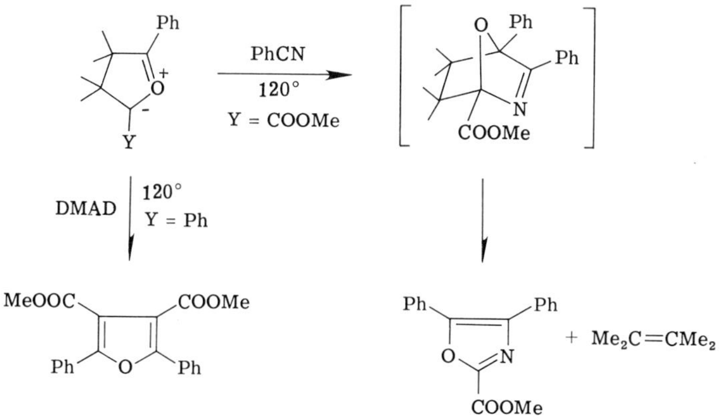

$$
\text{(4-90)}
$$

not necessary to bring about cleavage of the C–C bond in competition with C–O scission.[164]

The ring opening and subsequent cycloaddition reactions of tetracyanoethylene oxide have been studied in considerable detail.[165-167] The tetracyanocarbonyl ylide is a very reactive substance, which will even add to benzene [Eq. (4-91)]. However, addition to a Schiff base was shown to occur

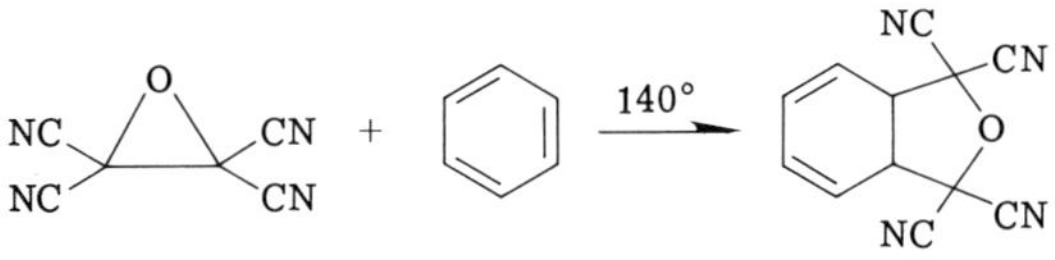

$$
\text{(4-91)}
$$

in a nonconcerted fashion [Eq. (4-92)]. In the absence of a dipolarophile, the

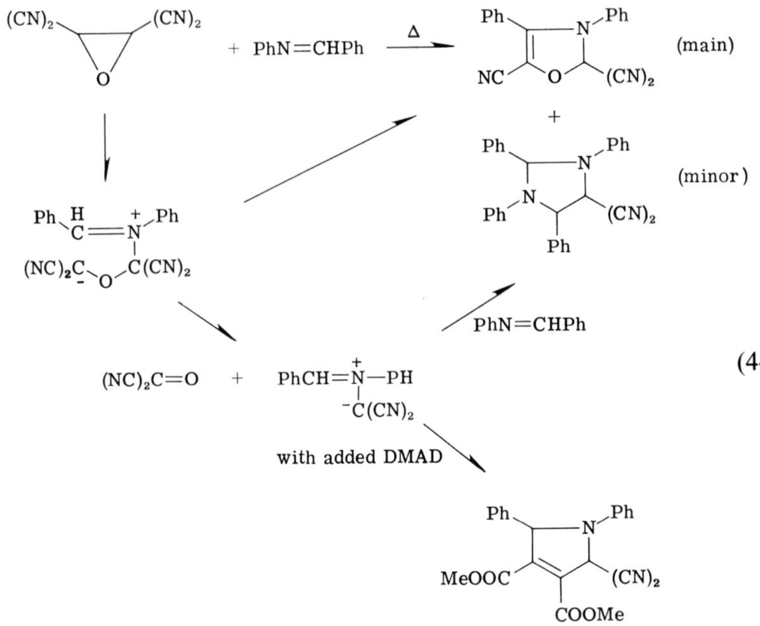

(CN)₂ (CN)₂ epoxide   + PhN=CHPh —Δ→   structure (main)

Ph, N–Ph ring with NC, O, (CN)₂   (main)

+

Ph, N–Ph ring with Ph, N, Ph, (CN)₂   (minor)

(NC)₂C=O + PhCH=N⁺—PH with ⁻C(CN)₂

(4-92)

with added DMAD

Ph, N–Ph ring with MeOOC, COOMe, (CN)₂

carbonyl ylide may decompose unimolecularly with formation of a carbene [Eq. (4-93)].[168] Robert and co-workers[168–172] have examined a series of

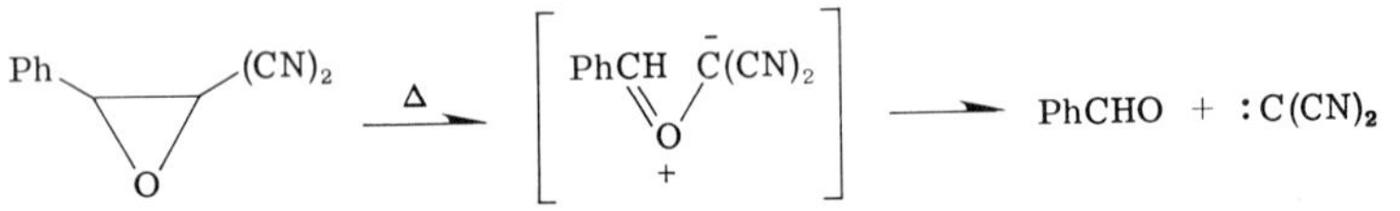

Ph, (CN)₂ epoxide —Δ→ [PhCH⁻ C(CN)₂ over O, +] ⟶ PhCHO + :C(CN)₂ (4-93)

2-substituted 1,1-dicyanoethylene oxides. Rates of the reactions of 2-*p*-Y-phenyl-1,1-dicyanoethylene oxide with DMAD, aldehydes and Schiff bases were determined. In all cases the rates varied with Y, decreasing in the order MeO,H,Cl,NO₂. Slow formation of the carbonyl ylide followed by a rapid cycloaddition is indicated by this rate sequence.

The Cope rearrangement of *cis*-divinylcyclopropane[173] and conversion of the trans isomer to the cis prior to Cope rearrangement,[174] is mimicked, at least formally, by divinylethylene oxides. Chuche and his co-workers have studied the oxide system in some detail to show that carbonyl ylides are the intermediates in this latter series.[175–177] Most of their work was done with 1-phenyl-2-vinylethylene oxides. The general pattern of their results is illustrated in Eq. (4-94). Their interpretation is supported strongly by the

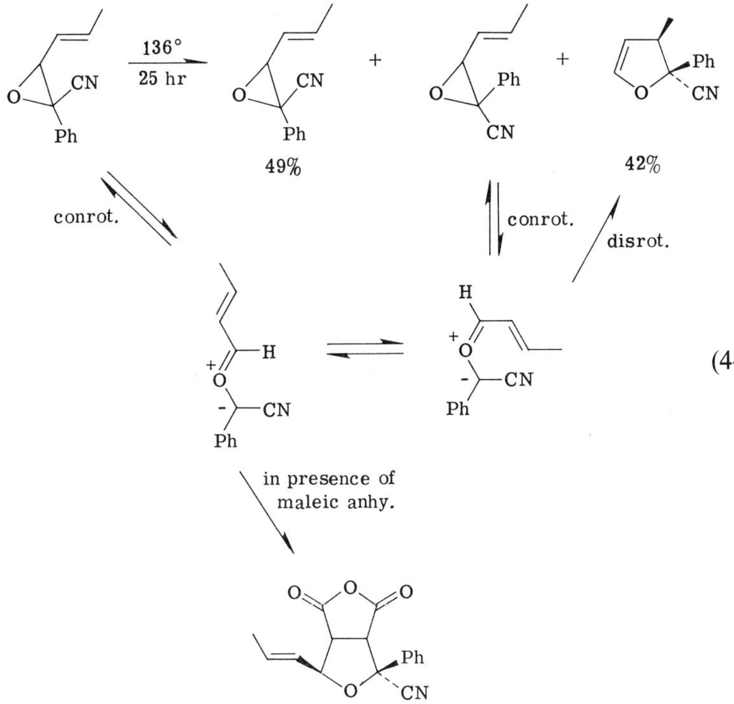

(4-94)

kinetic studies of Crawford and his students,[178,223] who showed that racemization of optically active 2-vinyloxirane occurred 40 times faster than isomerization of *cis*- and *trans*-3-deuterio-2-vinyloxiranes, and optically active *trans*-divinyloxirane gives 2-vinyl-1,2-dihydrofuran which is 99.6% racemic.[223]

C. Episulfides

By analogy with the epoxides, episulfides would be expected to undergo reversible ring opening to thiocarbonyl ylides [Eq. (4-95)]. The reaction

(4-95)

should be facilitated by groups which would stabilize the negative charge. However, there appears to be no evidence that ring opening can occur, though the ring closure of the thiocarbonyl ylides is a well established and very facile reaction. Work in this area has been reviewed by Kellogg,[179] who has contributed a large portion of the stereochemical studies of the ring closure.

The most common route to thiocarbonyl ylides proceeds via a thiadiazoline which may or may not be isolated. The thiadiazoline is thermally unstable, losing nitrogen readily and, in the absence of any reactive dipolarophile or appropriate proton source, the thiocarbonyl ylide closes to the episulfide. Early work on the reaction of thiocarbonyl compounds with diazo compounds gave the episulfide directly.[180,181] Middleton was the first to observe the thiadiazoline [Eq. (4-96)].[182,182a] Evidence for the inter-

$$(CF_3)_2C=S \ + \ (CF_3)_2CN_2 \quad \longrightarrow \qquad \qquad \xrightarrow{\Delta} \qquad \qquad (4\text{-}96)$$

mediate formation of a thiocarbonyl ylide was presented by Kellogg and Wassenaar [Eq. (4-97)].[183] Kinetic studies showed that an intermediate was

$$(4\text{-}97)$$

present and that in the presence of excess dimethyl azodicarboxylate the cycloaddition rate was 35 times faster than closure to the episulfide.[189]

Advantage was taken of the cycloaddition capture to demonstrate the conrotatory ring closure to the episulfide.[184,185,186] The stereoselectivity is maintained even in the face of serious steric repulsion [Eq. (4-98). R = tert-butyl].

R = Et, t-Bu

$$(4\text{-}98)$$

Desulfurization of episulfides by thiophiles permits the above route to episulfides to be converted to a synthesis of hindered olefins.[183,187] Where the thiadiazoline is formed from a ketone [Eq. (4-99)], the olefin will be a

$$(4\text{-}99)$$

symmetric one. The method has been used to prepare a variety of interesting alkenes[224,225] including compounds **69**,[188] **70**,[189] and **71**.[190] A modified

form of the scheme can be used to prepare unsymmetrical alkenes having aromatic groups [Eq. (4-100)].[191,191a] This route was based on a synthesis for

$$(4\text{-}100)$$

episulfides developed by Pederson.[192] The synthesis has been used to prepare triphenylethylene as well as **72**. Unsymmetrical olefins may also be prepared from thioketones with diazoalkanes [Eq. (4-101)].[193,226] This procedure was

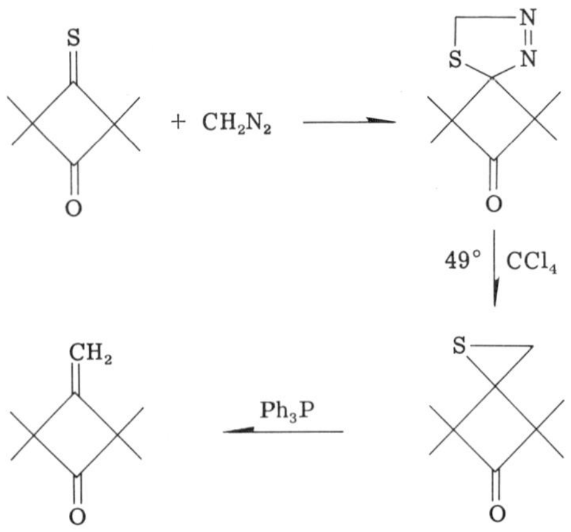

(4-101)

employed to prepare 1,3-dimethylene-2,2,4,4-tetramethylcyclobutane and several homologs.[194] [194a] It provides a specially valuable approach for the preparation of hindered alkenes [Eq. (4-102)].[195] Selenium may replace the

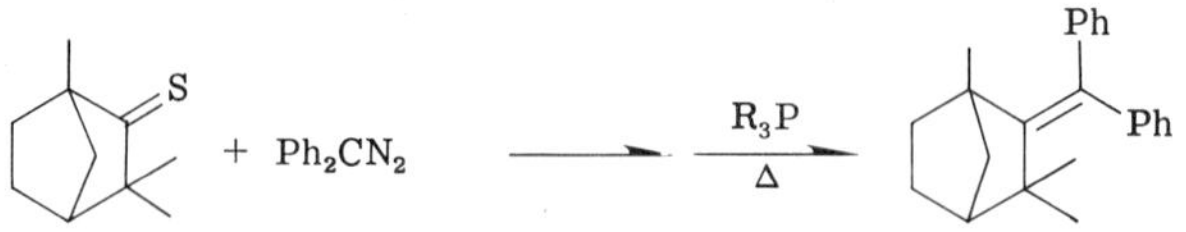

(4-102)

sulfur without deteriment to the procedure.[227]

D. Three-Membered Rings with Two or Three Heteroatoms

With two heteroatoms (N,O,S) present in the ring there exist six potential candidates for the electrocyclic reaction. Of these only two appear to have been studied, the oxaziridenes and thioxiranes. No experimental study of the stereochemistry is possible in either case [Eqs. (4-103) and (4-104)], but

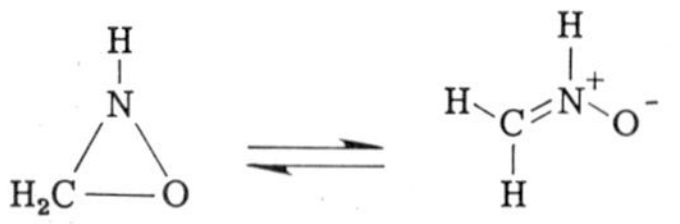

(4-103)

(4-104)

Shilling and Snyder[106] have calculated that for the oxaziridines both conrotatory and disrotatory thermal reactions are allowed, and Snyder[196,219]
has found that for the thioxiranes the reaction is allowed thermally but no
stereochemical conclusions were drawn. Protonated on oxygen the thioxir-
ane was predicted to undergo an allowed concerted electrocyclic reaction
with either rotation.

Study of the oxaziridine–nitrone electrocyclic reaction commences with
the discovery by Emmons[197] that imines can be oxidized to oxaziridines by
peracids. The oxaziridine structure was confirmed by obtaining an optically
active example. Emmons observed the facile thermal ring opening to form
a nitrone [Eq. (4-105)].[197] When all of the substituents are alkyl groups, a

$$\text{PhHC}\underset{\text{N}}{\overset{\text{O}}{\triangle}}\text{CMe}_3 \quad \xrightarrow[\text{MeCN}]{\Delta} \quad \overset{\text{O}^-}{\underset{+}{\text{PhCH}=\text{N}}}-\text{CMe}_3 \qquad (4\text{-}105)$$

different cleavage is observed at higher temperature [Eq. (4-106)].[198] This

$$(4\text{-}106)$$

reaction bears a close formal resemblance to the ketone formation observed
with certain oxiranes [Eq. (4-93)].[168] However, in view of the presence of the
weak N-O bond, a free radical process [Eq. (4-107)] seems more likely.

$$(4\text{-}107)$$

Splitter and Calvin[199,199a] have indicated that nitrone formation is favored
by aryl substitution on carbon, and if aryl groups are present on carbon and
nitrogen the reaction will proceed only with electron releasing substitution
on the C-aryl group and electron attracting substituents on the N-aryl group.

The rate of the nitrone formation from 3-phenyl-N-*tert*-butyloxaziridine
was reported as 5.37×10^{-5} sec^{-1} at $100°$ ($\Delta H^{\ddagger} = 28$ kcal/mol, $\Delta S^{\ddagger} = 3$
eu).[200] Substituents on the ring have a very great influence on the rate of
this reaction. Thus, for example, **73** gives a mixture of nitrones [Eq. (4-108)]

$$(4\text{-}108)$$

$$Ar = Me_2N-\langle\!\langle\ \rangle\!\rangle-$$

with $k = 6.8 \times 10^{-4}$ sec^{-1} at $-8°$[201] The *cis* and *trans*-nitrones interconvert at $-8°$ with $k = 2.9 \times 10^{-5}$ sec^{-1}. Those authors also noted that the
nitrone can be converted to the oxaziridine photochemically. At $-60°$ the
photochemical process is stereospecific with *trans*-nitrone giving *trans*-
oxaziridine, hence as the methyl group swings out-of-place, the bulky aryl
group rotates away from the methyl. The result is reasonable on steric
grounds but does not relate to the conservation of orbital symmetry theory.
Unfortunately, the interconversion rates of both oxaziridine and nitrone at
$-8°$ prevented any stereochemical conclusions with respect to the thermal
process.

Snyder[196] has reported that **74**, the lachrymatory compound of onions,
undergoes a ring closure to 3-ethylthioxirane which decomposes to give
propionaldehyde [Eq. (4-109)]. The last step is reminiscent of the ketone

$$(4\text{-}109)$$

formation from oxaziridines and probably occurs by a related mechanism.

The two isomeric compound diazirines and diazoalkanes could conceivably
interconvert by an electrocyclic process. Both isomers are known[228,229] and
while the interconversion has been considered, it has generally been rejected
because the isomers often lead to different products [Eqs. (4-110) and

$$(4\text{-}110)$$

$$(4\text{-}111)$$

(4-111)].[230] However, a clean example of the electrocyclic behavior has been
uncovered using a diazoketone [Eq. (4-112)].[231] It has also been shown that

$$\text{(4-112)}$$

3-phenyl-3-propyldiazirine is converted thermally to the diazoalkane.[234] However, 3-methoxy-3-chlorodiazirine has been suggested to decompose thermally via a chelotropic process.[235]

Of the various three membered rings with all heteroatoms, only one has been reported to participate in a retro-electrocyclization. An azoxy compound was found to be converted photochemically to an oxadiaziridine which reverts thermally to the azoxy reactant [Eq. (4-113)].[202] Where $R_1 = R_2 =$

$$R_1 - \underset{+}{N} = N - R_2 \quad \underset{\Delta}{\overset{h\nu}{\rightleftharpoons}} \quad R_1 N \overset{O}{\underset{\triangle}{\qquad}} NR_2 \qquad \text{(4-113)}$$

tert-butyl the half-life at 28° in carbon tetrachloride is 290 min.

REFERENCES

1. W. v. E. Doering and D. M. Laflamme, *Tetrahedron* **2**, 75 (1958).
2. For consideration of this point see. W. Kirmse, "Carbene Chemistry," 2nd ed., Academic Press, New York, 1971.
3. W. T. Borden, *Tetrahedron Lett.* p. 447 (1967).
4. M. J. S. Dewar, E. Haselbach, and M. Shanshal, *J. Am. Chem. Soc.* **92**, 3505 (1970).
5. R. Hoffmann, private communication cited by J. M. Wallbrick, J. W. Wilson, Jr., and W. M. Jones, *J. Am. Chem. Soc.* **90**, 2895 (1968).
6. Compare a footnote in W. M. Jones and D. L. Krause, *J. Am. Chem. Soc.* **93**, 551 (1971).
7. P. S. Skell and R. R. Engel, *J. Am. Chem. Soc.* **88**, 3749 (1966).
8. P. S. Skell and R. R. Engel, *J. Am. Chem. Soc.* **89**, 2912 (1967).
9. W. M. Jones, *J. Am. Chem. Soc.* **82**, 6200 (1960).
10. W. J. Jones, M. H. Grasley, and W. S. Brey, Jr., *J. Am. Chem. Soc.* **85**, 2754 (1963).
11. W. M. Jones, M. H. Grasley, and D. G. Baarda, *J. Am. Chem. Soc.* **86**, 912 (1964).
12. W. M. Jones, J. W. Wilson, Jr., and F. B. Tutwiler, *J. Am. Chem. Soc.* **85**, 3309 (1963).
13. W. M. Jones and J. W. Wilson, Jr., *Tetrahedron Lett.* p. 1587 (1965).
14. S. M. Mason and G. W. Vance, *Tetrahedron Lett.* p. 1593 (1965).
15. W. M. Jones and J. M. Walbrick, *Tetrahedron Lett.* p. 5229 (1968).
16. W. R. Moore and R. D. Bach, *J. Am. Chem. Soc.* **94**, 3148 (1972).
17. K. Kleveland and L. Skattebøl, *Acta Chem. Scand.* **29**, 191 (1975).
17a. K. Kleveland and L. Skattebøl, *Chem. Commun.* p. 432 (1973).
18. A. C. Cope and S. A. Mehta, *J. Am. Chem. Soc.* **86**, 5626 (1964).
19. A. C. Cope, W. R. Moore, R. D. Bach, and H. J. S. Winkler, *J. Am. Chem. Soc.* **92**, 1243 (1970).
20. W. R. Moore, H. W. Anderson, S. D. Clark, and T. M. Ozretich, *J. Am. Chem. Soc.* **93**, 4932 (1971).

21. W. M. Jones and M. H. Grasley, *Tetrahedron Lett.* p. 927 (1962).
22. W. M. Jones and J. M. Walbrick, *J. Org. Chem.* **34,** 2217 (1969).
23. H. R. Ward, R. G. Lawler, and H. P. Loken, *J. Am. Chem. Soc.* **90,** 7359 (1968).
24. H. Nozaki, T. Aratani, and R. Noyori, *Tetrahedron Lett.* p. 2087 (1968).
25. L. Skattebøl, *Tetrahedron Lett.* p. 2361 (1970).
26. W. R. Moore, K. G. Taylor, P. Müller, S. S. Hall, and Z. L. F. Gaibel, *Tetrahedron Lett.* p. 2365 (1970).
27. W. R. Moore and J. B. Hill, *Tetrahedron Lett.* p. 4343 (1970).
28. W. R. Moore and J. B. Hill, *Tetrahedron Lett.* p. 4553 (1970).
29. L. Skattebøl, *Tetrahedron Lett.* p. 167 (1961).
30. L. Skattebøl, *Acta Chem. Scand.* **17,** 1683 (1963).
31. M. S. Baird, *Chem. Commun.* p. 1145 (1971).
32. L. Skattebøl, *Tetrahedron* **23,** 1107 (1967).
33. L. Skattebøl, *J. Org. Chem.* **31,** 2789 (1966).
34. M. S. Baird and C. B. Reese, *J. Chem. Soc. Chem. Commun.* p. 523 (1972).
35. W. R. Moore, H. R. Ward, and R. F. Merritt, *J. Am. Chem. Soc.* **83,** 2019 (1961).
36. W. R. Moore and H. R. Ward, *J. Org. Chem.* **25,** 2073 (1960).
37. E. T. Marquis and P. D. Gardner, *Chem. Commun.* p. 726 (1966).
38. E. T. Marquis and P. D. Gardner, *Tetrahedron Lett.* p. 2793 (1966).
39. C. G. Cardenas, B. A. Shoulders, and P. D. Gardner, *J. Org. Chem.* **32,** 1220 (1967).
40. M. S. Baird and C. B. Reese, *Tetrahedron* **32,** 2153 (1976).
40a. M. S. Baird and C. B. Reese, *Chem. Commun.* p. 1519 (1970).
41. W. R. Moore and W. R. Moser, *J. Am. Chem. Soc.* **92,** 5469 (1970).
42. W. R. Moore and W. R. Moser, *J. Org. Chem.* **35,** 908 (1970).
43. J. Sonnenberg and S. Winstein, *J. Org. Chem.* **27,** 748 (1962).
44. H. J. J. Loozen, W. A. Castenmiller, E. J. M. Buter, and H. M. Buck, *J. Org. Chem.* **41,** 2965 (1976).
45. T. J. Logan, *Tetrahedron Lett.* p. 173 (1961).
46. W. R. Moore and H. R. Ward, *J. Org. Chem.* **27,** 4179 (1962).
47. R. Maurin and M. Bertrand, *Bull. Soc. Chim. Fr.* p. 2349 (1972).
48. P. D. Gardner and M. Narayana, *J. Org. Chem.* **26,** 3518 (1965).
49. K. G. Untch, D. J. Martin, and N. T. Catellucci, *J. Org. Chem.* **30,** 3572 (1965).
50. M. Bertrand and C. Santelli-Rouvier, *Bull. Soc. Chim. Fr.* p. 2775 (1972).
51. M. Bertrand, M. Santelli, and R. Maurin, *Bull. Soc. Chim. Fr.* p. 998 (1967).
52. T. Hujama, T. Mishima, K. Kitatani, and H. Nozaki, *Tetrahedron Lett.* p. 3297 (1974).
53. L. Skattebøl, *Tetrahedron Lett.* p. 2175 (1965).
54. W. J. Ball, S. R. Landor, and N. Punja, *J. Chem. Soc. C* p. 194 (1967).
55. W. R. Moore and T. M. Ozretich, *Tetrahedron Lett.* p. 3205 (1967).
56. G. Maier, *Tetrahedron Lett.* p. 3603 (1965).
57. F. T. Bond and D. E. Bradway, *J. Am. Chem. Soc.* **87,** 4977 (1965).
58. P. J. Garratt, K. C. Nicholaou, and F. Sondheimer, *J. Org. Chem.* **38,** 865 (1973).
59. P. J. Garratt, K. C. Nicholaou, and F. Sondheimer, *Chem. Commun.* p. 1219 (1970).
59a. P. J. Garratt, K. C. Nicholaou, and F. Sondheimer, *J. Am. Chem. Soc.* **95,** 9582 (1973).
60. P. G. Gassman, *Accts. Chem. Res.* **3,** 26 (1970).
61. P. G. Gassman and D. K. Dygos, *J. Am. Chem. Soc.* **91,** 1543 (1969).
61a. P. G. Gassman, D. K. Dygos, and J. E. Trent, *J. Am. Chem. Soc.* **92,** 2084 (1970).
62. R. G. Weiss, *Tetrahedron* **27,** 271 (1971).
63. D. C. Horwell and C. W. Rees, *Chem. Commun.* p. 1428 (1969).
64. A. D. Walsh, *J. Chem. Soc.* pp. 2260, 2266, 2288, 2296, 2301 (1953). See also Y. Takahata,

G. W. Schnuelle, and R. G. Parr, *J. Am. Chem. Soc.* **93**, 784 (1971); and H. B. Thompson, *J. Am. Chem. Soc.* **93**, 4609 (1971).

65. W. Kirmse, "Carbene, Carbenoide and Carbenanaloge," pp. 174–176, Verlag Chemie, Weinheim, 1969.
66. G. L. Closs and L. E. Closs, *J. Am. Chem. Soc.* **83**, 1003 (1961).
66a. G. L. Closs and L. E. Closs, *J. Am. Chem. Soc.* **85**, 99 (1963).
67. G. L. Closs and L. E. Closs, *J. Am. Chem. Soc.* **83**, 2015 (1961).
67a. G. L. Closs, L. E. Closs, and W. A. Böll, *J. Am. Chem. Soc.* **85**, 3796 (1963).
68. G. L. Closs and W. A. Böll, *Angew. Chem. J. Am. Chem. Soc.* **75**, 640 (1963); G. L. Closs and W. A. Böll, *J. Am. Chem. Soc.* **85**, 3904 (1963); G. L. Closs, W. A. Böll, H. Heyn, and V. Dev, *J. Am. Chem. Soc.* **90**, 173 (1968).
69. G. L. Closs and K. D. Krantz, *J. Org. Chem.* **31**, 638 (1966).
70. F. Fisher and D. E. Applequist, *J. Am. Chem. Soc.* **30**, 2089 (1965).
71. A. C. Day and M. C. Whiting, *J. Chem. Soc. C* p. 1719 (1966).
72. C. Dietrich-Buchecker and M. Franck-Neumann, *Tetrahedron* **33**, 751 (1977).
73. K. B. Wiberg and R. A. Fenoglio, *J. Am. Chem. Soc.* **90**, 3395 (1968).
74. H. M. Frey, *Adv. Phys. Org. Chem.* **4**, 170 (1966).
75. H. A. Stechl, *Chem. Ber.* **97**, 2681 (1964).
76. M. A. Battiste, B. Halton, and R. H. Grubbs, *Chem. Commun.* p. 907 (1967).
77. R. Srinivasan, *Chem. Commun.* p. 1041 (1971).
78. R. Srinivasan, *J. Am. Chem. Soc.* **91**, 6250 (1969).
79. E. J. York, W. Dittmar, J. R. Stevenson, and R. G. Bergman, *J. Am. Chem. Soc.* **94**, 2882 (1972).
79a. E. J. York, W. Dittmar, J. R. Stevenson, and R. G. Bergman, *J. Am. Chem. Soc.* **95**, 5680 (1973).
80. G. L. Closs, L. R. Kaplan, and V. I. Bendall, *J. Am. Chem. Soc.* **89**, 3378 (1967).
81. R. J. Crawford, J. Hamelin, and B. Strehlke, *J. Am. Chem. Soc.* **93**, 3810 (1971).
82. R. D. Steeper and P. D. Gardner, *Tetrahedron Lett.* p. 767 (1973).
83. J. M. Davis, K. J. Shea, and R. G. Bergman, *J. Am. Chem. Soc.* **99**, 1499 (1977).
83a. J. M. Davis, K. J. Shea, and R. G. Bergman, *Angew. Chem.* **88**, 254 (1976).
84. G. Smolinsky, *J. Am. Chem. Soc.* **83**, 4483 (1961).
84a. G. Smolinsky, *J. Org. Chem.* **27**, 3557 (1962).
85. G. Smolinsky and C. W. Pryde, *J. Org. Chem.* **32**, 2411 (1968).
86. K. Isomura, S. Kobayashi, and H. Taniguchi, *Tetrahedron Lett.* p. 3499 (1968).
87. K. Isomura, M. Okada, and H. Taniguchi, *Tetrahedron Lett.* p. 4073 (1969).
88. T. Nishiwaki, *Tetrahedron Lett.* p. 2049 (1969).
89. T. L. Gilchrist, G. E. Gymer, and C. W. Rees, *Chem. Commun.* p. 1519 (1971).
90. D. J. Anderson, T. L. Gilchrist, G. E. Gymer, and C. W. Rees, *Chem. Commun.* p. 1518 (1971).
91. L. A. Wendling and R. G. Bergman, *J. Org. Chem.* **41**, 831 (1976).
92. A. Demoulin, H. Gorissen, A.-M. Hesbain-Fresque, and L. Ghosez, *J. Am. Chem. Soc.* **97**, 4409 (1975).
93. T. Nishiwaki, A. Nakano, and H. Matsuoka, *J. Chem. Soc. C* p. 1825 (1970).
94. D. T. Clark and G. Smale, *Tetrahedron Lett.* p. 3673 (1968).
95. M. J. S. Dewar and S. Kirschner, *J. Am. Chem. Soc.* **93**, 4290, 4291, 4292 (1971).
96. P. Merlet, S. D. Peyerimhoff, R. J. Buenker, and S. Shih, *J. Am. Chem. Soc.* **96**, 959 (1974).
97. G. Wittig, V. Rauchenstrauch, and F. Wingler, *Tetrahedron Suppl.* **7**, 189 (1966) and also refer. 103.
98. M. E. Londrigan and J. E. Mulvaney, *J. Org. Chem.* **37**, 2823 (1972).

120   4. Four Electron–Three Atom Systems

99. P. Eberhard and R. Huisgen, *J. Am. Chem. Soc.* **94,** 1345, 1346 (1972).
100. E. N. Marvell and C. Lin, *Tetrahedron Lett.* p. 2697 (1973).
100a. E. N. Marvell and C. Lin, *J. Am. Chem. Soc.* **100,** 877 (1978).
101. W. T. Ford and M. Newcomb, *J. Am. Chem. Soc.* **95,** 6277 (1973).
102. W. T. Ford and M. Newcomb, *J. Am. Chem. Soc.* **95,** 7186 (1973).
103. M. Newcomb and W. T. Ford, *J. Am. Chem. Soc.* **96,** 2968 (1974).
104. G. Boche, D. Martens, and H.-U. Wagner, *J. Am. Chem. Soc.* **98,** 2668 (1976).
105. R. Huisgen, R. Grashey, and J. Sauer, "Chemistry of Alkenes" (S. Patai, ed.), pp. 806–878, Wiley (Interscience), New York, 1964.
106. B. Shilling and J. P. Snyder, *J. Am. Chem. Soc.* **97,** 4422 (1975).
107. H. Staudinger and K. Mieschler, *Helv. Chim. Acta* **2,** 554 (1919).
108. J. W. Lown, *Rec. Chem. Progress* **32,** 51 (1971).
109. H. W. Heine and R. Peavy, *Tetrahedron Lett.* p. 3123 (1965).
110. A. Padwa and L. Hamilton, *Tetrahedron Lett.* p. 4362 (1965).
111. R. Huisgen, W. Scheer, G. Szeimies, and H. Huber, *Tetrahedron Lett.* p. 397 (1966).
112. R. Huisgen, *Angew. Chem. Internl. Edit.* **2,** 633 (1963).
113. R. Huisgen, W. Scheer, and H. Huber, *J. Am. Chem. Soc.* **89,** 1753 (1967).
114. R. Huisgen, W. Scheer, and H. Huber, *Angew. Chem. Internl. Edit.* **8,** 602 (1969).
115. H. Hermann, R. Huisgen, and H. Mäder, *J. Am. Chem. Soc.* **93,** 1779 (1971).
116. R. Huisgen, W. Scheer, H. Mäder, and E. Brunn, *Angew. Chem. Internl. Edit.* **8,** 604 (1969).
117. R. Huisgen and H. Mäder, *Angew. Chem. Internl. Edit.* **8,** 604 (1969).
117a. R. Huisgen and H. Mäder, *J. Am. Chem. Soc.* **93,** 1777 (1971).
118. H. W. Heine, A. B. Smith III, and J. D. Bower, *J. Org. Chem.* **33,** 1097 (1968).
119. H. W. Heine and R. P. Henzel, *J. Org. Chem.* **34,** 171 (1969).
120. J. W. Lown and K. Matsumoto, *J. Chem. Soc. D* p. 692 (1970).
121. J. W. Lown and K. Matsumoto, *J. Org. Chem.* **36,** 1405 (1971).
122. D. L. Garling and N. H. Cromwell, *J. Org. Chem.* **38,** 654 (1973).
123. E. Hansen and K. Undheim, *J. Chem. Soc. Perkin Trans. 1* p. 305 (1975).
124. H. W. Heine, R. Peavy, and A. J. Durbetaki, *J. Org. Chem.* **31,** 3924 (1966).
125. J. H. Hall and R. Huisgen, *J. Chem. Soc. Chem. Commun.* p. 1187 (1971).
125a. J. H. Hall, R. Huisgen, C. H. Ross, and W. Scheer, *J. Chem. Soc. Chem. Commun.* p. 1188 (1971).
126. J. W. Lown, J. P. Moser, and R. Westwood, *Can. J. Chem.* **47,** 4335 (1969).
127. J. W. Lown and M. H. Akhtar, *Can. J. Chem.* **50,** 2236 (1972).
128. Y. Gelas-Mialhe, R. Hierle, and R. Vessière, *Bull. Soc. Chem. Fr.* p. 709 (1974).
129. A. Padwa and L. Hamilton, *J. Heterocycl. Chem.* **4,** 118 (1967).
130. T. Kauffmann, H. Berg, and E. Kauffmann, *Angew. Chem. Internl. Edit.* **9,** 380 (1970).
131. T. Kauffmann, H. Berg, E. Ludorff, and A. Woltermann, *Angew. Chem. Internl. Edit.* **9,** 960 (1970).
132. T. Kauffmann, D. Berger, B. Scheerer, and A. Woltermann, *Angew. Chem. Internl. Edit.* **9,** 961 (1970).
133. T. Kauffmann and R. Eidenschink, *Angew. Chem. Internl. Edit.* **10,** 739 (1971).
134. T. Kauffmann and R. Eidenschink, *Angew. Chem. Internl. Edit.* **11,** 290 (1972).
135. T. Kauffmann, K. Habersaat, and E. Köppelmann, *Angew. Chem. Internl. Edit.* **11,** 291 (1972).
136. R. Eidenschink and T. Kauffmann, *Angew. Chem. Internl. Edit.* **11,** 292 (1972).
137. T. Kauffmann and R. Eidenschink, *Angew. Chem. Internl. Edit.* **12,** 568 (1973).
138. T. Kauffmann, *Angew. Chem. Internl. Edit.* **13,** 827 (1974).
139. P. B. Woller and N. H. Cromwell, *J. Heterocycl. Chem.* **5,** 579 (1968).

140. F. Texier and R. Carrié, *Bull. Soc. Chim. Fr.* p. 2373 (1972).
141. F. Texier and R. Carrié, *Bull. Soc. Chim. Fr.* p. 2381 (1972).
142. F. Texier and R. Carrié, *Bull. Soc. Chim. Fr.* p. 3437 (1973).
143. F. Texier, R. Carrié, and J. Jaz, *Chem. Commun.* p. 199 (1972).
144. J. W. Lown, R. K. Smalley, G. Dallas, and T. W. Maloney, *Can. J. Chem.* **48,** 89 (1970).
145. G. Dallas, J. W. Lown, and J. P. Moser, *Chem. Commun.* p. 278 (1970).
146. A. Padwa and W. Eisenhardt, *Chem. Commun.* p. 380 (1968).
147. J. E. Baldwin, R. G. Pudussery, A. K. Quershi, and B. Sklartz, *J. Am. Chem. Soc.* **90,** 5325 (1968).
148. J. W. Lown, G. Dallas, and T. W. Maloney, *Can. J. Chem.* **47,** 3557 (1969).
149. J. W. Lown and J. P. Moser, *Chem. Commun.* p. 247 (1970).
150. A. Padwa and W. Eisenhardt, *J. Org. Chem.* **35,** 2472 (1970).
151. W. J. Linn, O. W. Webster, and R. E. Benson, *J. Am. Chem. Soc.* **85,** 2032 (1963).
152. E. F. Ullman and J. E. Milks, *J. Am. Chem. Soc.* **84,** 1315 (1962).
153. E. F. Ullman and J. E. Milks, *J. Am. Chem. Soc.* **86,** 3814 (1964).
154. E. F. Hayes and A. K. Q. Siu, *J. Am. Chem. Soc.* **93,** 2090 (1971).
155. H. C. Longuet-Higgins and E. W. Abrahamson, *J. Am. Chem. Soc.* **87,** 2045 (1965).
156. H. Hamberger and R. Huisgen, *Chem. Commun.* p. 1190 (1971).
157. A. Dahmen, H. Hamberger, R. Huisgen, and V. Morkowski, *Chem. Commun.* p. 1192 (1971).
158. H. H. J. MacDonald and R. J. Crawford, *Can. J. Chem.* **50,** 428 (1972).
159. E. Ullmann, *J. Am. Chem. Soc.* **85,** 3529 (1963).
160. J. M. Dunston and P. Yates, *Tetrahedron Lett.* p. 505 (1964).
161. D. R. Arnold and L. A. Karnishky, *J. Am. Chem. Soc.* **92,** 1404 (1970).
162. D. R. Arnold and Y. C. Chang, *J. Heterocycl. Chem.* **8,** 1097 (1971).
163. J. W. Lown and K. Matsumoto, *Can. J. Chem.* **49,** 3443 (1971).
164. D. L. Garin, *J. Org. Chem.* **34,** 2355 (1969).
165. W. J. Linn and R. E. Benson, *J. Am. Chem. Soc.* **87,** 3657 (1965).
166. W. J. Linn, *J. Am. Chem. Soc.* **87,** 3665 (1965).
167. W. J. Linn and E. Ciganek, *J. Org. Chem.* **34,** 2147 (1969).
168. A. Robert and A. Foucaud, *Bull. Soc. Chim. Fr.* p. 2531 (1969).
169. A. Robert, J.-J. Pommeret, and A. Foucaud, *Tetrahedron Lett.* p. 231 (1971).
170. J.-J. Pommeret and A. Robert, *Tetrahedron* **27,** 2977 (1971).
171. A. Robert, J.-J. Pommeret, and A. Foucaud, *Tetrahedron* **28,** 2085 (1972).
172. A. Robert, J.-J. Pommeret, E. Marchand, and A. Foucaud, *Tetrahedron* **29,** 463 (1973).
173. J. M. Brown, B. T. Golding, and J. J. Stofko, Jr., *Chem. Commun.* p. 319 (1973).
174. J. E. Baldwin and C. Ullenius, *J. Am. Chem. Soc.* **96,** 1542 (1974).
175. J. C. Pommelet, N. Manisse, and J. Chuche, *Tetrahedron* **28,** 3929 (1972).
176. J. C. Palladini and J. Chuche, *Tetrahedron Lett.* p. 4383 (1971).
176a. J. C. Palladini and J. Chuche, *Bull. Soc. Chim. Fr.* p. 197 (1974).
177. M. S. Medinagh and J. Chuche, *Tetrahedron Lett.* p. 793 (1977).
178. R. J. Crawford, S. B. Lutener, and R. D. Cockroft, *Can. J. Chem.* **54,** 3364 (1976).
179. R. M. Kellogg, *Tetrahedron* **32,** 2165 (1976).
180. H. Staudinger and J. Siegwart, *Helv. Chim. Acta* **3,** 833 (1920).
181. A. Schönberg and S. Nickel, *Chem. Ber.* **64,** 2323 (1931).
182. W. J. Middleton and W. H. Sharkey, *J. Org. Chem.* **30,** 1384 (1965).
182a. W. H. Middleton, *J. Org. Chem.* **34,** 3201 (1969).
183. R. M. Kellogg and S. Wassenaar, *Tetrahedron Lett.* p. 1987 (1970).
184. J. Buter, S. Wassenaar, and R. M. Kellogg, *J. Org. Chem.* **37,** 4045 (1972).
185. R. M. Kellogg, S. Wassenaar, and J. Buter, *Tetrahedron Lett.* p. 4689 (1970).

186. R. M. Kellogg, M. Noteboom, and J. K. Kaiser, *J. Org. Chem.* **40,** 2573 (1975).
187. D. H. R. Barton, E. H. Smith, and B. J. Willis, *Chem. Commun.* p. 1225 (1970).
188. A. P. Schaap and G. R. Faler, *J. Org. Chem.* **38,** 3061 (1973).
189. H. Sauter, H.-G. Hörster, and H. Prinzbach, *Angew. Chem. Internl. Edit.* **12,** 991 (1973).
190. L. K. Bee, J. Beeby, J. W. Everett, and P. J. Garratt, *J. Org. Chem.* **40,** 2212 (1975).
191. D. H. R. Barton and B. J. Willis, *Chem. Commun.* p. 1225 (1970).
191a. D. H. R. Barton and B. J. Willis, *J. Chem. Soc. Perkin Trans. 1* p. 305 (1972).
192. C. T. Pederson, *Acta. Chem. Scand.* **22,** 247 (1968).
193. C. E. Diebert, *J. Org. Chem.* **35,** 1501 (1970).
194. A. P. Krapcho, D. R. Rao, M. P. Silvon, and B. Abegaz, *J. Org. Chem.* **36,** 3885 (1971).
194a. A. D. Krapcho, D. E. Horn, D. R. Rao, and B. Abegaz, *J. Org. Chem.* **37,** 1575 (1972).
195. D. H. R. Barton, F. S. Guziec, Jr., and I. Shahak, *J. Chem. Soc. Perkin Trans. 1* p. 1794 (1974).
196. J. P. Snyder, *J. Am. Chem. Soc.* **96,** 5005 (1974).
197. W. D. Emmons, *J. Am. Chem. Soc.* **78,** 6208 (1956).
198. W. D. Emmons, *J. Am. Chem. Soc.* **79,** 5739 (1957).
199. J. S. Splitter and M. Calvin, *J. Org. Chem.* **23,** 65 (1958).
199a. J. S. Splitter and M. Calvin, *J. Org. Chem.* **30,** 3427 (1965).
200. M. F. Hawthorne and R. D. Strahm, *J. Org. Chem.* **22,** 1263 (1957).
201. J. S. Splitter, T.-M. Su, H. Ohno, and M. Calvin, *J. Am. Chem. Soc.* **93,** 4075 (1971).
202. F. D. Greene and S. S. Hecht, *J. Org. Chem.* **35,** 2482 (1970).
203. J. H. Davis, W. A. Goddard II, and R. G. Bergman, *J. Am. Chem. Soc.* **98,** 4015 (1976).
204. A. Padwa, J. Smolanoff, and A. Tremper, *J. Org. Chem.* **41,** 543 (1976).
205. T. Nishiwaki and T. Saito, *J. Chem. Soc. C.* p. 3021 (1971).
206. G. H. Coleman and G. P. Waugh, *Proc. Iowa Acad. Sci.* **49,** 115 (1933).
207. G. H. Coleman and C. S. Nicholopolous, *Proc. Iowa Acad. Sci.* **49,** 286 (1942).
208. T. W. J. Taylor, J. S. Owen, and D. Whittacker, *J. Chem. Soc.* p. 206 (1938).
209. B. K. Campbell and K. N. Campbell, *J. Org. Chem.* **9,** 178 (1944).
210. C. G. Stuckwisch, *Synthesis* p. 469 (1973).
211. R. M. Kellogg, *Tetrahedron* **32,** 2165 (1976).
212. W. J. Jones and U. H. Brinker, "Pericyclic Reactions" (A. P. Marchard and R. E. Lehr, eds.), Vol. I, pp. 169–189, Academic Press, New York, 1977.
213. P. W. Dillon and G. R. Underwood, *J. Am. Chem. Soc.* **99,** 2435 (1977).
214. P. Blickle and H. Hopf, *Tetrahedron Lett.* p. 449 (1978).
215. J. Ripoll and A. Thullier, *Tetrahedron* **33,** 1333 (1977).
216. J. H. Davis, W. A. Goddard III, and R. G. Bergman, *J. Am. Chem. Soc.* **99,** 2427 (1977).
217. See ref. 212, pp. 131–159.
218. R. G. Ford, *J. Am. Chem. Soc.* **99,** 2389 (1977).
219. L. Carlsen and J. P. Snyder, *J. Org. Chem.* **43,** 2216 (1978).
220. V. Markowski and R. Huisgen, *Chem. Commun.* p. 439 (1977).
221. V. Markowski and R. Huisgen, *Chem. Commun.* p. 440 (1977).
222. R. Huisgen, *Angew. Chem.* **89,** 589 (1977).
223. R. J. Crawford, V. Vukov, and H. Tokunaga, *Can. J. Chem.* **51,** 3718 (1973).
224. D. J. Humphreys, C. E. Newhall, G. H. Phillips, and G. A. S. Smith, *J. Chem. Soc. Perkin Trans. 1* p. 45 (1978).
225. A. Krebs and W. Rüger, *Tetrahedron Lett.* p. 1305 (1979).
226. F. Cordt, R. M. Frank, and D. Lenoir, *Tetrahedron Lett.* p. 505 (1979).
227. T. G. Back, D. H. R. Barton, M. R. Britten-Kelly, and F. S. Guziec, *J. Chem. Soc. Perkin Trans. 1* p. 2079 (1976).

228. E. Schmitz, *Angew. Chem. Int. Ed. Engl.* **3,** 333 (1964).

229. E. Schmitz, A. Stark, and C. Hörig, *Chem. Ber.* **98,** 2509 (1965).

230. C. G. Overberger and J.-P. Anselme, *J. Org. Chem.* **29,** 1188 (1964).

231. E. Voigt and H. Meier, *Angew. Chem. Int. Ed. Engl.* **14,** 103 (1977).

232. See refer. 212, pp. 136–159.

233. M. J. S. Dewar and D. Landman, *J. Am. Chem. Soc.* **99,** 6179 (1977).

234. M. T. H. Liu and B. M. Jennings, *Can. J. Chem.* **55,** 3596 (1977).

235. N. P. Smith and I. D. R. Stevens, *J. Chem. Soc. Perkin Trans. 2* p. 213 (1979).

CHAPTER 5

FOUR ELECTRON–FOUR AND
FIVE ATOM SYSTEMS

I. CYCLOBUTENE-BUTADIENE

A. Introduction

The key molecule in this classic case of an electrocyclic reaction is cyclo-butene, because its strain energy of 28.5 kcal/mol[1] is sufficient to render it thermodynamically less stable than butadiene. Cyclobutene was prepared first in the early years of this century,[2] and during a study of its properties it was noted that a sample prepared by Hofmann elimination was con-taminated with butadiene.[3] Attempted preparations by heating cyclobutyl-ammonium phosphate or bromocyclobutane with quinoline gave only butadiene. Roberts and Sauer[4] noted that pure cyclobutene could be ob-tained by elimination if the olefin was removed from the reaction mixture immediately. Pyrolyses of the xanthate of cyclobutanol produced solely butadiene. In retrospect, the evidence pointing to a thermal instability of cyclobutene seems quite convincing, but no attempt was made prior to the 1950s to establish this rigorously.

Early in the 1950s the thermal instability of a number of substituted cyclo-butenes was recognized.[5–9] However, the first careful mechanistic study appeared in 1958.[10] Conversion of cyclobutene to butadiene was shown to be a homogeneous unimolecular reaction. At 60% conversion the sole product was butadiene and a material balance of 99.3% was attained. The rate was independent of the surface area, and reaction was not inhibited by nitric oxide nor was an induction period observed. The activation energy was 32.5 kcal/mol, $\Delta S^{\ddagger} \simeq 0$. The data are completely consistent with a simple intramolecular reaction, i.e., a retro-electrocyclic reaction.

B. Theoretical Studies

The early qualitative approaches to the electrocyclic reaction of buta-diene–cyclobutene described in Chapter 2 were followed by more quantita-tive calculations of the potential energy surface for the reaction. In almost all these later studies simplifying assumptions proved necessary to moderate the computational problems. How much these simplifications have influenced the results is uncertain, but the calculations have led to very different conclusions. This diversity was illustrated dramatically by the first pair of studies published.[11,12,12a]

Filer[11] used the extended Hückel (EH) method to determine the potential surface for the ground state reaction via conrotatory, disrotatory, and asymmetric paths. The four carbon atoms were confined to a plane, i.e., no rotation about the C_2-C_3 bond (butadiene numbering) was permitted. Asymmetry provided no energetic advantage, and only the two symmetric routes need be discussed. On the calculated surface butadiene (s-cis) lay ca. 19 kcal/mol lower than cyclobutene and the energy barriers for conrotatory and disrotatory routes were almost identical. However, the conrotatory route was considered preferable since in the region of the transition state a drastic change in molecular geometry was required on the disrotatory path. The transition state occurred rather late (toward cyclobutene), i.e., the methylene terminii had rotated from 0° (butadiene) to 60° at the transition state.

The second study,[12,12a] a valence bond procedure, varied three parameters—methylene rotation, rotation about C_2-C_3, and bending of the $C_1C_2C_3$ or $C_2C_3C_4$ bond angles. No bond length changes were permitted and the hybridization of C_1 and C_4 was maintained as Sp^2. The calculation included electronic, bond angle strain and non-bonded interaction energies. Within the restrictions imposed, butadiene (s-cis) is non-planar with a 20° twist about the central bond, and butadiene is more stable than cyclobutene by about 36.5 kcal/mol. Thermal ring closure via the conrotatory route required no activation energy, while the forbidden disrotatory closure had an activation energy of about 100 kcal/mol. On the preferred route as the bond angles ($C_1C_2C_3$ and $C_2C_3C_4$) decrease from 120° to 90° the methylene groups slowly conrotate and the twist about the C_2-C_3 bond increases to 40°. When the bond angles reach 95° methylene rotation suddenly accelerates and the twist drops rapidly toward 0°. Since no barrier to conrotatory closure was found, the nature of the transition state cannot be specified.

Dewar and his students have made two separate studies of the butadiene electrocyclization, the first using MINDO/2[13] and the second MINDO/3.[14,15] All of the reports are brief so only sketchy results are available. The earlier study[13] showed that the conrotatory cyclization was a normal process, i.e., both forward and reverse reactions follow the same path, and if the four carbon atoms were restricted to coplanarity, E_a was 90 kcal/mol. MINDO/2 calculation of the antiaromatic disrotatory closure gave the unanticipated result that neither ring closure nor its reverse could occur. Essentially the reactant and product lie in two potential valleys separated by a ridge which does not contain the lowest energy route out of either valley. The result was attributed to the tendency of each calculation (forward and reverse) to lead toward an excited state of the reactant. Non-crossing was presumed to result from retention of the symmetry plane throughout the reaction. This was confirmed by a MINDO/3 calculation[14] in which the symmetry restriction was raised. The forbidden stereochemistry was then achieved via a single methylene rotation at the transition state. Disrotation occurred after passage through the transition state. It is important to note that this model indicates the forbidden transition state lies on the path which would lead to the orthogonal diradical, and is thus energetically lower than the diradical. The MINDO/3 transition state for the conrotatory route is shown in Fig. 5-1. The allowed reaction had $E_a = 49.0$ kcal/mol, and $\Delta E = E_a^{\text{forb.}} - E_a^{\text{allow.}}$ was 16.6 kcal/mol.

A careful *ab initio* SCF calculation with configuration interaction has been carried out.[16] Again the results are surprising. While the predictions of the orbital symmetry analyses are confirmed, the mechanism of ring opening via either rotational process was found to be a disjointed proceeding with bond stretching preceding rotation of the methylene groups. After completion of the rotation, the CCC angles widened until s-*cis*-butadiene was formed. Coplanarity of the four carbons was assumed. The SCF calculation gave E_a for the conrotatory reaction as 53 kcal/mol (cyclobutene → butadiene), but with configuration interaction the value was reduced to 45 kcal/mol. As expected, configuration interaction plays a more important role with the forbidden reaction, and the best result obtained gave $\Delta E_{\text{disrot.-conrot.}} = 13.8$ kcal/mol. Further calculations which allowed for out-of-plane twisting about the C_2-C_3 bond[17] did not change the disjointed character of the stretch vs. rotation process, but did indicate a moderate reduction of E_a for the conrotatory reaction (to ca. 40 kcal/mol). At the transition state the dihedral angle was $45°$, but the authors found that s-*cis*-butadiene was still the initial product. An appraisal of the frontier orbital approach and the extent to which its predictions can be expected to match the results of quantitative calculations has been made,[18] and in general for electrocyclic reactions the prognosis was excellent.

A second MINDO/2 study has appeared.[19] The transition state geometry is essentially that of Fig. 5-1. However, the calculation indicates cyclobutene is 20.5 kcal/mol *more* stable than butadiene, a general problem of MINDO/2 which overestimates the stability of small rings. E_a for conrotatory reaction was calculated as 49.5 kcal/mol. This calculation suggested that the direct product of ring opening could be s-*trans*-butadiene.

Rastelli *et al.*[20] have used an "orbital following" procedure with Del Re's maximum localization criterion to develop geometries of minimum energy

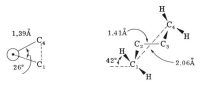

Fig. 5-1. Newman projection and front view of the conrotatory transition state (MINDO/3).

for cyclobutene and butadiene. Imposing the reaction conditions which specify the C_1-C_4 distance (R) and the rotation angle (θ) of the terminal methylene groups with values intermediate between those of the reactant and product, geometries for intermediate states were then determined by the same procedure. Energy values for the states were calculated from the Hückel approximation and non-bonded interaction energies. The resultant contour maps for the reactant surfaces permit development of reaction profiles for the conrotatory and disrotatory reactions. The conrotatory process appears to proceed normally with an activation energy (cyclobutene → butadiene) of 33.4 kcal/mol. No minimum energy path connects reactant and product in the disrotatory case.

Tee and Yates[21,21a] have used the principle of least motion to calculate whether cyclobutene opens in a conrotatory or disrotatory manner. The results agree with the orbital symmetry rule in that case. The HMO method was used to predict the effect of benzannelation which indicated a lowering of $\Delta H^{\ddagger}$ for the disrotatory route.[450]

The series of theoretical studies is a study in contrasts. The conrotatory stereochemistry is consistently better than the disrotatory, but the energy difference varies from nearly zero to 100 kcal/mol. The transition state varies from early to late, and the activation energy for the allowed reaction varied from zero to close to 50 kcal/mol. Without some experimental information to set up guidelines, it would be difficult to choose the best results at present. Obviously the energetics of the electrocyclic reaction is a very sensitive function of the molecular mechanics.

C. The Allowed Reaction

The prime requisite for the "symmetry allowed" reaction of cyclobutene is that it give butadiene by a conrotatory process. Thus, stereochemistry is one experimental criterion for the allowed reaction.

1. Stereochemistry

Orbital symmetry theory predictions for 3,4-disubstituted cyclobutenes are summarized in Eqs. (5-1) and (5-2). Further predictions can be added

$$(5\text{-}1)$$

$$(5\text{-}2)$$

if it is assumed that the largest substituent will rotate outwards. In many respects this assumption about the inward or outward rotation can be more important than the conrotatory rule, though certainly less stereorestrictive. Thus, the product composition from 3-monosubstituted, 3,4-disubstituted (in most cases), most 3,3,4-trisubstituted and 3,3,4,4-tetrasubstituted molecules will depend on which groups turn outward whether conrotatory rotation can be observed or not.

The simplest clear-cut test of the conrotational rule is provided by the *cis*-3,4-disubstituted cyclobutenes. A single product with the thermodynamically less stable cis, trans configuration is predicted [Eq. (5-1) $a = b$]. The first example of this type is shown in Eq. (5-3). Synthesis insured the cis

$$(5\text{-}3)$$

configuration and the sole product (79%) was dimethyl cis, *trans*-muconate.[22,22a]

Similar results have been obtained for *cis*-3,4-dichlorocyclobutene,[23] *cis*-3,4-dimethylcyclobutene [Eq. (5-4)][24,25] and *cis*-3,4-diphenylcyclobu-

$$(5\text{-}4)$$

tene.[238] Phenyl substituents should increase the rate of a diradical reaction.[27] If the transition state for the forbidden reaction should possess considerable radical character, that process might compete with the allowed reaction. However, the phenyl substituents stabilized the allowed transition state and again the allowed process was the sole reaction observed. In this case the stereoselectivity exceeded 99%.[26] A recent study of *cis*-3,4-dimethylcyclobutene[28] has extended the stereoselectivity of the allowed process to 99.995%.

In view of these results it is clear that *trans*-3,4-disubstituted cyclobutenes should produce exclusively the more stable *trans, trans*-1,4-disubstituted butadienes. There appear to have been only two of these simple cases subjected to careful examination. Winter[24] showed that *trans*-3,4-dimethylcyclobutene gives at least 98% of *trans, trans*-2,4-hexadiene. While *trans*-3,4-dichlorocyclobutadiene was not isolated, decomposition of **75** gave pure

$$(5\text{-}5)$$

85%
m.p. 37–38°

trans, trans-1,4-dichlorobutadiene.[7]

Additional substituents at C_1 or C_1 and C_2 would be expected to introduce steric interactions with the outward rotating groups and thus in some cases to alter the stereoselectivity. With *cis*-3,4-disubstituted cyclobutenes having like substituents no alteration of stereoselectivity has been observed, though the analytical sensitivity was undoubtedly insufficient to uncover the minor changes which might have been expected. Thus, **76**,[30,31,32] **77**,[33] and dimethyl 1,2-diphenyl-*cis*-3,4-cyclobutene dicarboxylate[451] gave only the *E,Z*-butadienes on thermolysis [Eqs. (5-6) and (5-7)]. For *trans*-3,4-disub-

$$(5\text{-}6)$$

$$(5\text{-}7)$$

stituted examples with equivalent substituents the ratio of trans, trans and cis, cis products should show an increase in the cis, cis isomer. However, only *trans, trans*-3,4-dimethyl-2,4-hexadiene (>99% purity) was obtained from *trans*-1,2,3,4-tetramethylcyclobutene.[32] Though *trans*-1,2,3,4-tetraphenylcyclobutene could not be obtained in pure form because of its thermal instability, rearrangement at 25° gave E,E-1,2,3,4-tetraphenylbutadiene as the main product.[33] A most intriguing result was reported by Winter and Honig[34] when **78** was subjected to thermolysis [Eq. (5-8)]. The apparent

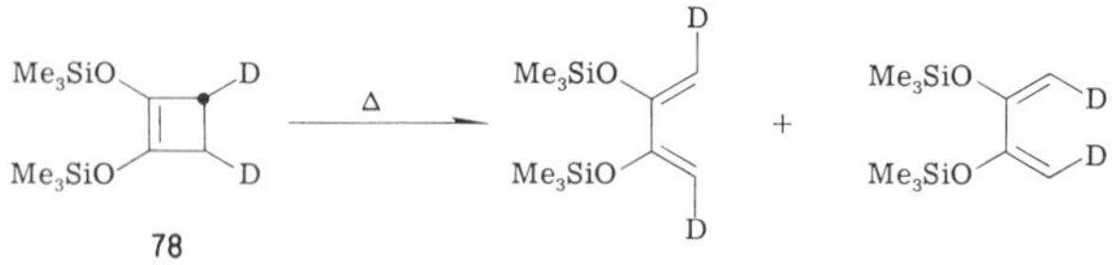

$$\tag{5-8}$$

78   47.7%   52.3%

sensitivity to small steric factors exhibited there would suggest that with larger groups at C_3 and C_4, the E,E product might be formed with very high selectivity. However, when **79** was heated, the kinetic results correspond to a

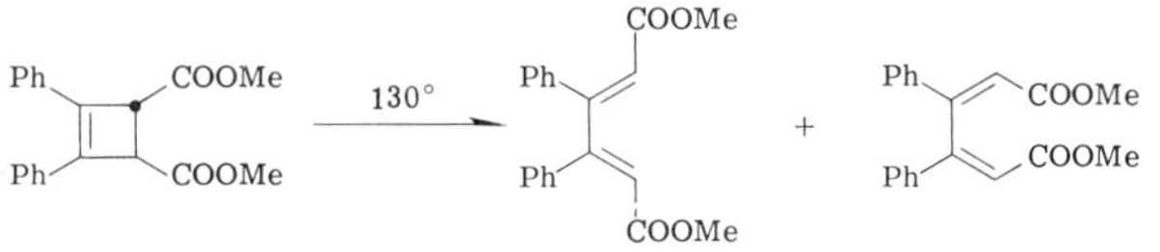

$$\tag{5-9}$$

79   76%   24%

ratio of 76% E, E and 24% Z, Z forms of the product [Eq. (5-9)].[35] The Z,Z-butadiene is thermodynamically the most stable isomer, and since the reactions of Eq. (5-9) are reversible, the final product ratio is not kinetically determined.[451]

Maier and Wiessler[36] made the surprising observation that **80** gives a 9:1 mixture of E,E and Z,Z products [Eq. (5-10)]. They also reported that

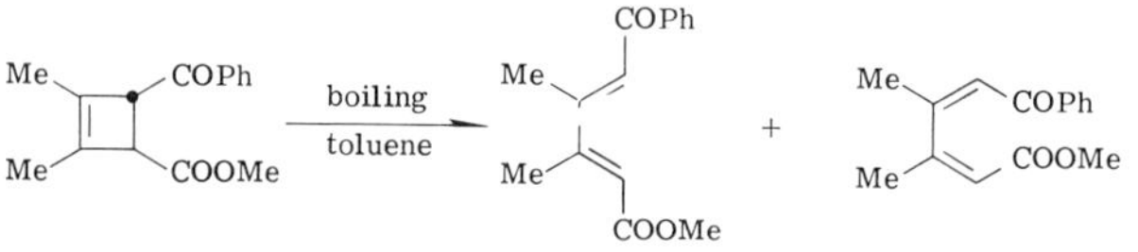

$$\tag{5-10}$$

80   90%   10%

81 gave a 2:1 mixture of two E,Z products [Eq. (5-11)]. The major isomer in

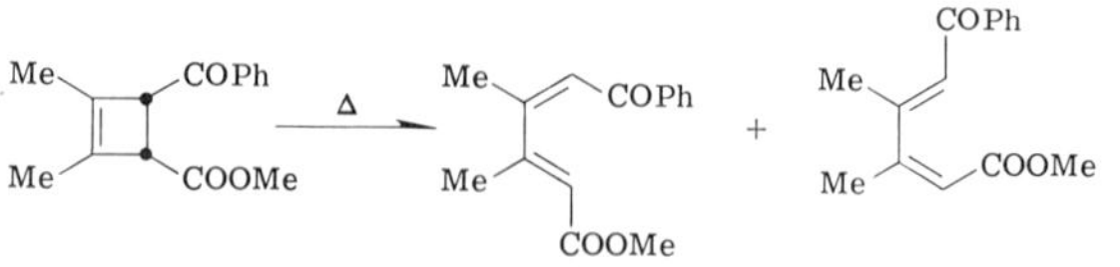

$$\tag{5-11}$$

81   67%   33%

this latter case can be transformed into the minor isomer of Eq. (5-10) via an intermediate pyran (see Chapter 7, Section II).

$$(5\text{-}12)$$

Reactions of 3-substituted cyclobutenes leading to either *trans-* or *cis*-1-substituted butadienes are thermally allowed, but inward rotation is sterically more congested than the converse. In fact the steric problem appears to be serious enough to prevent formation of any *cis*-butadiene in the examples studied to date. Thus, 3-methylcyclobutene gives only *trans*-1,3-pentadiene, though ca. 0.2% of the cis isomer would not have been found by the analysis used.[37] Not surprisingly only *trans*-2-methyl-1,3-pentadiene was identified in the product from 1,3-dimethyl-cyclobutene. Only *trans*-1-phenyl-1,3-butadiene was obtained from 3-phenylcyclobutene,[39] and both 1,4-dimethyl[38] and 1,2,3-trimethylcyclobutenes[40] gave only trans products. Attempted solvolysis of 3-(β-tosyloxyethyl) cyclobutene in acetic acid gave *trans*-1-acetoxy-3,5-hexadiene, and ring opening of 3-(β-acetoxyethyl) cyclobutene in boiling xylene gave the same product.[41] Thermolysis of 2-methyl-3-trimethylsiloxycyclobutene gave the trimethylsilyl enol ether of 2-vinylpropionaldehyde in trans configuration.[42] Careful thermolysis of 2-methyl-2-cyclobutenol-OD at 100° permitted isolation of the deuteriated enol of 2-vinylpropionaldehyde.[42] A *trans*-triene is obtained on ring opening of 1-methyl-3-(2-methylpropenyl) cyclobutene.[43]

Obviously with 3,3-disubstituted cyclobutenes one substituent must turn inward. Main interest attaches to molecules with two different substituents, but with monocyclic systems very little appears to have been done. Pyrolysis of **82** gave three products [Eq. (5-13)].[44] one of which was not identified.

$$(5\text{-}13)$$

Clearly the methyl group rotates inward at least 83% of the time. The third product could conceivably have been Z-2-methyl-4-phenyl-1,3-pentadiene. Conversely, when **83** was heated one single product was isolated [Eq. (5-14)].[44] In this case the larger group appears to have turned inward

$$(5\text{-}14)$$

exclusively. It does not seem likely that the alternative E-triene would revert
to cyclobutene under these conditions. The behavior of 3-ethyl-3-methyl-
cyclobutene is also surprising in that both *cis*- and *trans*-4-methyl-1,3-
hexadienes are found [Eq. (5-15)], but the *cis*-isomer is the main product.[45]

$$\text{(5-15)}$$

68%   32%

Trisubstituted cyclobutenes with a 3,3,4-substitution pattern should
undergo a conrotatory ring opening dominated by outward rotation of two
of the substituents. All cases known [see for example Eq. (5-16)] follow the
expected pattern.[36,452,453]

$$\text{(5-16)}$$

R = Me Y = COPh
R = Ph Y = COOMe

All of the known examples of tetrasubstituted (3,3,4,4) systems are sym-
metrically substituted with a plane of symmetry perpendicular to the ring.
All give rise to the single butadiene expected on the basis of the conrotatory
rule [Eq. (5-17)]. Examples with A=B=Me D=COOMe,[31] A=Cl, B=Me,

$$\text{(5-17)}$$

D=COOMe,[46] and A=B=Ph, D=Me[47] A=Ph, B=Me, D=COOMe[452]
have been studied.

The stereochemical results with monocyclic cyclobutenes are also valid
for benzocyclobutenes, but the very ready reversibility introduces a new
feature which can influence the apparent stereochemistry. Monosubstituted
benzocyclobutenes favor outward rotation of the single substituent. Generally
the stereochemical result now must be determined by trapping the *o*-xylylene
via a Diels–Alder reaction [Eq. (5-18)].[48] Stereochemistry of the ring open-

$$\text{(5-18)}$$

ing was kinetically controlled since optically active benzocyclobutenol was not racemized during the reaction. The adduct was optically inactive as expected. A similar result was obtained with the methyl ether of cyclobutenol. Preferential outward rotation was also observed by Oppolzer[49] with an amide substituent [Eq. (5-19)].

$$(5\text{-}19)$$

However, it can be demonstrated that inward rotation must occur in certain cases, since internal trapping via a twin electrocyclization was observed. Thus, for example, vinylbenzocyclobutene gives 1,2-dihydronaphthalene as

$$(5\text{-}20)$$

the sole isolable product.[50] An analogous reaction leads to an isochromene [Eq. (5-21)].[51,52]

$$(5\text{-}21)$$

R = Me, Et, i-Pr, Ph

sym-Diphenylbenzocyclobutenes, both cis and trans, have been carefully studied. The two isomers are interconverted slowly even at 65°.[53] However, each isomer undergoes ring opening to an intermediate which can be trapped by a reactive dienophile.[54] Kinetic analysis showed the presence of an inter-

$$(5\text{-}22)$$

mediate.[54] The adducts were identified by their symmetry properties, the trans reactant with a twofold axis gave an adduct with a plane of symmetry, and the cis reactant with a symmetry plane produced an axially symmetric adduct.[54,55,56] In view of the symmetric nature of the Diels–Alder reaction, these results show that formation of the intermediate involves a conrotatory process. *trans*-Dimethoxybenzocyclobutene reacts readily in the presence of maleic anhydride to give a single stereoisomeric adduct which was assigned the configuration shown [Eq. (5-23)].[57] The assignment appears to be based

$$(5\text{-}23)$$

on theory. Conversely, *cis*-dimethoxybenzocyclobutene does not undergo ring opening even at 110° and eventually decomposes.[57] The methoxyl group apparently exhibits a strong inhibition of an inward rotation.

This conclusion is supported by the failure of the asym-dimethoxybenzo-cyclobutene to undergo a thermal ring opening.[57] The same sort of behavior is exhibited by **84** [Eq. (5-29)].[58] No explanation or rationale for this special

$$(5\text{-}24)$$

84

resistance to an inward rotation has been advanced. It is moreover very interesting to note that that reluctance does not carry over to monocyclic cyclobutenes. Even a 3,3,4,4-tetraalkoxycyclobutene reacts quite readily under normal thermolysis [Eq. (5-25)].[59] Other examples with bicyclic

$$(5\text{-}25)$$

$$R = Me, Et$$
$$Ar = Ph, \textit{p}\text{-Tol, An,}$$

molecules are illustrated by Eqs. (5-26) and (5-27).[60] Thus, the unexpected behavior is apparently restricted to benzocyclobutenes.

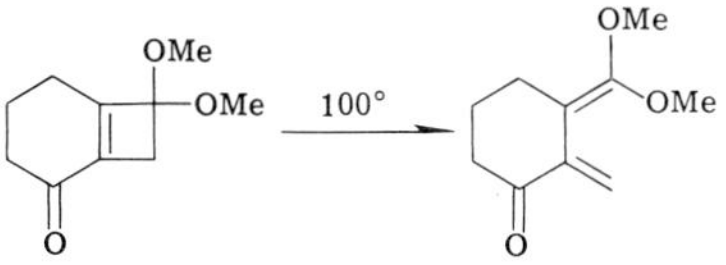

(5-26)

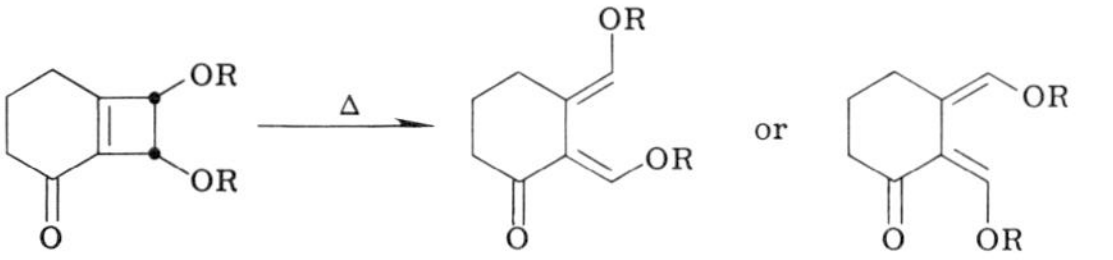

(5-27)

Disubstituted benzocyclobutenes with both substituents on one carbon normally show steric rotational control.[61] The results shown here were based on trapping with maleic anhydride or by a study of the products derived

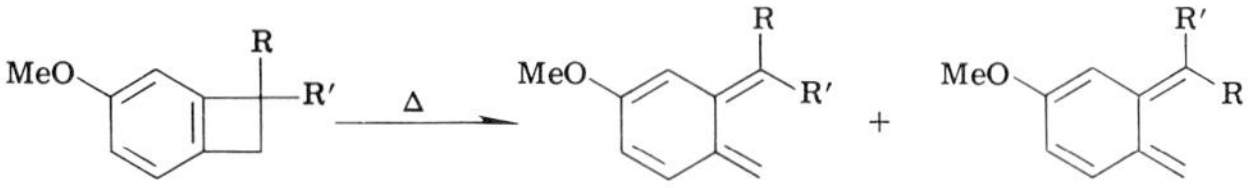

(5-28)

R	R'		
Me	CN	54%	46%
(CH$_2$)$_4$CH=CH$_2$	CN	100%	
(CH$_2$)$_4$CH=CH$_2$	CH$_2$CN	60%	40%
(CH$_2$)$_5$Me	Me	50%	50%

from a 1,5-hydrogen shift [Eq. (5-29)]. No evidence that the product ratios

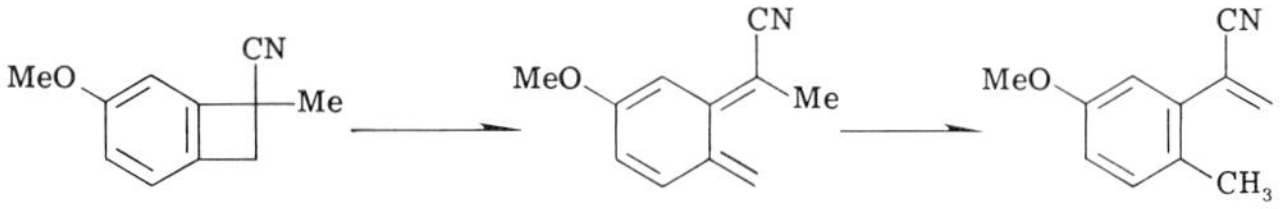

(5-29)

were kinetically determined was provided. When one of the two groups is hydroxyl or alkoxyl that group never turns inward [Eqs. (5-30) and (5-31)].[57]

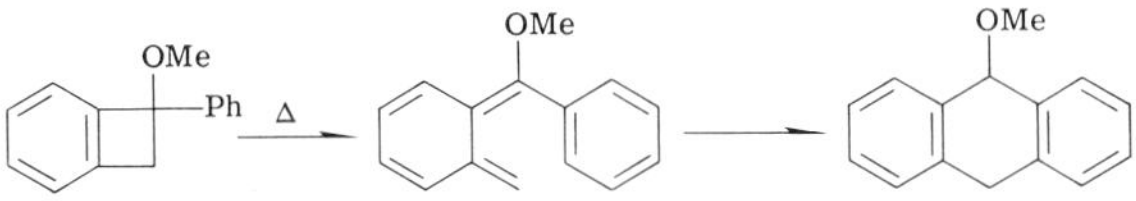

(5-30)

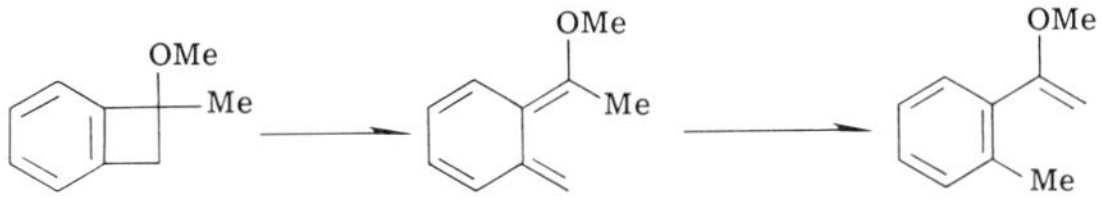

(5-31)

2. Structural Effects on Rates

Ring openings of cyclobutenes free to adopt the conrotatory path vary in rate from molecules which open so readily at room temperature that isolation of the pure reactant is very difficult[33] to compounds which do not react at 300°.[80] Cyclobutenes for which adequate rate data have been reported are listed in Table 5-1 along with representative rates and activation parameters. Perhaps the most striking feature of the table is the consistent pattern of a small positive $\Delta S^{\ddagger}$. This leads to the conclusion that the transition state is quite similar in structure to the reactant. Also notable is the fact that with the exception of phenyl and fluoro substituents, the variation in rate with structure is modest, overall a factor of less than one hundred. Generally, substitution in the 1- or 1,2-positions lowers the rate via an increase of $\Delta H^{\ddagger}$

TABLE 5-1

Rate Data for Cyclobutenes Undergoing Conrotatory Ring Opening

Compound	Temperature (°)	$k \times 10^9$ sec^{-1}	E_a	log A	$\Delta H^{\ddagger}$	$\Delta S^{\ddagger a}$	Reference
Cyclobutene	150	2.50	32.2	13.06	31.4	0.50	62
1-Methylcyclobutene	152	0.654	35.1	13.79	34.3	3.7	63
1-Ethylcyclobutene	152	0.71	34.6	13.29	33.8	1.5	64
1-Propylcyclobutene	150	0.63	34.6	13.55	33.8	2.7	65
1-Isopropylcyclobutene	150	0.46	34.7	13.55	33.9	2.7	65
1-Allylcyclobutene	150	0.66	34.2	13.48	33.4	2.4	65
1-Cyclopropylcyclobutene	152	0.88	34.1	13.48	33.3	2.4	65
1-Chlorocyclobutene	150	0.79	33.6	13.26	32.8	1.4	66
1-Bromocyclobutene	150	1.36	33.8	13.53	33.0	2.6	66
1-Cyanocyclobutene	—	—	33.6	13.38	32.8	2.0	67
3-Methylcyclobutene	150	16.7	31.5	13.52	30.7	2.6	37
3-Phenylcyclobutene	97.6	15	26.0	12.4	25.2	−2.5	39
3-β-Acetoxyethylcyclobutene	100	0.122	—	—	29.1	−3.2	41
1,2-Dimethylcyclobutene[b]	159	0.435	36.0	13.84	35.2	4.1	68
1,2-Dimethylcyclobutene[c]	155	0.42	34.6	13.29	33.8	1.5	69
1,2-Diphenylcyclobutene	155	2.79	32.0	12.8	31.2	−0.69	70
1-Ethyl-2-vinylcyclobutene	161	0.58	35.7	13.73	34.9	3.6	40
1,3-Dimethylcyclobutene	153	5.06	33.0	13.65	32.2	3.2	38
1,4-Dimethylcyclobutene	153	2.40	33.4	13.52	32.6	2.6	38
3,3-Dimethylcyclobutene	159	0.48	36.1	13.93	35.3	4.5	71
3,3-Diethylcyclobutene	155	0.64	34.7	13.53	33.9	2.6	45
3-Ethyl-3-methylcyclobutene[d]	160	0.239	35.9	13.53	35.1	2.6	45
3-Ethyl-3-methylcyclobutene[e]	160	0.522	35.2	13.50	34.4	2.5	45
cis-3,4-Dimethyl	—	—	34.3	13.88	33.5	4.2	25
cis-3,4-Diphenyl	26.8	0.180	24.5	13.1	23.7	0.7	26

by about 1–2 kcal/mol. Conversely, substitution in the 3-position increases the rate via a decrease in $\Delta H^{\ddagger}$ of similar magnitude.

Considerable interest attaches to the question of how severe is the steric inhibition to inward rotation of groups during the conrotatory process. For a methyl group this can be evaluated by consideration of 2- methyl- and 3,3-dimethylcyclobutenes. Following the lead of Frey et al.[71] we will use $\Delta G^{\ddagger}$ for this evaluation. If one assumes that the second methyl group would augment the rate to the same extent as the first in the absence of the steric effect, then the inward turning methyl raises $\Delta G^{\ddagger}$ by 5.4 kcal/mol. The same effect can also be evaluated from the cis-3,4-dimethylcyclobutene rate, giving a value of 3.3 kcal/mol. Again, we can estimate the same parameter from the rate data for cis- and trans-1,2,3,4-tetramethylcyclobutenes, giving again the 3.3 kcal/mol value. Finally, from 1-methylcyclobutene and 1,3,3-trimethyl-cyclobutene, a value of 4.8 kcal/mol is obtained. One rationale as to why the

TABLE 5-1 (Continued)

Compound	Temperature (°)	$k \times 10^9$ sec^{-1}	E	log A	$\Delta H^{\ddagger}$	$\Delta S^{\ddagger a}$	Reference
1,3,3-Trimethyl	173	0.597	37.0	13.90	36.2	4.3	71
1,2,3-Trimethyl	151	0.766	33.7	13.28	32.9	1.5	40
cis-1,2,3,4-Tetramethyl	158	0.135	37.4	14.10	36.6	5.3	32
trans-1,2,3,4-Tetramethyl	150	3.23	33.6	13.85	32.8	4.1	32
cis-1,2,3,4-Tetraphenyl	50	0.8	25	12.8	24.2	−0.7	33
trans-1,2,3,4-Tetraphenyl	24	0.5	21	11.1	20.2	−8.2	23
1,2,3,4-Pentamethyl	175	2.6	34.0	—	—	—	23
1,2,3,3,4,4-Hexamethyl	175	0.45	40.	—	—	—	23
3-Bromo-cis-1,2,3,4-Tetraphenyl	—	—	—	—	24.5	1.3	72
cis-1,2,3,4-Tetraphenyl-3,4-Dimethyl	25	0.189	—	—	29.1	0.5	47
3,3,4,4-Tetrafluoro	306	1.03	47.9	14.09	47.1	5.3	73
1,2-bistrifluoromethyl-3,3,4,4-tetrafluoro	—	—	46.0	13.64	38.0	3.2	74
Hexafluoro	—	—	47.1	14.12	46.3	5.3	75
Hexadeuterio	155	2.51	33.8	13.66	33.0	3.2	76
trans-Diphenyl-benzocyclobutene	50	26.0	—	—	20.6	−7.0	77
cis-Diphenyl-benzocyclobutene	50	0.37	—	—	25.2	0.0	77
Bicyclo[3.2.0]hept-1(5)-ene	126	1.03	31.6	13.3	30.8	1.5	78
Bicyclo[4.2.0]oct-1(6)-ene	—	—	32.7	—	—	—	79
Tricyclo[4.2.1.0^{2,5}]non-2(5)-ene	—	—	26.6	—	—	—	79

^{a} Calculated at 150°.
^{b} Gas phase study.
^{c} Rate in dodecene solution.
^{d} Methyl group turning inward.
^{e} Ethyl group turning inward.

value is larger for geminal substitution than for vicinal is considered below. The $\Delta\Delta H^{\ddagger}$ values for the same set are 5.3, 3.5, 3.8, and 3.3, and the averages are $\Delta\Delta G^{\ddagger} = 4.2$ kcal/mol and $\Delta\Delta H^{\ddagger} = 4.0$. The less detailed kinetic study of tetraphenylcyclobutenes leads to a value of $\Delta\Delta H^{\ddagger} = 4.0$ kcal/mol when one phenyl turns inward.

Frey et al.[72] have analyzed the $\Delta\Delta G^{\ddagger}$ for methyl groups in varying positions and have suggested there exists an excellent additivity relationship. However, a broader survey indicates that that relation is limited in extent. The data used are shown in Table 5-2. Assuming that the information derived from monosubstituted cyclobutenes can be used as the primary source of $\Delta G^{\ddagger}$ values, the additivity relation holds reasonably for 1,3- and 1,4-dimethyl-cyclobutenes. The result is poorer for 1,2-dimethylcyclobutene, but steric interactions between the two methyls may be a cause of the deviation. If so it suggests that the interaction decreases in going from reactant to transition state, which points to a transition state twisted about the C_1-C_2 bond. The overestimate of the $\Delta\Delta G^{\ddagger}$ for cis-3,4-dimethylcyclobutene based on the $\Delta\Delta G^{\ddagger}$ increase for an inward rotating methyl derived from the 3,3-dimethylcyclo-butene has been noted above. Again this could be a ground state interaction between the methyls in the cis-3,4 isomer which is relieved at the transition state.

Several examples can be treated in more than one way. For example, trans-1,2,3,4-tetramethylcyclobutene could be considered as two 1-Me sub-situtents and two 3-Me groups, i.e., $\Delta\Delta G^{\ddagger} = 0$. Alternatively it could be treated as a combination of 1,2-dimethyl and two 3-Me groups, where $\Delta\Delta G^{\ddagger} = -0.9$. Similarly 1,2,3,3,4-pentamethylcyclobutene can be considered as two 1-Me, two 3-Me, and one inward twisting methyl ($\Delta\Delta G^{\ddagger} = 2.6$), or as a 1,2-dimethyl, 3,3-dimethyl and a 3-methyl ($\Delta\Delta G^{\ddagger} = 2.9$), or again as a 1,2-dimethyl, cis-2,4-dimethyl and a 3-methyl ($\Delta\Delta G^{\ddagger} = 0.8$). Generally the agreement is only modest.

One of the most interesting results is that less energy is expended to twist an ethyl group inward than a methyl group. The same result appears with both 3-ethyl-3-methylcyclobutene and when comparing 3,3-dimethyl and 3,3-diethylcyclobutenes. One possible explanation of this unexpected observation could be that the added chain length permits atoms in the ethyl group to be positioned in the transition state so that van der Waals attractive forces stabilize that transition state. The free energy difference between axial and equatorial groups has been attributed largely to differences in the van der Waals attractions between the group and other atoms in the cyclohexane ring.[81,82] Though the information is less accurate, the relatively small $\Delta\Delta H^{\ddagger}$ for the inward moving phenyl may involve both repulsive and attractive van der Waals forces giving a balance smaller than one might have expected. Alternatively the transition state for the 3,4-diphenylcyclobutene may be

TABLE 5-2

Changes in Free Energy of Activation with Substitution

Compound	$\Delta\Delta G^{\ddagger}$	$\Delta\Delta G^{\ddagger}$ (calc.)[a]
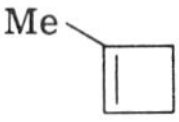 Me	1.6	
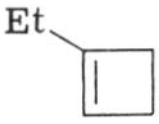 Et	2.0	
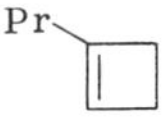 Pr	1.5	
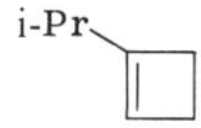 i-Pr	1.6	
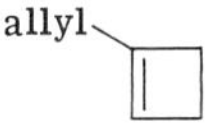 allyl	1.2	
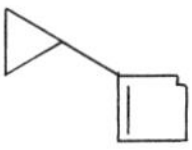	1.0	
Cl	1.0	
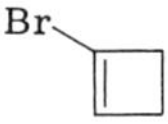 Br	0.7	
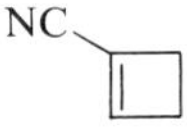 NC	0.8	
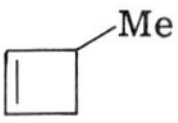 Me	−1.6	
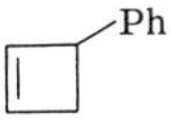 Ph	−4.9	
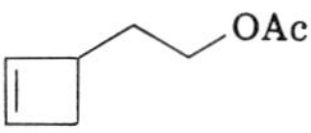 OAc	−0.7	

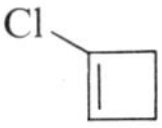

(*continued*)

TABLE 5-2 (Continued)

Compound	$\Delta\Delta G^{\ddagger}$	$\Delta\Delta G^{\ddagger}$ (calc.)[a]
(cyclobutene, Me, Me)	2.3	$2 \times 1.6 = 3.2$
(cyclobutene, Ph, Ph)	0.3	
(cyclobutene, Et, =CH₂)	2.2	
(cyclobutane ring, Me, Me)	−0.4	$1.6 - 1.6 = 0.0$
(Me, Me)	+0.3	$1.6 - 1.6 = 0.0$
(Me, Me)	+2.2	
(Et, Et)	+1.6	
(Et, Me) (Me in)	+2.8	$-1.6 + 3.8 = 2.2$
(Et in)	+2.1	$-1.6 + 3.2 = 1.6$
(Me, Me)	+0.1	$-1.6 + 3.8 = 2.2$
(Ph, Ph)	−7.8	
(Me, Me, Me)	+1.1	$+2.3 - 1.6 = 0.7$
(Me, Me, Me)	+3.2	$+1.6 + 2.2 = 3.8$

TABLE 5-2 (Continued)

Compound	$\Delta\Delta G^{\ddagger}$	$\Delta\Delta G^{\ddagger}$ (calc.)^a
Me–Me / Me–Me (cyclobutene)	+3.2	+2.3 + 0.1 = 2.4
Me–Me / Me–Me (cyclobutene)	−0.1	+2.3 − 2(1.6) = −0.9
Me–Me, Me / Me–Me (cyclobutene)	+2.8* (* E_a used directly)	+2.3 + 2.2 − 1.6 = 2.9
Me, Me–Me / Me–Me, Me (cyclobutene)	+8.8*	+2.3 + 2(2.2) = 6.7
Ph–Ph / Ph–Ph (cyclobutene)	−6.7	+0.3 − 7.8 = −7.5
Ph–Ph / Ph–Ph (cyclobutene)	−7.5	+0.3 + 2(−4.9) = −9.5

^{a} See text for the calculations.

further toward the ring opened form as is suggested by the small negative $\Delta S^{\ddagger}$. Conjugative requirements for the phenyls reduce their rotational freedom and lower $\Delta S^{\ddagger}$ despite the greater degree of ring opening.

One unusual example of a ring opening which indicates that fusion of a cyclohexene ring trans about a cyclobutene does not greatly enhance the retro-electrocyclization has been published.[459] Thus, *trans*-1-methyl-7,8-bis(trimethylsiloxy)bicyclo[4.2.0]octa-3,7-diene has a half life of 150 min at 100°.

3. Structural Effects on Equilibrium

Cyclobutene is thermodynamically less stable than s-*trans*-butadiene, and at 150° the ring opening reaction leads to butadiene as the sole detectable product. Presumably, s-*cis*-butadiene is the initial product of the ring opening, but it is rapidly converted to the more stable s-trans conformer. The difference in energy between the s-cis and s-trans conformers of butadiene is 2.4 kcal/mol. Wiberg and Fenoglio[83] measured the heats of combustion of cyclobutene and butadiene, giving $\Delta\Delta H^\circ$ as 11.4 kcal/mol. The increase in entropy from cyclobutene to butadiene is 4.5 cal/degree/mol,[84,85] which gives $\Delta G^\circ = 12.9$ kcal/mol at 298°K for conversion of cyclobutene to s-*trans*-butadiene. For the retro-electrocyclic reaction $\Delta H^\ddagger = 31.4$ kcal/mol and $\Delta S^\ddagger = +0.50$ eu, so $\Delta G^\ddagger = 31.2$ kcal/mol at 298°K. Thus, one can construct the reaction profile, Fig. 5-2, for the reaction.

The free energy difference of 12.9 kcal/mol is large enough that small perturbations in the stability, principally of butadiene, would produce no observable influence on the equilibrium. However, relatively strong perturbations of either steric or electronic origin could bring about measurable influence, particularly at high temperatures. In fact the earliest observation

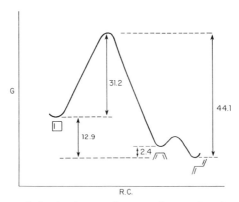

Fig. 5-2. Reaction profile for the electrocyclic reaction between butadiene and cyclobutene.

of such an equilibrium disturbance was made prior to the earliest mechanistic studies of the cyclobutene–butadiene reaction. Prober and Miller[86] found that hexafluorobutadiene cyclizes to hexafluorocyclobutene at 500°. Actually, the reaction proceeds to equilibrium,[87] and a careful study[75] gave $\Delta H^\circ = 11.7$ kcal/mol and $\Delta S^\circ = 9.56$ for the equilibrium ($K = $ buta/cyclo). The forward (k_1) and reverse (k_2) reactions [Eq. (5-32)] have E_a and log A of 35.4, 12.03 and 47.0, 14.12, respectively. It has also been shown that hexafluorobutadiene is non-planar and exists as a cisoid conformer with a dihedral angle near 48°.[88]

$$\text{(5-32)}$$

Alteration of the equilibrium in this fashion appears to be very sensitive to the substitution pattern for the fluorines. Thus, 3,3,4,4-tetrafluorocyclo-butene gives 1,1,4,4-tetrafluorobutadiene when heated at 700°.[89] The equilibrium constant ($K = $ buta/cyclo) is greater than 200° at 600°K.[73] At 700° 1-trifluoromethyl-2,3,3-trifluorocyclobutene is in equilibrium with 3-trifluoromethyl-1,1,2-trifluorobutadiene with $K = $ buto/cyclo $\simeq 1.5$.[90] At 300°K the equilibrium constant for 1,2-bistrifluoromethyl-3,3,4,4-tetrafluorocyclobutene and its ring opened valence isomer is 0.17.[74] In Table 5-3 the yields of substituted butadienes found on pyrolysis of a number of fluorine containing cyclobutenes are listed. The yields are not accurate measures of the equilibrium content of the butadienes since some polymerization does occur, but there is a rough relation to the equilibrium value. At 600° **85** undergoes ring opening followed by a series of reactions leading to formation of an acyl fluoride [Eq. (5-33)]. In view of these results the observa-

$$(5\text{-}33)$$

tions of Brune and Schwab[91] seem surprising. They obtained the following equilibrium data for halosubstituted cyclobutenes, and attributed the influence of the substituents on the equilibria to a mixture of steric and electronic factors. To alter the equilibrium from that of cyclobutene $\rightleftarrows$ butadiene to the present case requires a $\Delta\Delta G°$ of 12.7 kcal/mol (at 150°). For comparison of steric vs. electronic contributions it would be very helpful to have the free energy data for *trans*-1,2,3,4-tetramethylcyclobutene and *E,E*-3,4-dimethyl-2,4-hexadiene.

R	X	Temp.°	$K = \dfrac{buta}{cyclo}$	
H	Cl	150	95.5/4.5	= 21
H	Br	150	67.4/32.6	= 2.1
Me	Cl	175	90/10	= 9*
Me	Br	175	70/30	= 2.3*

* some decomposition reactions lowered the accuracy.

5. Four Electron–Four and Five Atom Systems

TABLE 5-3

Yields of Substituted Fluorobutadienes Obtained on Pyrolysis of Some Fluorocyclobutenes[89]

Compound	Temperature (°)	% Butadiene
(cyclobutene: F, F_2 / F, F_2)	700	10
(cyclobutene: F_2 / F_2)	700	~100
(cyclobutene: Cl, F_2 / F_2)	700	38
(cyclobutene: HOH_2C, F_2 / F_2)	650	50
(cyclobutene: Ph, F_2 / F_2)	700	60
(cyclobutene: ClH_2C, F_2 / F_2)	550	100
(cyclobutene: MeOOC, F_2 / F_2)	525	55
(cyclobutene: F_2 / Ph, Cl_2)	700	40
(cyclobutene: F_2 / MeOOC, Cl_2)	550	90
(cyclobutene: F_2 / MeOOC, F, Cl)	550	80
(cyclobutene: F_2 / Ph, F, Cl)	750	60

Polysubstitution also appears to have a profound effect on the equilibrium constant. In methylene chloride the equilibrium of Eq. (5-34) has $K = 2.50$ at 37°, $\Delta H° = -5.56$ and $\Delta S° = -16.3$ eu.[72] Again, $\Delta\Delta G°$ to convert K for cyclobutene to that for this example is 12.2 kcal/mol (at 300°K). In view

$$(5\text{-}34)$$

of the greater freedom of the butadiene molecule to avoid steric congestion, the origins of the required energy changes are uncertain.

A more obvious basis for biasing the equilibrium in favor of the cyclic molecule involves incorporation of one of the diene double bonds into an allene. The heat of formation of 1,2-butadiene is 38.78 kcal/mol,[92] and that of cyclobutene is 37.45 kcal/mol. Thus, the equilibrium between vinylallene and methylenecyclobutene can be expected to favor the cyclic molecule. This expectation is clearly born out with 3,7-dimethyl-2,3,5,6-octatetraene which forms 3,4-diisopropylidenecyclobutene at 250°.[92] Biallenyl undergoes ring closure in 15 min at 200° [Eq. (5-35)].[93,93a] Ring closure proceeds with the

$$(5\text{-}35)$$

expected conrotation when **86** is heated [Eq. (5-36)].[94] This reaction appar-

$$(5\text{-}36)$$

86

ently plays a role in the thermal conversion of 1,5-diynes to dimethylene–cyclobutenes.[95,96,97] In accord with this mechanism **87** gives **88** [Eq.

$$(5\text{-}37)$$

(5-38)].[97] Huntsman has proposed an alternative mechanism via an inter-

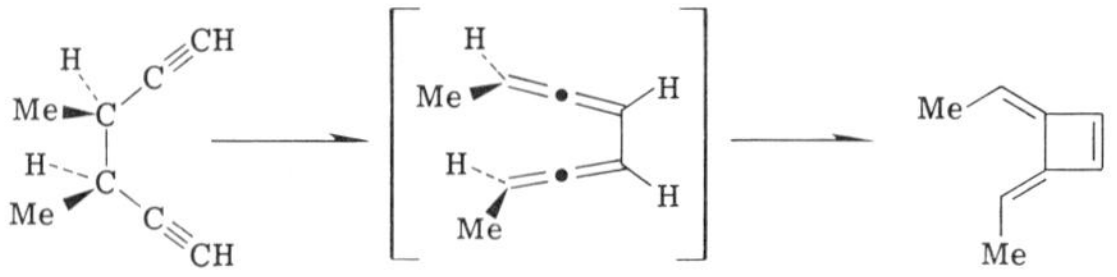

(5-38)

87   88

mediate with bonding between C_2 and C_5.[97] It is interesting that a single allene moiety is not sufficient to make the ring closed product the main constituent at equilibrium, at least at temperatures above 175°.[98]

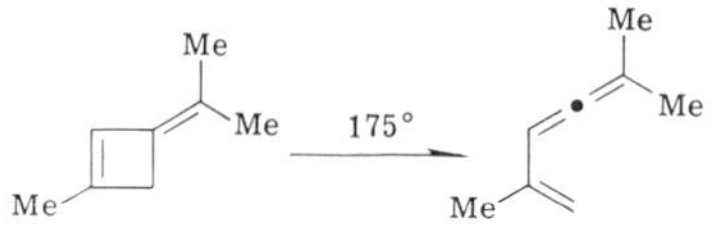

(5-39)

Rather similar results might be expected for cyclobutenones where one double bond in the acyclic valence isomer becomes part of a ketene group. Though no equilibrium constant data are known, it is clear that the reaction can proceed in both directions to a measurable extent. Jenney and Roberts[99] observed that optically active **89** racemizes readily in chloroform at 100°. The rate of racemization in $CHCl_3$ at 100° was 1.7×10^{-4} sec^{-1}, and in ethanol at 100° the loss of optical activity and rate of formation of inactive

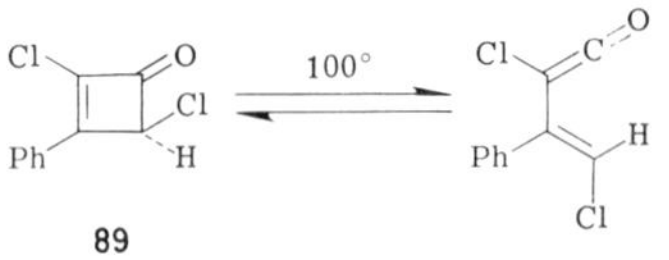

(5-40)

89

ethyl 2,4-dichloro-3-phenyl-3-butenoate were equal. When 2,4-dichloro-3-phenyl-3-butenoic acid was heated in acetic anhydride and the solvent evaporated the cyclobutenone was obtained. Ring opening of 2-chloro-3-phenyl-4-piperidinocyclobutenone occurs much more rapidly than **89**, $t_{1/2} \sim 30$ min at 20°.[100] Ring opening has also been postulated as a means of rationalizing the reactions of 3-phenylcyclobutenedione with methanol [Eq. (5-41)].[101] Further examples of thermal ring openings of substituted

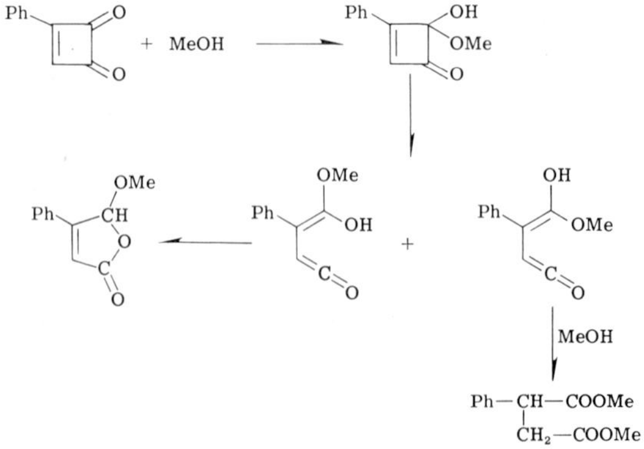

(5-41)

phenylcyclobutenones have been noted.[102,454-456] An attempt to utilize the reverse reaction as an electrocyclic synthesis for cyclobutenones was most disappointing.[102] Only 2-chloro-3-phenyl-3-butenoic acid gave a cyclized product (20–30%) when treated with acetic anhydride.

$$(5\text{-}42)$$

More recently Mayr[104] reported that 2,4,4-trimethylcyclobutenone converts thermally to methyl isobutenyl ketene, but the ketene does not revert to the cyclobutenone. Schiess and Radimerski[105,105a] have found that vinylketenes with terminal substituents undergo cis-trans isomerization during pyrolyses and the intermediacy of a cyclobutenone was proposed. Direct

$$(5\text{-}43)$$

ring opening of a cyclobutenedione to a bisketene has been proposed,[101,106] but no adequate evidence for the reaction has been obtained. By analogy with the bisallene, ring closure would seem to be a more likely reaction.

Benzocyclobutenes represent a prime example of reversal of the equilibrium favoring the butadiene moiety. Reversibility has been demonstrated

$$(5\text{-}44)$$

via separate experiments which show that o-xylylene can be generated from a benzocyclobutene, and that a benzocyclobutene can be formed from an alternative source of the o-xylylene [Eq. (5-45)]. The presence of the unstable

$$(5\text{-}45)$$

o-xylylene entity has normally been shown by trapping with a dienophile, but o-xylylene itself has been isolated at −78°.[107] When warmed to 0° it dimerizes as shown in Eq. (5-46). The o-xylylene was generated by pyrolysis

$$(5\text{-}46)$$

of o-methylbenzyltrimethyl ammonium hydroxide at 200°–250° under reduced pressure. The o-xylylene survives long enough to be condensed at

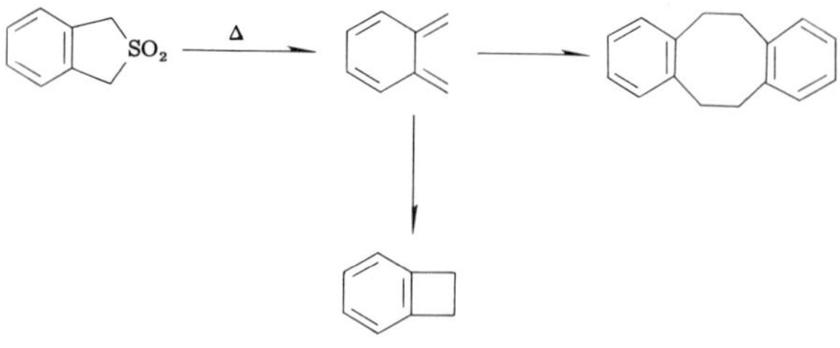

(5-47)

−78°. When generated by heating a sulfone in diethylphthalate the o-xylylene dimerizes to dibenzcyclooctadiene, but in the vapor phase the o-xylylene ring closes to give benzocyclobutene.[108] When benzocyclobutene is heated in the presence of maleic anhydride, the Diels–Alder adduct of o-xylylene is obtained in good yield.[109]

Studies of the same sort have been made for 1,2-naphthocyclobutene,[110] 2,3-naphthocyclobutene,[111] trans-dibromobenzocyclobutene,[112,109] cis-diiodobenzocyclobutene,[113] and cis- and trans-diphenylbenzocyclobutene.[54,55,56] Both cis- and trans-diphenylbenzocyclobutenes have been synthesized by formation of the diphenyl-o-xylylenes.[114,115] An early report[115] of the stereochemistry of the synthesis was later clarified when the conservation of orbital symmetry theory made revisions in order.[116] The ready thermal

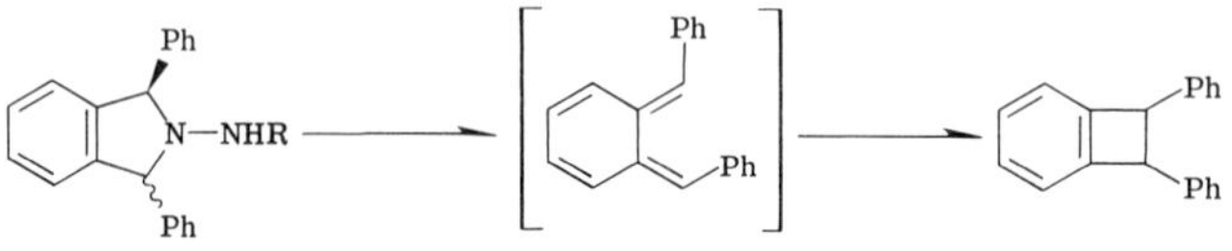

(5-48)

cis to trans conversion has not been provided with a mechanism. Hart has generated a very interesting synthesis of dichlorobenzocyclobutenes from sterically hindered o-methylbenzotrichlorides [Eq. (5-49)].[117]

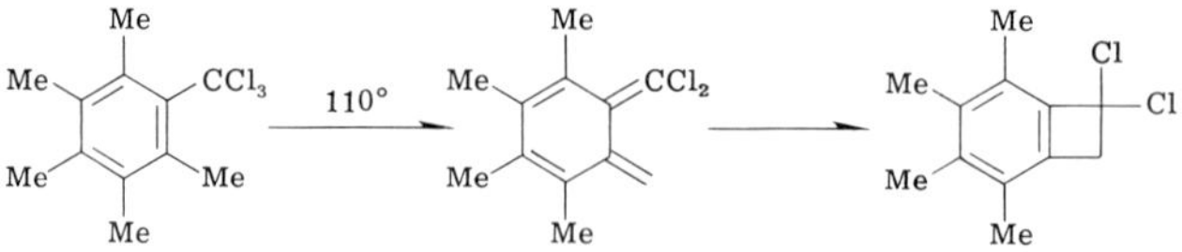

(5-49)

Benzocyclobutenone can also be prepared from the relevant ketene [Eq. (5-50)].[119]

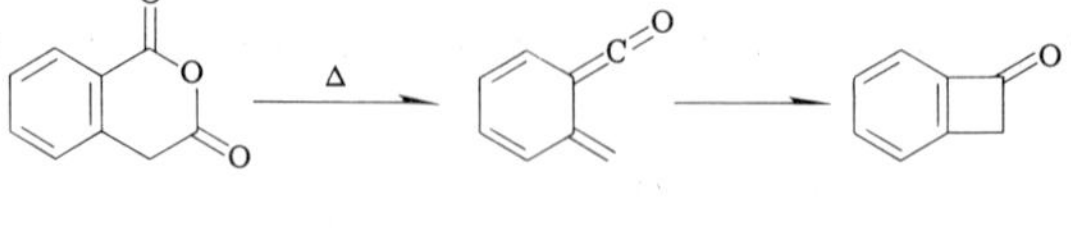

(5-50)

Perhaps the most convincing evidence of the reversible nature of the ring opening of benzocyclobutenes comes from *trans*-diphenylbenzocyclobutene which exhibits reversible thermochromism.[54] The racemization of optically active methoxybenzocyclobutene also indicates the ring opening is reversible.[48] Formation of an isochromene from acetobenzocyclobutene[51] must also be attributed to the reversibility of the electrocyclic process since *trans*-aceto-*o*-xylylene must be formed but cannot generate the product and must revert to reactant. Reversion is, however, prevented where some other reaction can intervene. Thus, for example, optically active benzocyclobutenol loses optical activity at the same rate it is converted to *o*-tolualdehyde.[48] The benzocyclobutene **90** shows no isomerization, apparently because the

$$(5\text{-}51)$$

hydrogen shift is faster than electrocyclization.[118] An interesting *o*-xylylene has been prepared[124] which shows no tendency to revert to the benzocyclobutene form. The stability of this compound is attributable to push–pull stabilization of the diene system.

$$(5\text{-}52)$$

Trans double bonds develop considerable strain when incorporated in medium rings. Thus, *cis, trans*-1,3-cyclooctadiene undergoes an electrocyclic closure at ca. 80°, while the cyclononadiene requires 185° to react.[120,121] The more highly strained *trans*-2-*cis*-4-*cis*-6-cyclooctatrienone gives 5,7-

$$(5\text{-}53)$$

bicyclo[4.2.0]octadiene-2-one at room temperature.[122] It is also possible that 6-bicyclo[3.2.0]heptene-3-one is formed by thermal electrocyclization of *cis, trans*-3,5-cycloheptadienone generated photochemically by sensitized irradiation of 3,5-cycloheptadienone.[123]

4. *Cis-Trans* Isomerization of Conjugated Dienes

Electrocyclization of an appropriately substituted diene and reversion provide a route to geometric isomerization of the dienes [Eqs. (5-54) and (5-55)]. Both cases are characterized by the necessary concurrent isomeriza-

$$(5\text{-}54)$$

$$(5\text{-}55)$$

tion of both double bonds. That this route can be expected to play a significant role in many cases is evident on energetic grounds. The data of Section I,C,3 show that $\Delta H^{\ddagger}$ for conversion of butadiene to cyclobutene is 42.8 kcal/mol. Thermal stereomutation of *cis*-2-butene requires $\Delta H^{\ddagger} = 62.8$ kcal/mol.[125] and subtraction of the allylic resonance energy of ca. 13 kcal/mol[126,127] gives $\Delta H^{\ddagger} \sim 50$ kcal/mol for isomerization of butadiene by the singlet radical route.

The first examples of this isomerization process involved highly substituted dienes [Eqs. (5-56) and (5-57)].[47] Note that in both cases the double mutation

$$\Delta H^{\ddagger} = 25.3 \quad \Delta S^{\ddagger} = -14.5$$

$$(5\text{-}56)$$

process occurs without intervention of any mono cis-trans conversion. The reaction of Eq. (5-57) was run for 51 days and produced no cis-cis product,

$$\Delta H^{\ddagger} = 29.2 \quad \Delta S^{\ddagger} = -6.5$$

$$(5\text{-}57)$$

which indicates that the singlet radical process does not compete with this route.

Frey *et al.*[128] suggested that *cis*-2,3-dimethyl-1,3-pentadiene is converted to the trans isomer via 1,2,3-trimethylcyclobutene. The equilibrium concentration of the cyclobutene was measured at 600°K, and since the rates of reversion to the two diene isomers were known the rate for the cyclobutene route could be calculated. The agreement between calculated and measured rates was reasonable. For the isomerization a $\Delta H^{\ddagger}$ of 43.2 kcal/mol was obtained.

Analysis of the isomerizations of 1,4-dideuteriobutadiene in the gas phase at 910° revealed a more complex set of reactions.[129] The cis, cis isomer converts to the trans, trans form much faster than the cis, trans isomer appears. However, the latter form does appear indicating that intervention of the singlet radical process can occur at this temperature. At lower temperatures the route via cyclobutene would dominate since it has a lower enthalpy of activation. Thus the electrocyclic mechanism may be expected to be general unless severe steric factors are present. Another example has been uncovered by Dalrymple and Russo[35] [Eq. (5-58)], and in that case the intermediate is sufficiently stable to be isolated and identified.

$$(5\text{-}58)$$

Arguments have been presented by Marvell to suggest that the same electrocyclic mechanism can be applied to the geometric isomerizations of 1,8-diphenyloctatetraenes.[130] The trans, cis, cis, trans isomer is converted to the all trans isomer to a greater extent than could be accommodated by a singlet radical process.

$$(5\text{-}59)$$

D. The "Forbidden" Process

Given the structural freedom to adopt either a conrotatory or disrotatory ring opening, cyclobutenes inevitably adopt the conrotation route. It is thus valuable to study molecules where conrotation is structurally denied. This can be accomplished by incorporating the cyclobutene into a bicyclic (or polycyclic) molecule of cis-[n.2.0] type where the monocycle (n + 4) cannot

$$\text{(5-60)}$$

accommodate a stable trans double bond. These systems have been quite extensively investigated. Aside from special mechanisms which could pertain in certain cases (see below), three general routes need to be considered. First, conrotatory retro-electrocyclization forms a strained trans double bond, which is rapidly converted to a stable cis double bond. Second, the "forbidden" concerted ring opening occurs in disrotatory fashion giving two cis double bonds directly. Third, bond fission occurs via a single rotation leading to an orthogonal diradical with allylic stabilization, followed very rapidly by a second rotation leading to the dicis diene. Probabilities for the activation of these routes vary with the value of n.

1. Bicyclo[2.1.0]Pent-2-Enes

The interesting story of 2-bicyclo[2.1.0]pentene begins in 1965 with the initial synthesis of this hitherto elusive molecule.[131] In solution, rapid reversion of the bicyclic molecule to cyclopentadiene (and other products)

$$\text{(5-61)}$$

($t_{1/2} \sim 2$ hrs) was observed. The gas phase reaction was studied kinetically and $E_a = 26.9$ kcal/mol, log $A = 14.2$ were reported.[132] The low $t_{1/2}$ in solution was attributed to polymerization. The obvious conclusion is that this is an example of a retroelectrocyclization which, if concerted, occurs via the forbidden disrotatory route. If not concerted, then a diradical reaction is a most plausible result. The $\Delta S^{\ddagger} = +6$ suggests that the strained transition state of a conrotatory reaction plays no role here. However, it is quite possible that the molecule might adopt an alternative route if a symmetry allowed one were available.

At least four routes other than a direct retro-electrocyclization can be devised for this reaction. Two involve hydrogen shifts [Eqs. (5-62) and (5-63)], and the other two necessitate carbon migrations [Eqs. (5-64) and (5-65)].

$$\text{(5-62)}$$

$$\text{(5-63)}$$

$$\text{(5-64)}$$

John Baldwin was the first to perceive and investigate these alternatives.

$$\text{(5-65)}$$

Of these the reactions of Eqs. (5-62) and (5-64) are symmetry allowed if concerted and his studies were confined to these two. The more appealing of these [Eq. (5-62)] can be considered as a retro-ene reaction rendered facile by relief of the ca. 57 kcal/mol of strain in the molecule.[133] However, rearrangement of a 5-D labeled 2-bicyclo[2.1.0]pentene showed no deuterium migration, and a fully deuterated molecule showed no primary isotope effect.[134] This eliminates both mechanisms involving hydrogen migrations.

To differentiate between the electrocyclic and the 1,3-C migration route [Eq. (5-64)] Baldwin and Andrist[135] prepared and rearranged 2-methyl-2-bicyclo[2.1.0]pentene. The product was trapped with N-phenylmaleimide to circumvent possible interconversion of 1- and 2-methylcyclopentadienes by the known internal 1,5-hydrogen shifts.[136] Only the adduct of 1-methylcyclopentadiene was obtained, a result which appears to eliminate the electro-

$$\text{(5-66)}$$

cyclic route, but also raises the question of why migration proceeded exclusively toward the methyl substituted carbon. McLean and Findlay[137] had obtained both 1- and 2-methylcyclopentadienes in the absence of a trapping reagent, and a further study by that group with N-phenylmaleimide gave at least 95% of the adduct of 2-methylcyclopentadiene![138] Finally a thorough kinetic analysis of the reaction of both 1-methyl and 2-methyl-2-bicyclo[2.1.0]pentenes[139] gave the data of Eq. (5-67). The result appeared to provide

$$\text{(5-67)}$$

support for the C-migration route, and incidentally answered the question raised above, since both expected migrations were indeed active.

This result did not still the controversy since Flowers and Frey[140] argued that if direct rupture of the internal bond gave a vibrationally activated cyclopentadiene, both 1- and 2-methyl-cyclopentadienes could come directly from this "hot" molecule. Some calculations by those authors indicated that the kinetic results were quantitatively interpretable by such a scheme. Further data in support of this suggestion were accumulated by Brauman et al.[141] In particular the ratio of 2-methyl to 1-methylcyclopentadienes arising from 2-methyl-2-bicyclo[2.1.0]pentene was pressure and deactivator dependent. Both Brauman et al.[141] and Andrews et al.[142] found that the main initial product from 2-methyl-2-bicyclo[2.1.0]pentene in solution was 2-methylcyclopentadiene, $K \simeq 10$ in THF[141] and 13.9 in hexane.[142] The case for the specially activated intermediate was summarized by Farneth et al.[143]

The existence of any carbon shift mechanism has finally been eliminated by a double labelling experiment.[144] Rearrangement of a vicinal double ^{13}C-labeled 2-bicyclo[2.1.0]pentene, labeled at C_1 and C_5, showed that no separation of the labels occurred during ring opening. At the same time use of a 5-methyl-2-bicyclo[2.1.0]pentene showed that an activated species does certainly intervene in the gas phase. Thus, the sole route to ring opening is via fission of the transannular bond. Now the question of interest which remains is whether the bond cleavage is via a biradical or a concerted forbidden process. There exists no convincing experimental evidence which would permit a choice between these, but MINDO/3 calculations[145] support the forbidden concerted route with a symmetric transition state of C_s symmetry and a transannular bond length of 1.85 Å. The calculated activation energy is 27.1 kcal/mol in good agreement with the experimental value.

Franck-Neumann[146] has suggested that a bicyclo[2.1.0]pentene is an intermediate in the reaction between diphenylcyclopropenone and N,N-diethylaminophenylacetylene [Eq. (5-68)]. If this is the case then the substitu-

(5-68)

ents must enhance the rate of ring opening by a considerable amount.

A stable *o*-xylylene results when a three membered ring is fused to a benzocyclobutene.[460] Thus, 8,8-dimethyltricyclo[4.3.0.0^{6,9}]nona-1(6),2,4-triene gives the bicyclic *o*-xylylene at room temperature ($E_a = 18.9$ kcal/mol, log $A = 11.5$). This is obviously a very interesting reaction.

2. 2-Bicyclo[2.2.0]Hexenes

The chemistry of 2-bicyclo[2.2.0]hexenes has been developed mainly as a result of interest in cyclobutadiene. Bicyclo[2.2.0]hex-2-ene has been prepared and shown to undergo an electrocyclic reversion to 1,3-cyclohexadiene, relatively slowly at 130° and rapidly at 330°.[147] Unfortunately, no kinetic studies of this parent molecule have been made, so an important reference point is missing. It has been reported that 2-bicyclo[2.2.0]hexene is unrearranged after 48 hrs at room temperature.[148]

It is unfortunate that of the rather large number of bicyclo[2.2.0]hexenes which have been prepared,[151–153] few kinetic studies of the ring opening have been made. Seebach[149] found that **91** had a half-life of 4.2 hrs at 80.1

(5-69)

91

($E_a = 30.6$ kcal/mol). An $E_a = 42$ kcal/mol was reported for hexamethyl-2-bicyclo[2.2.0]hexene.[23] Study of a series of substituted 1,4-*endo*-5-*endo*-6-tetramethyl-2-bicyclo[2.2.0]hexenes showed that thermal ring opening is very sensitive to the substituents at C_2 and C_3.[150] The results are listed in Table 5-4. The extremely large effect of a carbomethoxy group on $\Delta H^{\ddagger}$

TABLE 5-4

Activation Enthalpies for Thermal Ring Opening
of Some 2-Bicyclo[2.2.0]Hexenes

Y	Z	$\Delta H^{\ddagger}$
Me	Me	43.7
COOMe	Me	32.3
COOMe	COOMe	28.4
COOMe	Ph	29.6

would not have been anticipated. Compared directly with the results of Seebach,[149] it shows that substitution of a carbomethoxy group for a methyl group is at least as effective in stabilizing the transition state at C_3 as at C_1. The stabilization, ca. 11 kcal/mol, is far larger than a carbomethoxy group exerts on a radical.[154] Bauld et al.[155] has found that remote methylene groups also produce a very profound rate enhancement. At 66.5° the rate [Eq. (5-70)] equals 7.61×10^{-5} sec^{-1} and $\Delta H^{\ddagger} = 17.2$ kcal/mol, $\Delta S^{\ddagger} =$

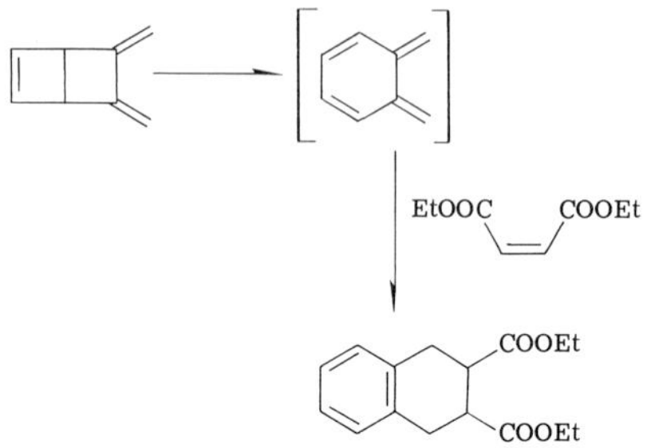

$$(5\text{-}70)$$

-27.2 eu. Rates are independent of the dienophile concentration.

Most of the remaining studies of 2-bicyclo[2.2.0]hexenes without added small rings have been of a qualitative nature. When heated to about 200° a variety of methylated 2-bicyclo[2.2.0]hexene-5,6-dicarboxylic acid anhydrides [Eqs. (5-71), (5-72), and (5-73)] undergo clean ring opening. Photo

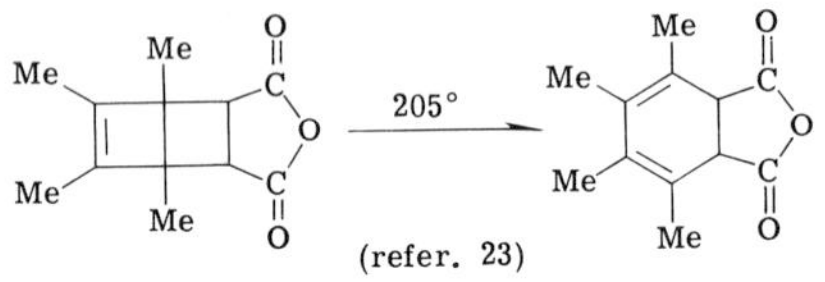

$$(5\text{-}71)$$

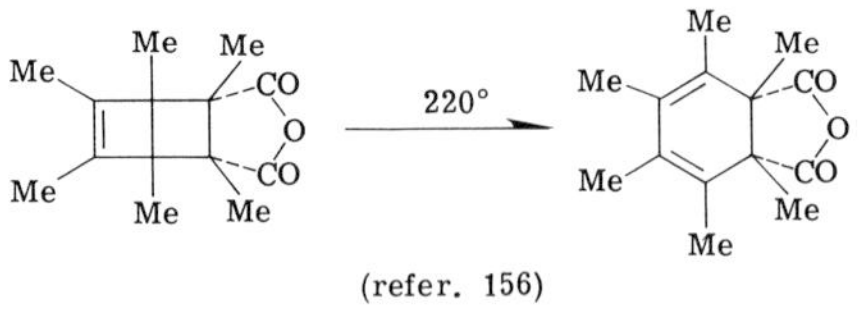

(refer. 23)

$$(5\text{-}72)$$

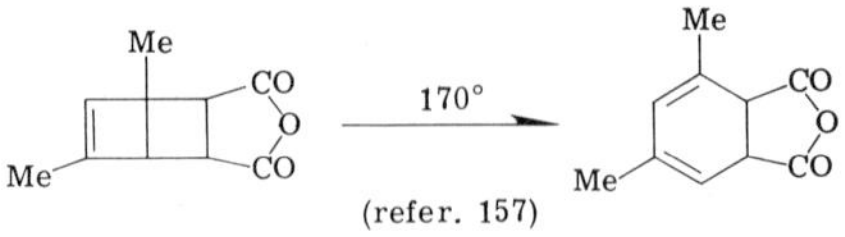

(refer. 156)

$$(5\text{-}73)$$

(refer. 157)

products from pyro- and isopyrocalciferols also undergo thermal reversion to the original reactants [Eqs. (5-74) and (5-75)].[158] Similarly the photo-

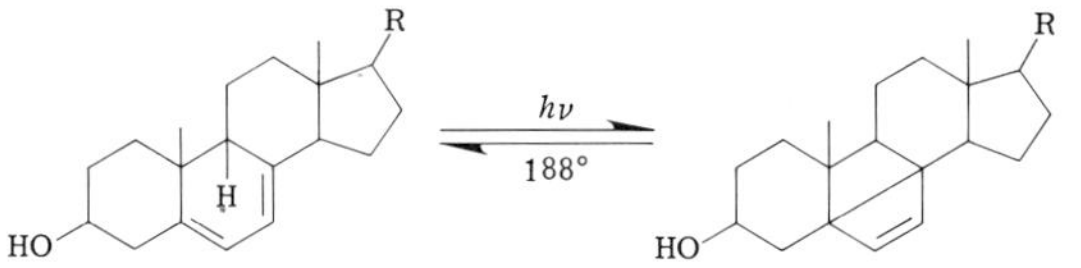

$$\text{(5-74)}$$

$E_a = 28.8$ (refer. 159)

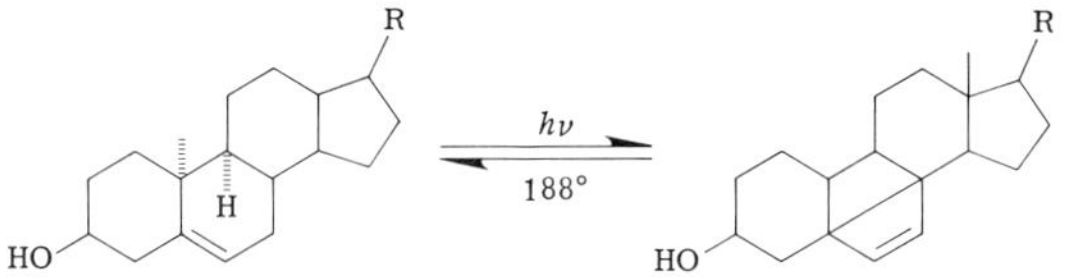

$$\text{(5-75)}$$

$E_a = 19.8$ (refer. 159)

isomer of levopimaric acid reverts to levopimaric acid at about 120°.[160,161] The data of Lawrence[161] convert to a rate $k = 4.5 \times 10^{-4} \text{ sec}^{-1}$ or $t_{1/2} \sim 25$ min. This rate is about ten times slower than the rate for 2,3-dimethyl-1,4-dicarbomethoxybicyclo[2.2.0]hex-2-ene.[149] This suggests that steric effects associated with fusing the cyclohexane ring across the 1,*exo*-6 positions must be responsible for the rate reduction.

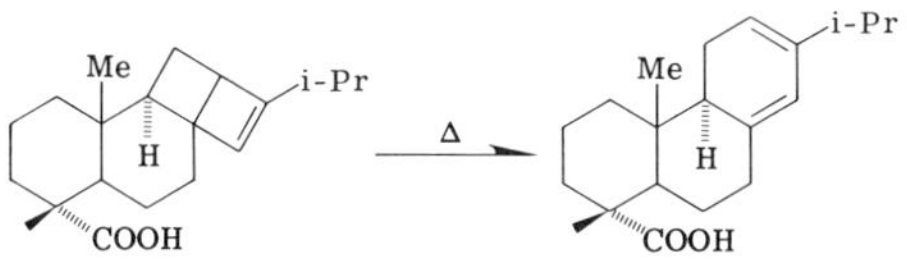

$$\text{(5-76)}$$

A very interesting rate effect has been observed by Ficini *et al.*[162] Bicyclohexenes with a diethylamino group at the bridgehead undergo rapid ring opening below 40°, while one having the diethylamino group at C_2 is stable under these conditions. However, the relation of the diethylamino group to one of the carbomethoxy groups may be the important feature permitting this easy reaction [Eq. (5-77)].

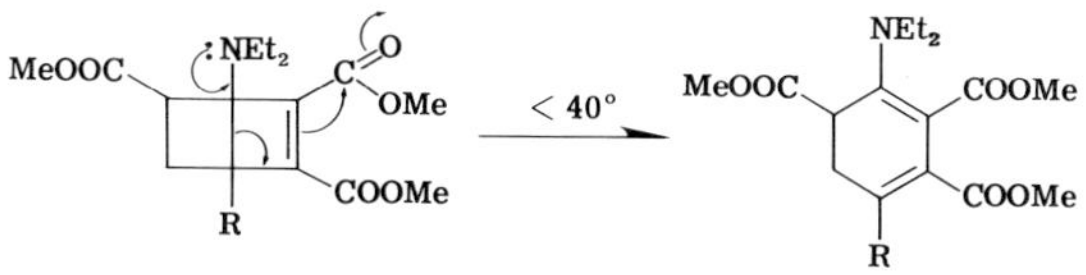

$$\text{(5-77)}$$

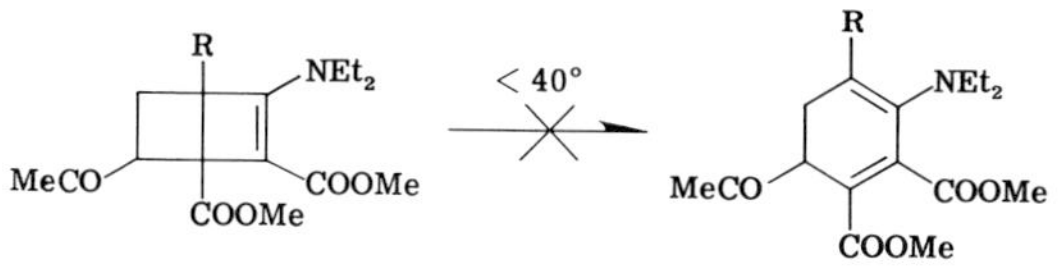

$$\text{(5-78)}$$

During the search for cyclobutadienes a number of Diels–Alder dimers of these unstable entities were uncovered. These dimers are usually tricyclo-[4.2.0.0^{2,5}]octa-3,7-dienes [Eq. (5-79)], and thus contain two separate

$$\square + \square \longrightarrow \square\square\square \qquad (5\text{-}79)$$

bicyclo[2.2.0]hexene groupings. These should undergo ring cleavage in two steps [Eq. (5-80)]. The second step is interesting since it involves a 6π retro-

$$\square\square\square \xrightarrow{\Delta} \bigcirc\square \longrightarrow \bigcirc \qquad (5\text{-}80)$$

electrocyclization on one side and a 4π on the other. The former is allowed, the latter forbidden but the allowed process takes precedence. Consequently in almost all cases the first step is rate determining and the product is a cyclooctatetraene derivative.

Both **92** and **93** have been obtained,[163] and both were observed to form cyclooctatetraene above 100°. The syn isomer reacts somewhat faster than

$$\square\square\square \xrightarrow{\Delta} \bigcirc \xleftarrow{\Delta} \square\square\square \qquad (5\text{-}81)$$

$$\text{92} \qquad\qquad\qquad \text{93}$$

the anti form, with $E_a(\text{syn}) = 30.49$ kcal/mol (log $A = 14.22$), and $E_a(\text{anti}) = 32.59$ kcal/mol (log $A = 14.01$).[164] The difference has been rationalized[164] as a ground state energy difference and not the result of the large repulsive interaction between the π electrons in the syn isomer.[165] A second study of the rates of the **92** and **93** gave a slightly different set of activation parameters,[166] and a MINDO/3 calculation gave $E_a(\text{syn}) = 57.7$ kcal/mol and $E_a(\text{anti}) = 56.1$ kcal/mol. Conversely, calculation for a route via the triplet state gave $E_a(\text{syn}) = 36.0$ and $E_a(\text{anti}) = 39.3$ kcal/mol. Frey[164] has suggested that the singlet diradical **94** is an intermediate [Eq. (5-82)], and since

$$\square\square\square \longrightarrow \bigcirc\square \longrightarrow \bigcirc\square \qquad (5\text{-}82)$$

$$\text{94}$$

the singlet-triplet splitting will be small, some intervention of the triplet diradical might occur. That the triplet can participate was indicated by the generation of light when **95** was heated in the presence of 9,10-dibromoanthracene.[166]

$$\bigcirc\square\square \xrightarrow{\Delta} \bigcirc\bigcirc \qquad (5\text{-}83)$$

$$\text{95}$$

Addition of an extra small ring to lengthen the ladder does not alter the reaction course appreciably. The epoxide of the anti isomer reacts readily at

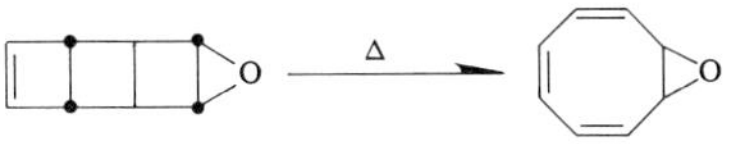

$$
\text{(5-84)}
$$

$$
\begin{aligned}
E_a &= 32.09 & \log A &= 13.76 \\
\Delta H^{\ddagger} &= 31.2 & \Delta S^{\ddagger} &= +1.4
\end{aligned}
$$

at 120°, and both **96** and **97** [Eq. (5-85)] proceed at rates comparable with their

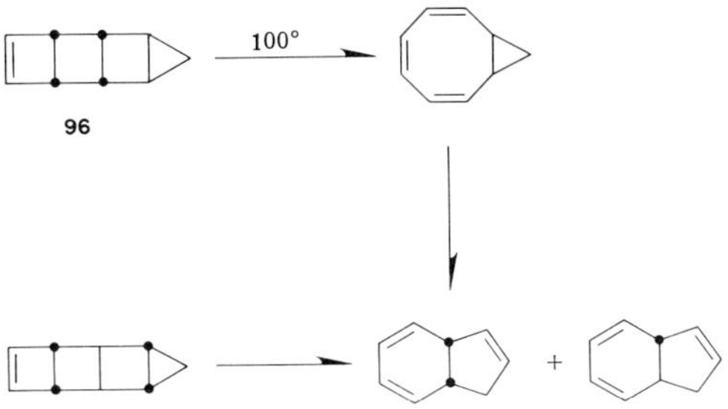

96

$$
\text{(5-85)}
$$

97 $\Delta H^{\ddagger} = 33.6$ $\Delta S^{\ddagger} = 6.1$

tricyclic parents.[167,167a] However, Dewar benzeneoxide has a half-life of 18

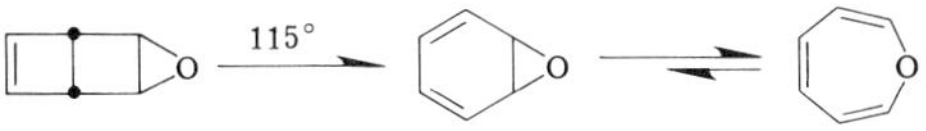

$$
\text{(5-86)}
$$

min at 115°, considerably shorter than that of tricyclo[4.2.0.0^{2,5}]octadiene (20 min at 140°).[168] The epoxide rearrangement does not involve any skeletal rearrangement, and $\Delta G^{\ddagger} = 29$ kcal/mol.[169] The ladder compound **98** appar-

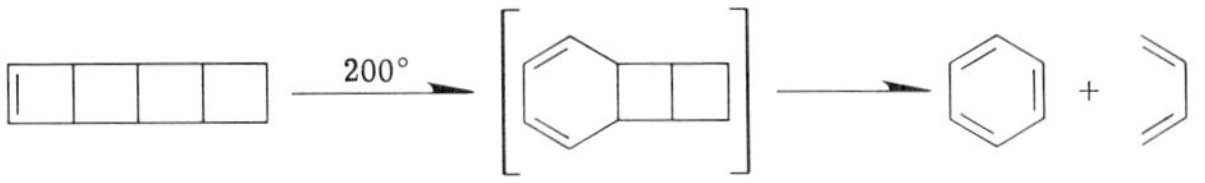

98

$$
\text{(5-87)}
$$

ently undergoes initial ring expansion but the resultant molecule fragments.[170]

Substitution can produce quite notable effects on the course of ring opening, but unfortunately little study has been made of the kinetics of these reactions. Dimers of dimethylcyclobutadiene react normally [Eq. (5-88)].[171]

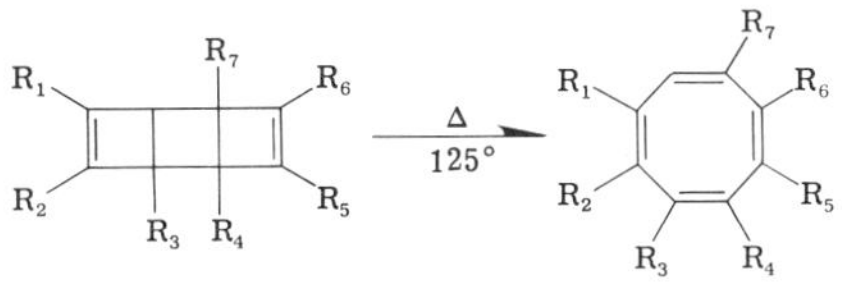

$$
\text{(5-88)}
$$

$$
\begin{aligned}
R_1 = R_2 = R_4 = R_7 = Me \qquad & R_3 = R_5 = R_6 = H \\
R_2 = R_3 = R_6 = R_7 = Me \qquad & R_1 = R_4 = R_5 = H \\
R_1 = R_2 = R_6 = R_7 = Me \qquad & R_3 = R_4 = R_5 = H
\end{aligned}
$$

Apparently all of these dimers were syn isomers. On the other hand, dimers of tetramethylcyclobutadiene behave abnormally [Eq. (5-89)]. Both syn[172]

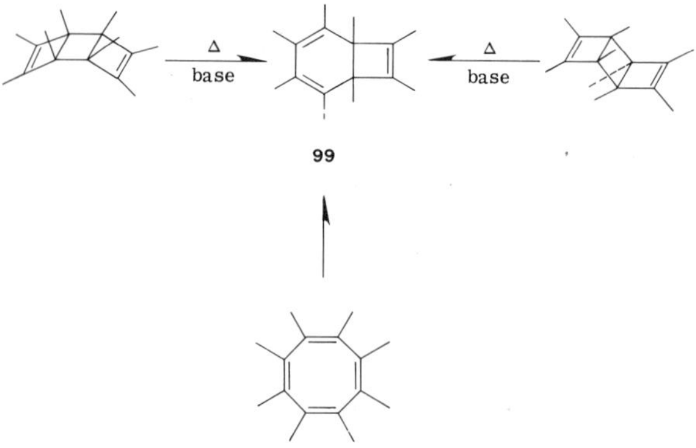

99

$$(5\text{-}89)$$

and anti[173] isomers, as well as octamethylcyclooctatetraene,[174] give **99**. Traces of acid produce mixtures containing isomers with terminal methylene groups which confused the early studies.[175]

By way of contrast, the fully brominated or chlorinated dimers react normally giving only the halogenated cyclooctatetraenes.[176] A complex case following the pattern of the tetramethyl dimers was uncovered by Wittig and Mayer.[177] A hexamethyltricyclo[4.1.0.0^{2,5}]hept-3-ene rearranged at 180°,[178] and Fonken and Moran[179] noted that tricyclo[5.2.0.0^{2,6}]nona-3,8-diene produced *cis*-dihydroindene on distillation. Maier[180] found that the epoxide **100** reacted thermally like its precursor diene [Eq. (5-90)].[180]

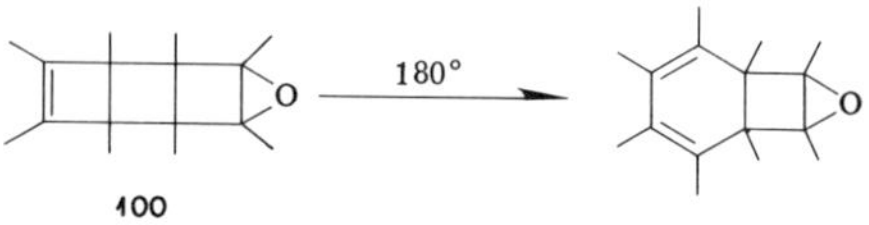

$$(5\text{-}90)$$

100

The presence of four phenyl groups accelerates ring cleavage, since the two dimers of diphenylcyclobutadiene form tetraphenylcyclooctatetraenes at 25°.[181] The half lives for these compounds are ca. 12 hrs at 25°. However,

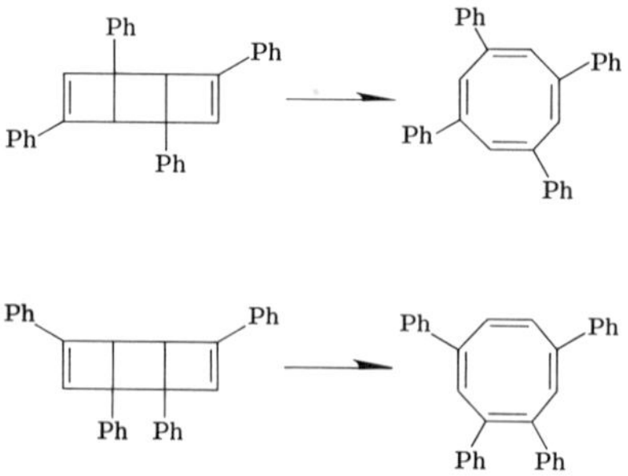

$$(5\text{-}91)$$

an excess of phenyl groups can prevent the reaction.[182] The presence of a number of fluorine atoms does not, however, deter the normal reaction.[183]

Attempts to generate benzocyclobutadienes gave dimers which rearranged prior to isolation. Nevertheless, the formation of the final products can be adequately rationalized by assuming the intermediacy of "ladder" type dimers. Cava and Napier[184] obtained a benzodihydrobiphenylene when zinc dust dehalogenation of 1,2-dibromo (or diiodo) benzocyclobutene was carried out under conditions inhibiting polymerization [Eq. (5-92)]. The

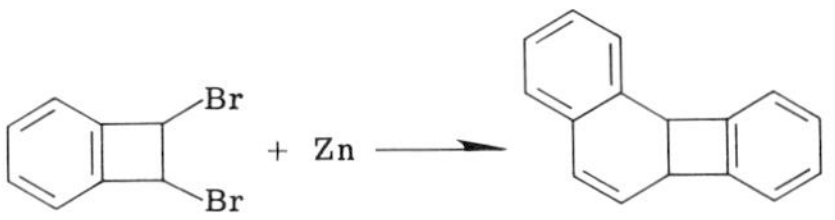

$$
\text{(5-92)}
$$

route assumed for formation of this product is shown in Eq. (5-93). A linear

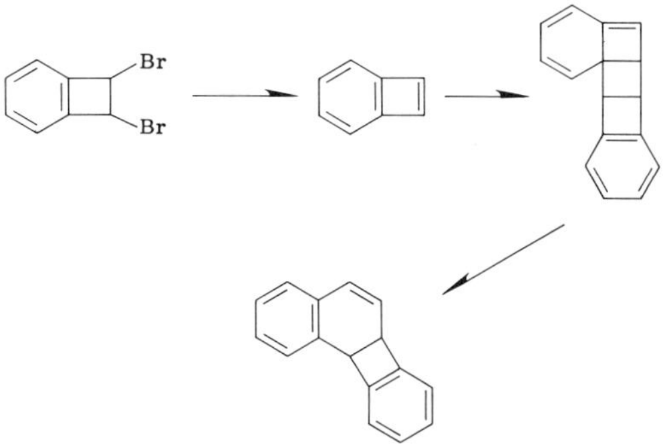

$$
\text{(5-93)}
$$

dimer was obtained when the dibromide was treated with lithium amalgam in the presence of nickel carbonyl.[185] That dimer formed dibenzocyclooctatetraene when heated to 135°. If the dibromide is dehydrohalogenated the angular dimer forms and rearranges prior to isolation. Loss of hydrogen bromide leads to an aromatized product [Eq. (5-94)].[186] Dehydrobromina-

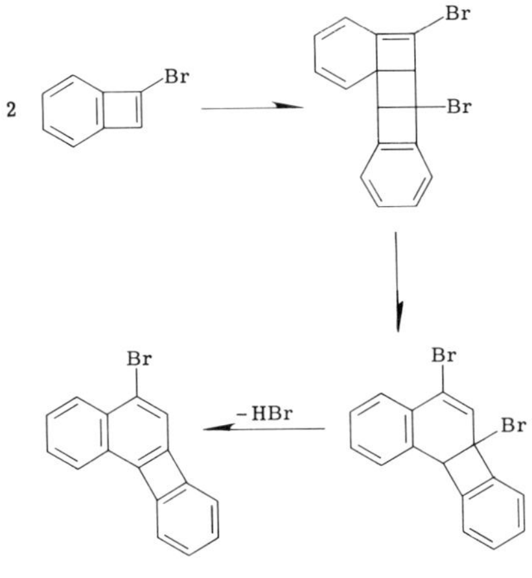

$$
\text{(5-94)}
$$

tion of 1,1-dibromobenzocyclobutene leads to the same product.[187] The position of the disappearing bromine atom is uncertain, but is shown here as suggested by Cava and Stucker.[186] However, where 1-*tert*-butylbenzocyclobutadiene was formed the dimeric product isolated had the second *tert*-butyl group on an α-carbon of the dihydronaphthalene ring.[188] A disubstituted benzocyclobutadiene, 1,2-dimethylbenzocyclobutadiene produced an angular dimer.[189] Hexachlorobenzocyclobutadiene on the other hand forms the linear dimer, which can be converted thermally to a dibenzocyclooctatetraene.[190]

The recent history of the valence isomers of benzene has added a fascinating chapter to the long story of aromaticity. Of these isomers only one, Dewar benzene or bicyclo[2.2.0]hexadiene, is related to benzene via an electrocyclic process. Prior to 1962 Dewar benzenes had been postulated as logical intermediates leading from cyclobutadienes to benzenes.[191,192] However, in 1962 van Tamelen and Pappas showed that a severely hindered

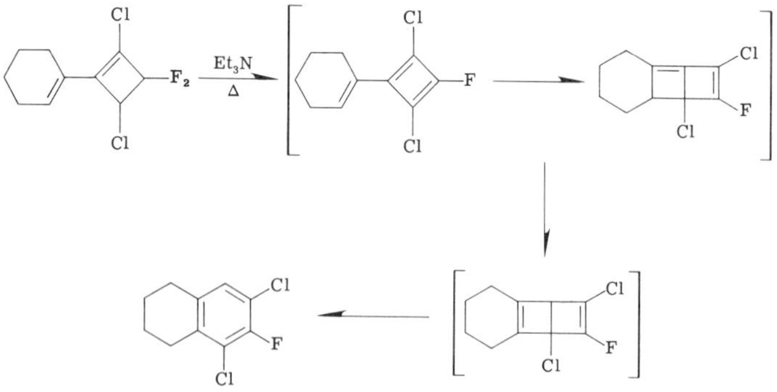

(5-95)

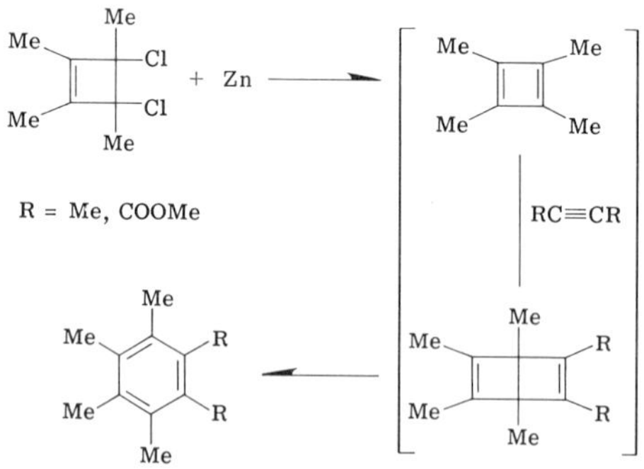

(5-96)

Dewar benzene could be isolated.[193] In fact, the 1,2,5-tri-*tert*-butylbicyclo-[2.2.0]hexadiene required 15 min at 200° for conversion to the benzenoid form. Shortly thereafter the parent Dewar benzene was isolated,[194] and proved to be rather more stable than was anticipated. At room temperature the half-life was estimated at 2 days, and 30 min at 90° converted it exclusively to benzene. Crieger and Zanker obtained the diester **101** and found its thermal

$$\text{(5-97)}$$

stability similar to that of the parent.[195] An activation energy of 27.7 kcal/mol for the ring cleavage was reported.[23]

Several kinetic studies of the isomerization of hexamethyl Dewar benzene have been reported.[196,197,197a,198] The earliest values[196] $E_a = 31.1$ kcal/mol and $\log A = 12$ have been revised[197,197a] to $E_a = 37.2$, $\log A = 15.03$ ($\Delta H^{\ddagger} = 36.4$, $\Delta S^{\ddagger} = 7.5$) and $\Delta H^{\ddagger} = 36.2$, $\Delta S^{\ddagger} = 7.7$.[198] Hexafluoro Dewar benzene reverts thermally to hexafluorobenzene with $E_a = 28.4$ kcal/mol and $\log A = 13.7$.[199,200] An unexpected influence of unsymmetrical substitution on rate was uncovered when Breslow et al.[201] found that 1-fluoro and 1-chloro-bicyclo[2.2.0]hexadienes reacted far faster than either the unsubstituted parent or 1,4-dichlorobicyclo[2.2.0]hexadiene. The results are given in Table 5-5. Lemal and Dunlap[202] showed that hexakis-trifluoromethyl Dewar benzene was of similar stability ($\Delta H^{\ddagger} = 37.4$, $\Delta S^{\ddagger} = 1.9$) to the hexamethyl derivative. Surprisingly, 1,4-hexamethylenebicyclo[2.2.0]hexadiene reacts about as readily as does the unsymmetrical 1-chloro compound.[204] The pentamethylene compound undergoes an alternate rearrangement.[205] Criegee and Huber gave $t_{1/2} = 2.3$ hrs at 90° for 2,3,5,6-tetrabromo-1,4-dimethylbicyclo[2.2.0]hexadiene.[203] This very large effect of substitution on the activation energy in this system is of considerable interest.

TABLE 5-5

Activation Parameters for Some Substituted Dewar Benzenes[201]

X	Y	$\Delta H^{\ddagger}$	$\Delta S^{\ddagger}$	Relative rate
H	H	23.0	−5.0	1.00
Cl	Cl	30.5	12.0	1.62×10^{-3}
Cl	H	19.1	−9.4	89.6
F	H	—	—	359.

Since a number of fluorinated benzenes will give Dewar benzenes on irradiation, a series of these fluorinated Dewar benzenes have been thermally converted to their benzenoid valence isomers. These include 1,2,5- and 2,3,5-trifluoro Dewar benzenes,[206] and the series given in Table 5-6. At temperatures above 550°K hexakis(trifluoromethyl) Dewar benzene is the favored valence isomer at equilibrium.[208]

When Dewar benzene, its 1-chloro or 1,4-dichloro derivatives are heated in the presence of 9,10-dibromoanthracene chemiluminescense is produced.[209] This important result shows that excited states are formed during the thermolysis of these Dewar benzenes. It has been shown that only the triplet state of benzene is accessible, and since the chlorinated derivatives are more efficient, the heavy atom effect on intersystem crossing operates as expected.[210] However, the efficiency for production of triplet benzene is relatively low in all cases (ca. 10^{-3}–10^{-4}). Generation of chemiluminescence during pyrolyses of bicyclopropenyls has been used as evidence for the intermediacy of Dewar benzenes in that reaction.[211] MINDO/3 calculations produced a transition state of C_s symmetry with one CH group having moved outward to a greater extent than the other.[212] When triplet formation was permitted, the calculation indicated that the crossing between ground and excited states occurs after the transition state.[213,458] This would account for the low efficiency of the formation of triplets since decay via the ground state surface competes effectively with intersystem crossing. Requirements for chemiexcitation include reaction via a forbidden route and a highly exothermic reaction.

TABLE 5-6

Temperatures Used to Rearrange a Series of Fluoro Dewar Benzenes[207]

Substituents	Temperature (°)
1,2,4-Trifluoro	80
1-Chloropentafluoro	80
1,4-Dichlorotetrafluoro	100
1-Iodopentafluoro	80
2,5-Dimethyltetrafluoro	100
2,6-Dimethyltetrafluoro	100
2,5-Diethoxytetrafluoro	190
2,6-Diethoxytetrafluoro	190
2,3-Diethoxytetrafluoro	195
1,2,3,4,5-Pentafluoro	80
1- and 2-Trifluoromethylpentafluoro	80

Reaction of Dewar benzenes as a four electron retro-electrocyclization raises the question of the role of the non-reacting double bond. An interaction between the two π bonds is to be expected, and a theoretical calculation[214] indicates a splitting between the symmetric and antisymmetric combination of 0.48 eV. Preparation of an optically active Dewar benzene and examination of its circular dichroism spectrum showed a transition at 218 nm, which might be assigned to the π–π interacting system.[215] Ignoring the small difference in ring strain, the $\Delta H^{\ddagger} = 43.7$ for reaction of hexamethylbicyclo[2.2.0]hexene[150] and 36.4 for hexamethyl Dewar benzene indicates the second double bond reduces the $\Delta H^{\ddagger}$ by about 7 kcal/mol.

3. Bicyclo[3.2.0]Hept-6-Enes

The thermal ring expansion of a bicyclo[3.2.0]hept-6-ene was uncovered by Chapman and Pasto in 1960.[216] Availability of the general photochemical ring closure [Eq. (5-98)] permitted both Dauben and Cargill[217] and Chapman

$$\text{(5-98)}$$

et al.[218] to establish that bicyclo[3.2.0]hept-6-ene itself will revert to cycloheptadiene at 400°–500°. Further examples of the thermal ring opening of substituted (methoxy and methyl substituents) 3,6-bicyclo[3.2.0]heptadiene-2-ones were added shortly.[219] Criegee and co-workers[23] prepared a series of substituted bicyclo[3.2.0]heptenes and determined qualitatively some relative rates. That the ring opening could be followed by migrations of the

$$\text{(5-99)}$$

double bonds (1,5-hydrogen shifts) was noted by Criegee and Furrer.[220] Additional bicyclo[3.2.0]heptenes were studied by Askani[221] and by Criegee et al.[222]

As a rough measure of the relative rates of rearrangement, Criegee used the temperature at which $t_{1/2}$ was about 2 hr.[22] The results are given in Table 5-7. Two rather rough conclusions can be drawn from this early work. First the presence of an unsaturated carbonyl moiety and/or a methoxyl group enormously enhances the rate of the ring opening. Second phenyl or carbalkoxy substitution at the bridgehead positions also enhances the rate considerably.

TABLE 5-7

Rough Relative Rates of Thermal Ring Opening of Some Bicyclo[3.2.0]Hept-6-Enes

Compound	Temperature (for $t_{1/2} \simeq 2$ hrs) (°)
	> 380
	> 380
	> 310
	365
	> 380
	270
	290

The number of representatives of this bicyclic system which have been given careful kinetic study is very limited. However, the basis for comparative studies of the effect of substitution has been provided, in that 6-bicyclo[3.2.0]-heptene has been studied.[223] Rearrangement proceeds readily in the gas phase at 275° ($k = 1.37 \times 10^{-4}$ sec^{-1}) and $E_a = 45.5$, log $A = 14.31$. In solution reaction is about twice as fast at the same temperature. The presence of a double bond in the five-membered ring accelerates the reaction markedly, the activation energy being lowered by 6 kcal/mol ($E_a = 39.5$ log $A -$ 14).[224] This is remarkably similar to the influence of a remote double bond on the rearrangement of bicyclo[2.2.0]hexene. Introduction of a bridgehead

$$(5\text{-}100)$$

102

methoxy and a C_2 carbonyl reduces the activation energy by an additional 6.0 kcal/mol.[225] The isopropyl group at C_4 probably has only a small effect. When the double bond in the cyclopentene ring is incorporated into an aro-

$$(5\text{-}101)$$

matic moiety the activation energy is essentially the same (39.3 kcal/mol) as for 2,6-bicyclo[2.2.0]heptadiene.[226] The possibility that a more remote double bond might intrude two electrons, converting this forbidden process to an allowed disrotatory six electron reaction has been considered, and for the reaction of Eq. (5-103), at least, has been discarded.[227] The differences

$$(5\text{-}102)$$

$$(5\text{-}103)$$

in activation energies between the compounds with and those without the remote double bond were attributed to differences in strain energy.

Some of the most amazing results were obtained by Nozoe and co-workers with a series of substituted phototropolones and aminotropones.[228] Their study was semiquantitative in nature in that they measured the temperature at which thermal reversion from the photo product was 50% complete in two minutes, i.e., $t_{1/2} = 120$ sec or $k = 6 \times 10^{-3}$ sec^{-1}. The results are shown in Table 5-8, and these demonstrate once again the extreme sensitivity of these reactions to substitution patterns. As was observed with Dewar benzenes, the unsymmetrical pattern of substitution at the ends of the bond being severed appears to provide the key to these extreme effects. However, use

TABLE 5-8

Temperatures at Which $t^{1/2} = 120$ Sec for Some 2,6-Bicyclo[3.2.0]Heptadiene-2-ones

X	Y	Temperature (°)
MeNH	CN	−75
NH$_2$	CN	−50
MeNH	COOH	−44
MeNH	COOMe	−33
NH$_2$	Ph	25
OH	CHO	25
OH	Ph	25
OH	CN	25
OH	COOH	25
OH	COOMe	25

of a methoxyl and a hydrogen did not result in such a large effect. The present pairs generally involve an electron attracting group paired with an electron donor. Though the replacement of a 1-methoxyl by a hydroxyl may introduce a special effect,[235] substitution of a phenyl or a carbonyl at C_5 reduces the free energy of activation by approximately 12–13 kcal/mol. That the change from methoxyl to a substituted amino group at C_1 can reduce the enthalpy of activation by about 9 kcal/mol has also been shown.[236] This reaction is

$$(5\text{-}104)$$

quite sensitive to the electron donating quality of the aromatic group ($\rho = -1.78$), and more polar solvents also increase the rate. For Ar = Ph in Eq. (5-104), the activation energy is 24.7 kcal/mol ($\Delta S^{\ddagger} = -8.9$ eu).

The importance of a 6-ene has been clearly illustrated [Eqs. (5-105) and

$$(5\text{-}105)$$

$$(5\text{-}106)$$

(5-106)].[234] Incidentally, these results show that an electron donor (Et_2N) and electron attractor $(C = 0)$ paired in this orientation give rise to no special accelerating effect. The unexpectedly large substituent effects do extend, however, to substitution of a heteroatom at C_2. Thus, N-carbethoxyazepine is formed readily from N-carbethoxy-2-azabicyclo[3.2.0]hepta-2,6-diene at 125°.[237] An oxygen has a similar effect [Eq. (5-107)].[237] A sulfur atom has

$$(5\text{-}107)$$

such a profound effect that ring opening occurs at 100°, even when the 3-ene

$$(5\text{-}108)$$

is missing.[238] Replace the 3-ene and the reaction rate increases as expected [Eq. (5-109)].[239] Further reaction leads to the formation of substituted

$$(5\text{-}109)$$

benzenes. Apparently the effect of selenium is more pronounced than that of sulfur.[240] Were it not for the behavior of the nitrogen and oxygen analogs, one might be tempted to suggest that a 6π electron disrotatory process was involved [Eq. (5-110)], but the general trend and the special influence of the sulfur in the absence of the 3-ene show this is not operative.

$$\text{(5-110)}$$

A number of additional examples of these electrocyclic reactions of 6-bicyclo[3.2.0]heptenes have been observed. Those which have not added any novel observations are recorded in Table 5-9.

In 1967, Miyashi *et al.*[242] uncovered a novel rearrangement of 3,6-bicyclo-[3.2.0]heptadien-2-ones[Eq. (5-111)], which is presently of uncertain mecha-

$$\text{(5-111)}$$

nism, but by no means clearly unrelated to the retro-electrocyclic reaction. Mukai has interpreted this new rearrangement as a concerted Cope rearrangement, the first example of an antara-antara [3,3] sigmatropic shift.[236] Baldwin found that the less strained but more mobile 3,6-bicyclo[3.3.0]-octadiene would not undergo an analogous rearrangement under much more

TABLE 5-9

Examples of Thermal Retro-Electrocyclic Reactions of 6-Bicyclo[3.2.0]Heptenes

Compound	Reaction temperature (°)	Reference
	300	229
	450	230
	430	241
	—	231

$$\text{(5-112)}$$

drastic conditions.[243] This evoked the suggestion that the rearrangement involved the intermediacy of a *cis, trans, cis*-cycloheptatriene. Miyashi *et al.*[225] measured the activation parameters for both the rearrangement and

$$\text{(5-113)}$$

its accompanying valence isomerization to the tropone. For **102** both processes have the same activation energy (33.5 kcal/mol) but log A for the formal Cope rearrangement is 12.7 while that for ring expansion is 12.0. Special need for the carbonyl group in the Cope rearrangement was suggested. Baldwin and Kaplan[244] countered by showing that 1-ethoxy-3,6-bicyclo[3.2.0]heptadiene will react normally in the Cope process. That the methoxyl group is also non-essential was shown by rearrangement of 1,4,4-trimethyl-2,6-bicyclo[3.2.0]heptadiene, albeit some 100× slower than those having a carbonyl group.[245] No final answer to the mechanism of this formal Cope rearrangement has been obtained.

Ring annelated bicyclo[3.2.0]heptenes show unusual behavior for benzo-cyclobutenes.[461,462] A dimethylbenzopleiadiene reverts to the cyclobutene isomer (E_a = 21.3, log A = 15.6) and the reaction has been considered a concerted disrotatory reaction.[461] A 2,3-naphtho analog does not revert to a cyclobutene form even at 100°.[462]

4. 7-Bicyclo[4.2.0]Octenes

The comprehensive study of the thermal reactions of cyclobutenes by Criegee and his students included a number of 7-bicyclo[4.2.0]octenes. Thus, Askani[221] prepared and pyrolyzed several bicyclooctenes. Both 1,6,7,8-tetramethyl and 7,8-dimethyl-1,6-dicarbomethoxy derivatives gave mixtures of monocyclic products at 300°–350°. A rough measure of the ease of ring opening was indicated by temperatures at which $t_{1/2} \simeq 2$ hrs (Table 5-10).[23] Clearly there exists no sharp demarcation between the rates of reaction for 6-bicyclo[3.2.0]heptenes and 7-bicyclo[4.2.0]octenes. This observation was quantitatively confirmed by Branton *et al.*[32] who showed that the parent ene has E_a = 43.2 and log A = 14.13. Thus, an increase in ring size by one

TABLE 5-10

Temperatures at Which $t^{1/2} \simeq 2$ hrs for Pyrolysis of Some 7-Bicyclo[4.2.0]Octenes

Compound	Temperature (°)	Reference
	250	23
	350	246, 247
COOMe ⟨structure⟩ COOMe	280	221, 23

carbon reduced E_a by only 2.3 kcal/mol while the ring strain in the reactants was lowered by about 5 kcal/mol.

However, the possibility that the rates might mask some novel route was suggested by the isolation of *cis*, *trans*-1,3-cyclooctadiene, and its conversion to 7-bicyclo[4.2.0]octene at 80°.[120] Liu[248] showed that the acetophenone sensitized photoconversion of *cis*, *cis*-1,3-cyclooctadiene to 7-bicyclo[4.2.0]-octene involved the intermediacy of the *cis*, *trans*-cyclocatadiene which ring closed via dark reaction. Schuster *et al.*[123] has questioned whether a similar route could be operative in the sensitized photolysis of 3,5-cyclohepta-dienone. Thermal reactions of 7-bicyclo[4.2.0]octene and the two cyclo-octadienes were carefully studied by McConaghy and Bloomfield[249] and Bloomfield *et al.*[250] The reaction between the bicyclic isomer and *cis*, *trans*-

$$\text{[structure]} \underset{k_{-1}}{\overset{k_1}{\rightleftharpoons}} \text{[structure]} \tag{5-114}$$

$$
\begin{array}{lll}
k_1 & \Delta H^\ddagger = 32.6 & \Delta S^\ddagger = \ 0.0 \\
k_{-1} & \Delta H^\ddagger = 26.9 & \Delta S^\ddagger = -1.0
\end{array}
$$

1,3-cyclooctadiene was shown to be reversible, and the rates and activation parameters for both forward and reverse steps were determined. At 250° the bicyclic compound gives only *cis*, *cis*-1,3-cyclooctadiene, and a concerted orbital symmetry allowed route for that reaction via a 1,5-hydrogen shift in the *cis*, *trans* diene [Eq. (5-115)] was suggested. The value for k_2 calculated

$$\text{(5-115)}$$

was reasonable. Cocks and Frey[78] used Bloomfield's data to obtain $E_a = 37.6$ kcal/mol and log $A = 14.0$ for the hydrogen shift step. Again the values appear quite reasonable, thus making the route permissable, if not operative. In view of these parameters the finding that cis, trans-1,3-cyclooctadiene gives 7% cis, cis-1,3-cyclooctadiene and only 2% 7-bicyclo[4.2.0]octene along with a series of dimers after standing 40 hrs at 25° in the dark seems rather surprising.[251]

Bicyclo[4.2.0]octa-2,7-diene is converted thermally to all-cis cycloocta-

$$\text{(5-116)}$$

triene.[252] That reaction was investigated further by Baldwin and Kaplan,[244] who found that at 180° the rate was 0.75×10^{-4} sec^{-1}. This compared with the rate for 7-bicyclo[4.2.0]octene[32] of 1.53×10^{-4} sec^{-1} at 253° indicates once again a substantial accelerating effect for the added double bond. In this study Baldwin and Kaplan also uncovered the degenerate rearrangement of Eq. (5-117). Generation of cis, trans, cis-cyclooctatriene via a retro

$$\text{(5-117)}$$

Diels–Alder reaction, and analysis of the kinetics showed that this strained triene closes to 2,7-bicyclo[4.2.0]octadiene, but does not go directly to all cis-cyclooctatriene.[253] This result, in conjunction with the acceleration caused by the 2-ene, raises some questions about the 1,5-hydrogen shift route to all cis-octadienes and trienes.

Thermal reactions of $(CH)_n$ compounds have been of considerable interest in recent years.[254] An important member of this group, "Nenitzescu's hydrocarbon" (103) contains the bicyclo[4.2.0]octene entity embedded in a tricyclic

103

system.[255] When heated the hydrocarbon gives a mixture containing naphthalene, 1,2-dihydronaphthalene, tetralin, cis-9,10-dihydronaphthalene, and 1-phenylbutadiene in varying proportions depending on the specific conditions used.[255-258] Rearrangement of the 7,8-dicarbomethoxy derivative

$$(5\text{-}118)$$

gave a mixture of two dihydronaphthalenes[259,260] [Eq. (5-119)] and not 1,5-

$$(5\text{-}119)$$

dicarbomethoxy-9,10-dihydronaphthalene as was originally suggested.[261] An analogous reaction [Eq. (5-120)] prompted Maier[262] to suggest that the

$$(5\text{-}120)$$

reaction was initiated by ring opening of the cyclobutene moiety. Pyrolysis of a benzo derivative led to a similar mechanistic proposal.[263] However,

$$(5\text{-}121)$$

104

Vedejs[258] showed that **104** cannot be an intermediate in that reaction, because it reacts too slowly. It was also shown that **105** gives a very different

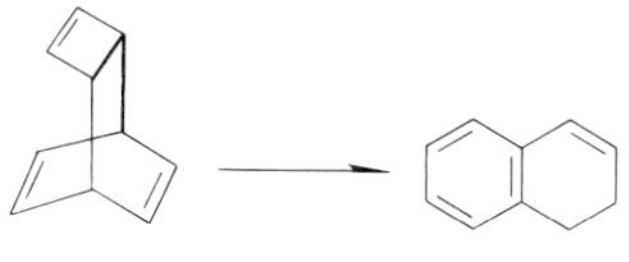

84%
E_a = 46.4
log A = 15.5

(5-122)

105

86%
E_a = 40.4
log A = 12.2

product ratio from Ninetzescu's hydrocarbon under the same conditions. The activation parameters were determined and a mechanism for the formation of 1,2-dihydronaphthalene was proposed [Eq. (5-123)].[264] If the scheme

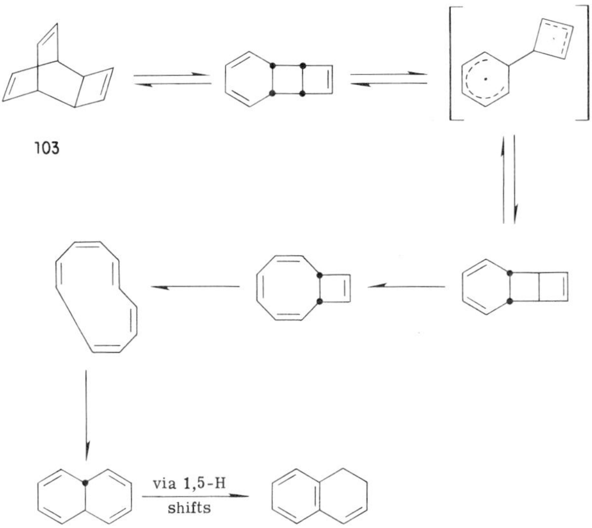

103

(5-123)

via 1,5-H shifts

is to apply to the carbomethoxy derivative, it will require some rather unexpectedly selective reactions. Whatever the problems with the entire mechanism, it seems clear that direct ring opening of the cyclobutene ring in Nenitzescu's hydrocarbon and its benzoanalogs occurs to only a minor extent.

Contrariwise, ring opening is the sole reaction of **106** and **107** at 500°.[265]

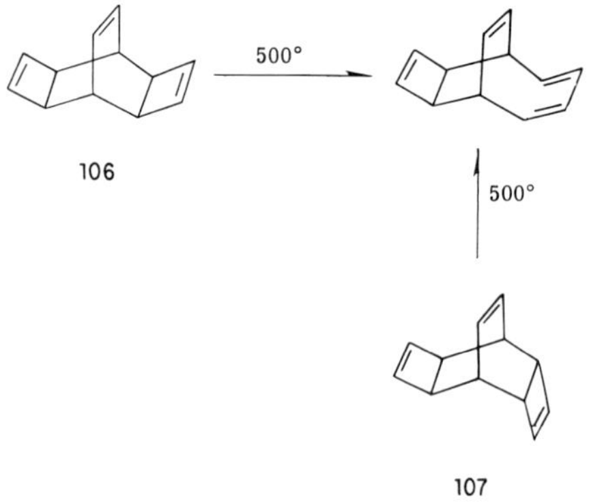

106

$$\text{(5-124)}$$

107

Monosubstitution of one cyclobutene ring leads to preferential cleavage of the non-substituted ring. This result would be expected based on the influence of substituents (Cl, Me) on the rate of the retroelectrocyclization of cyclobutene, assuming they exert a similar effect on the forbidden reaction. Collin and Sasse[266,266a] have studied several examples of the mechanistically ambiguous process of [Eq. (5-125)]. If the benzene ring participates the reaction

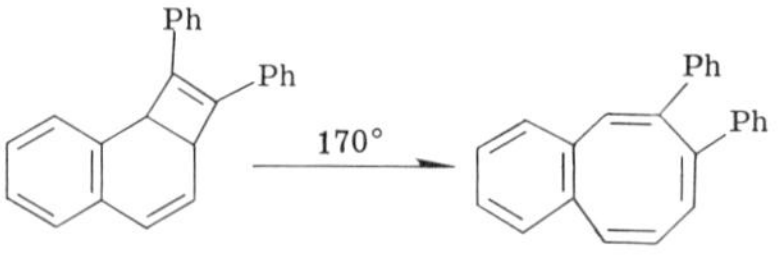

$$\text{(5-125)}$$

should be a facile 6π electron process. An unusual ring cleavage accompanied by loss of carbon monoxide has been observed.[267] Cyclooctatriendione has

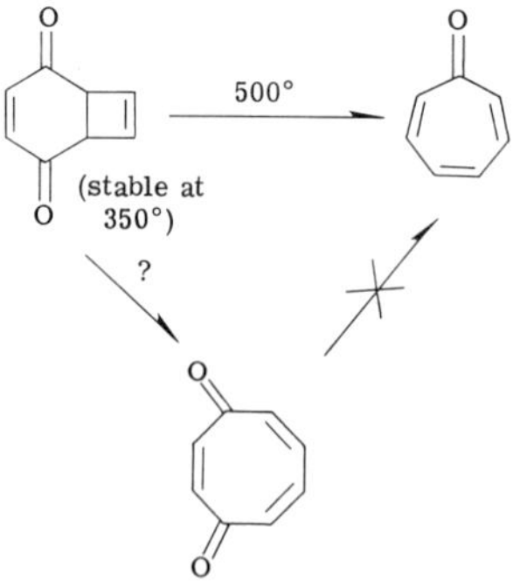

$$\text{(5-126)}$$

been shown not to be a participant in the process.[268] Replacement of the carbonyl groups by alkylidenes enhances the reaction rate and no fragmentation occurs.[269,270]

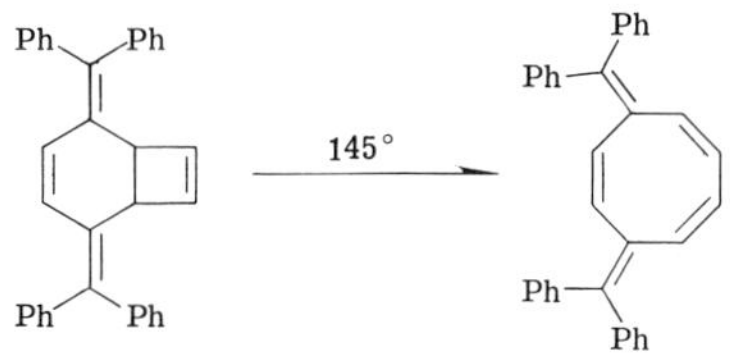

$$E_a = 36.5 \text{ kcal/mole}$$
$$\log A = 14.14$$

(5-127)

5. Bicyclo[*n*.2.0]Alkenes with *n* ≥ 5

The work of Bolz and Askani[246] with Criegee was reported in 1968,[247] and established that a rather considerable break in the plot of rate vs. *n* in the bicyclo[*n*.2.0]alkenes occurred between $n = 5$ and $n = 6$. Thus, **108** had $t_{1/2} \simeq 2$ hrs at 330°, whereas **109** had the same half-life at 180°. While the product from **108** was not described, **109** gave **110**. Radlick and Fenical[271]

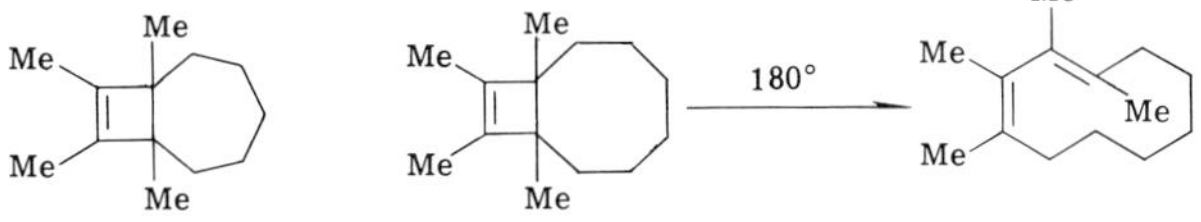

(5-128)

showed that 9-bicyclo[6.2.0]decene gave *cis, trans*-1,3-cyclodecadiene at 200°, while the 2,9-diene gave **111** [Eq. (5-129)], and the 4,9-diene gave

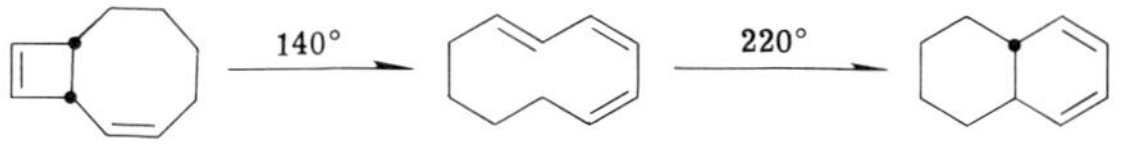

(5-129)

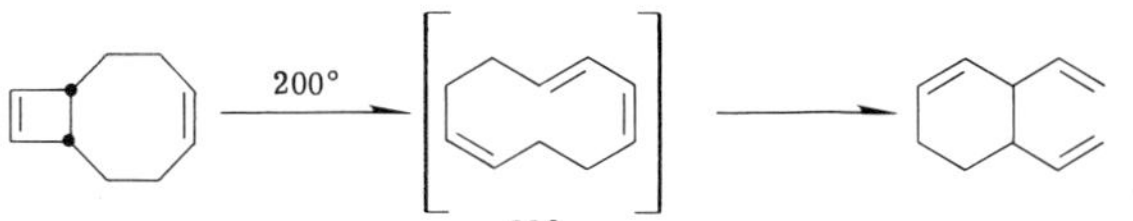

(5-130)

3,4-divinylcyclohexene. Since some cyclodecane was obtained along with diethylcyclohexane when this last product was reduced, the intermediacy of **112** was assumed. Ring opening of 9,10-dicarbomethoxy-2,9-bicyclo-[6.2.0]decadiene gave *trans, cis, cis*-2,3-dicarbomethoxycyclodecatriene accompanied by *trans*-2,3-dicarbomethoxy-2,4-bicyclo[4.4.0]decadiene.[249]

The apparent break in the rate curve noted by Criegee may be rather misleading. Thus, Schumate[272] found that bicyclo[5.2.0]nonene and bicyclo-[6.2.0]decene are in equilibrium with *cis, trans*-1,3-cycloalkadienes at 250° [Eqs. (5-131) and (5-132)] but rearrange more slowly to the *cis, cis*-dienes.

$$K = 0.5$$
$$\text{(at } 250°)$$

$$250°$$
$$k = 3.0 \times 10^{-4} \text{ sec}^{-1}$$

$$(5\text{-}131)$$

$$K > 100$$
$$\text{(at } 250°)$$

$$250°$$
$$k = 8.0 \times 10^{-4} \text{ sec}^{-1}$$

$$(5\text{-}132)$$

At 250° the rates of rearrangement to the *cis, cis*-dienes are 1.2×10^{-4} sec^{-1} for the [4.2.0] compound,[32] 3.0×10^{-4} sec^{-1} for the [5.2.0] and 8.0×10^{-4} for the [6.2.0]. Clearly for comparable reactions there exists no such sharp discontinuity in the rate curve.

Heating **113** gives **114** [Eq. (5-133)].[273] This reaction cannot proceed

$$\Delta$$

$$(5\text{-}133)$$

113   114

directly via either *cis, trans, cis, cis*- or *trans, cis, cis, cis*-cyclononatetraene which would have to give the trans-bicyclo[4.3.0]nonatriene. Ring expansion of **115** [Eq. (5-134)] gives a *cis, cis*-diene.[274] Fusion of an extra ring to a

$$\xrightarrow[200°]{\Delta}$$

$$(5\text{-}134)$$

115

benzocyclobutene does not prevent thermal rearrangement [Eqs. (5-135) and (5-136)]. Both primary products could arise via an initial allowed

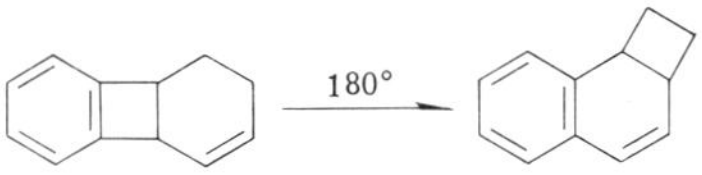

$$(5\text{-}135)$$

(refer. 275)

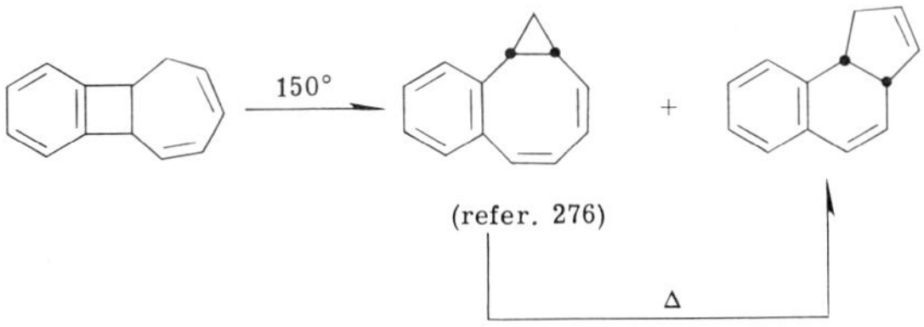

$$(5\text{-}136)$$

(refer. 276)

conrotatory ring opening followed by a second electrocyclic reaction [Eq.

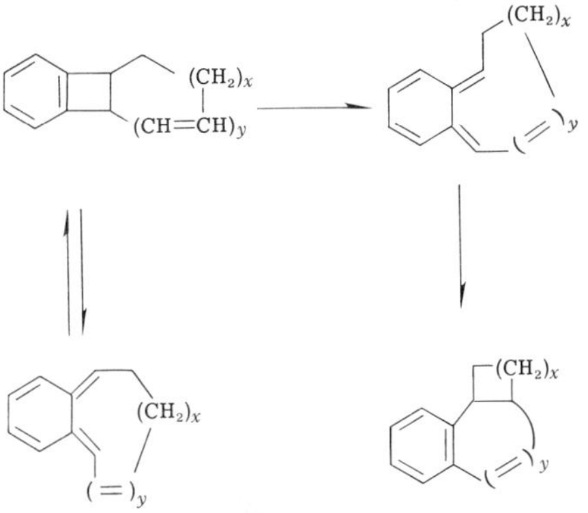

$$(5\text{-}137)$$

$$x = 1 \quad y = 1$$
$$x = 0 \quad y = 2$$

(5-137)]. Those cis products where $y = 1$ cannot come directly from the ring expanded intermediate via an allowed electrocyclic reaction.

Bicyclo[6.2.0]deca-2,4,6,9-tetraene has been prepared and is thermally un-stable with respect to *trans*-9,10-dihydronaphthalene.[277,278] Compound **116**

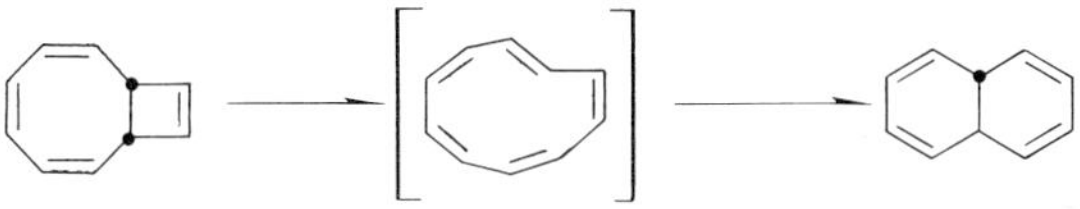

$$(5\text{-}138)$$

116

$$k_{(46.8°)} = 6.9 \times 10^{-5} \ \text{sec}^{-1}$$
$$\Delta H^{\ddagger} = 24.9 \ \text{kcal/mol}$$
$$\Delta S^{\ddagger} = -0.4 \ \text{eu}$$

is presumed as an intermediate, since it has been prepared and shown to give the trans-dihydronaphthalene in a very rapid reaction.[279] A halogenated bicyclo[6.2.0]decatetraene has been obtained and rearranges thermally at room temperature.[280] A dibenzoderivative also undergoes the same type of rearrangement, but at a higher temperature.[281]

$$\text{(5-139)}$$

6. Other Ways to Foil the Allowed Reaction

Since cis disubstituted cyclobutenes must open via conrotation to give one trans double bond, two laterally fused rings necessitate development of a trans double bond in one of the rings. If both rings contain less than eight

$$\text{(5-140)}$$

atoms, the conrotatory process is seriously inhibited. An early observation by Crowley[282,283] set the stage for examples of this type [Eq. (5-141)]. Liu[284]

$$\text{(5-141)}$$

suggested that sensitized irradiation of 1,1'-biscyclohexene gives tricyclo-[6.4.0²,⁷]dodec-1-ene via thermal ring closure of *cis, trans*-1,1'-bis-cyclo-hexene. Arguments for the cis geometry in the tricycle have been summarized by Dauben.[285]

Ball and Landor[286] noted that dimers of 1,2-cycloalkadienes would form diadducts with maleic anhydride. Dimers of both cycloheptadiene and cyclo-

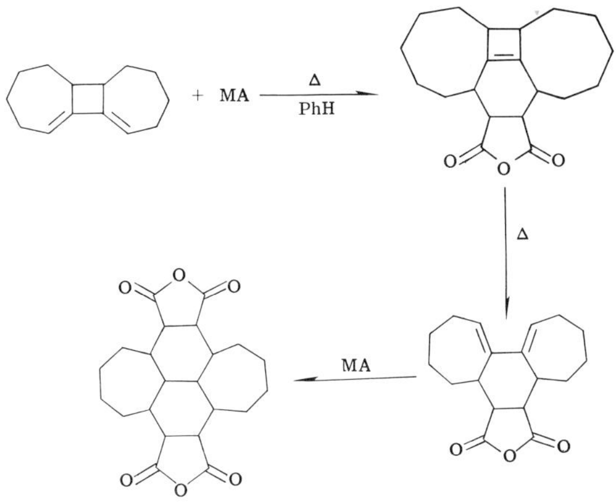

(5-142)

octadiene behave similarly. Yields of diadduct were modest and some evidence for the presence of the monoadduct was obtained. Cyclononadiene dimer also exhibits an analogous behavior.[287] Clarification was obtained when 1,2,6-cyclononatriene was found to give two stereoisomeric dimers, from which a diadduct and a monoadduct were obtained.[288] Interestingly

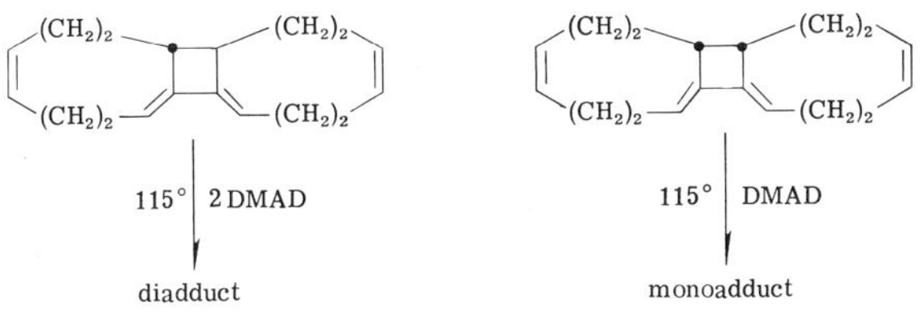

(5-143)

the cis dimer did not react to give a diadduct on prolonged heating even though the trans double bond would be in a nine carbon ring. Obviously a large fraction of the 3 kcal/mol difference between the cis and trans cyclononenes must be effective at the transition state.

A direct measure of the difference in activation energies for ring opening of several cis and trans tricyclic compounds was made by Criegee and Reinhardt.[289] Their results are given in Table 5-11. Rather surprisingly these E_a's compare quite accurately with those obtained from 3,4-fused rings of the same size. Thus, for example, for 7-bicyclo[4.2.0]octenes rearrangement to cis, cis-1,3-cyclooctadiene requires 43.2 kcal/mol, and to the cis, trans isomer, 32.6 kcal/mol. If Bloomfield's mechanism applies here, the 1,5-hydrogen shift cannot be operative and an intermediate trans-cyclohexene must fortuitously require about the same activation energy for its formation

TABLE 5-11

Relative Rates of Ring Opening of Stereo-Isomeric Tricyclic Cyclobutenes

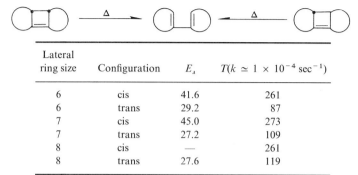

Lateral ring size	Configuration	E_a	$T(k \simeq 1 \times 10^{-4} \, sec^{-1})$
6	cis	41.6	261
6	trans	29.2	87
7	cis	45.0	273
7	trans	27.2	109
8	cis	—	261
8	trans	27.6	119

as does the H-shift in the eight-membered ring. Again, 8-bicyclo[5.2.0]nonene gives *cis*, *cis*-cyclononene with $E_a = 42.0$ kcal/mol, which compares quite reasonably with the 45.0 kcal/mol of the seven ring laterally fused system. Further study of these compounds could be very useful.

A rather specialized example of a forbidden electrocyclization to a cyclo-butene has been postulated to explain the results of Eq. (5-144).[290]

$$\text{Me} \underset{C\equiv CH}{\overset{C\equiv CH}{\triangleleft}} \xrightarrow[N_2]{350°} \left[\bigcirc_{Me} \right] \longrightarrow \bigcirc_{Me} \qquad (5\text{-}144)$$

E. Energetics and the "Forbidden" Mechanism

Conrotatory processes and the consequent ease with which they can be geometrically prevented have permitted the assessment of limits for the energy difference between allowed and forbidden concerted mechanisms. The actual mechanism of the reaction (or reactions) leading to the so-called forbidden product matters not a whit, since the forbidden concerted reaction is either being observed or is slower than the observed process, thus permitting a lower bound to the energy differential to be established. Perhaps the first estimate was that of Willcott and Overland,[224] (13 kcal/mol) which was based on the difference in activation energies for cyclobutene and bicyclo[3.2.0]hept-6-ene.

Brauman and Golden[291] used the same basic approach, but extended the basic set to include bicyclo[2.1.0], [3.2.0], and [4.2.0] enes. To do so they assumed that the strain energy present in the reactant was decreased to that of the product at the transition state. They concluded that the $\Delta\Delta H^{\ddagger}$ was about 15 kcal/mol. Lupton[292] used a different assumption to calculate $\Delta\Delta H^{\ddagger}$ from the same experimental data, thus reducing $\Delta\Delta H^{\ddagger}$ to about 10 kcal/mol. A more direct experimental test was applied by Doorakian and Freedman.[293] They found that **117** undergoes a degenerate rearrangement [Eq. (5-145)]

$$(5\text{-}145)$$

which was revealed by deuterium labeling. This reaction was allowed to run at 125° for 51 days without producing any measurable amount of cis, cis or trans, trans isomer. Calculation showed that $\Delta\Delta H^{\ddagger}$ must exceed 7.5 kcal/mol.

In 1972 Brauman[28] returned to this problem, confronting the experimental measure of $\Delta\Delta H^{\ddagger}$ directly. Thus, ring opening of *cis*-3,4-dimethylcyclobutene was shown by careful gas chromatographic analysis to give 0.005% of *trans, trans*-2,4-hexadiene. That value corresponds to a direct $\Delta\Delta G^{\ddagger}$ between conrotatory and disrotatory processes of 10.7 kcal/mol. Assuming $\Delta\Delta S^{\ddagger} \simeq 1 \pm 1$ eu, they give $\Delta\Delta H^{\ddagger} = 11 \pm 1$ kcal/mol as a direct difference, but since the conrotatory route moves one methyl inward at a cost of ca. 4 kcal/mol, $\Delta\Delta H^{\ddagger}$ must be greater than or equal to 15 kcal/mol in the absence of steric differences. The route which leads to the disrotational product is as yet unknown.

Comparison of the above result with that of the cis-trans isomerization of 1,4-dideuteriobutadiene[129] leads to some puzzling results. Stephenson, Gemmer, and Brauman presented data to support the conclusion that *cis, cis*-1,4-dideuteriobutadiene gives the trans-trans isomer rapidly via trans-3,4-dideuterio-cyclobutene, and the cis-trans isomer slowly via an orthogonal diradical. Using a value of 2.4 kcal/mol for the stabilization of a radical by a methyl group, derived from comparison of the rates of stereo-mutation of dideuterio- and dimethylcyclopropanes,[294,295] and values for the $\Delta H^{\ddagger}$ of cyclobutene and *cis*-3,4-dimethylcyclobutene taken from Table 5-1, the results of the two studies would require that *cis*-3,4-dimethylcyclo-butene be more stable than cyclobutene by about 7 kcal/mol. This does not seem to correspond to expectations.

F. Use in Synthesis

Electrocyclic ring openings of cyclobutenes have found two principal uses in synthesis. The first depends on the availability of cyclobutenes from cycloaddition reactions, and the ring opening then inserts a pair of carbons between the atoms originally joined by the double bond [Eq. (5-146)].

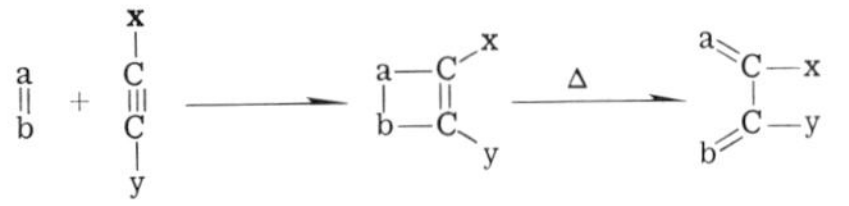

$$
\underset{b}{\overset{a}{\parallel}} + \underset{\underset{y}{\overset{|}{C}}}{\overset{\overset{x}{|}}{\underset{\parallel}{C}}} \longrightarrow \underset{b-C}{\overset{a-C}{\bigsqcup}}\overset{x}{\underset{y}{}} \xrightarrow{\Delta} \underset{b}{\overset{a}{\diagdown}}\overset{C-x}{\underset{C-y}{}} \tag{5-146}
$$

Brannock and co-workers[296] appear to have been the first to develop this route. The second broad usage which has developed exploits the high activity of o-quinodimethides as dienes in the Diels–Alder reaction [Eq. (5-147)]. Major use of this process has involved an intramolecular Diels–

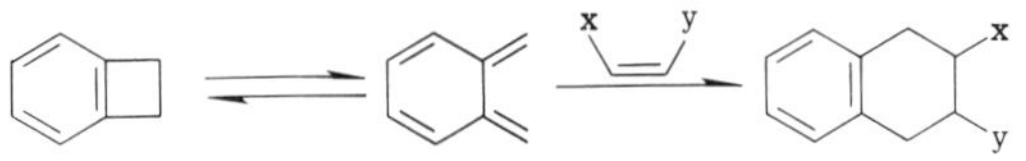

$$\tag{5-147}$$

Alder reaction, with Oppolzer and Kametani as the main proponents of the method.

During the early sixties Brannock,[296–298] Berchtold,[299] and Huebner and Dorfman,[300,301] all developed independently the same general reaction sequence to utilize Brannock's observation that electrophilic olefins will add to enamines to form cyclobutanes. Cyclobutenes were obtained by using electrophilic acetylenes such as dimethyl acetylenedicarboxylate or ethyl propiolate. These cyclobutenes were thermally labile and gave rise to some interesting ring-opened products. All three groups developed the route into

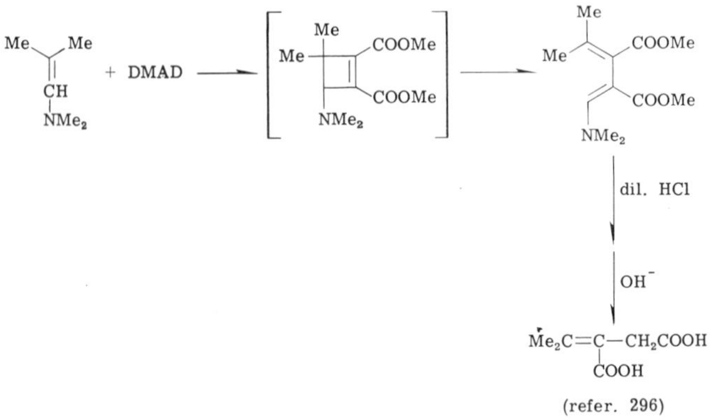

$$\tag{5-148}$$

$$Me_2C=\underset{\underset{COOH}{|}}{C}-CH_2COOH$$

(refer. 296)

$$\tag{5-149}$$

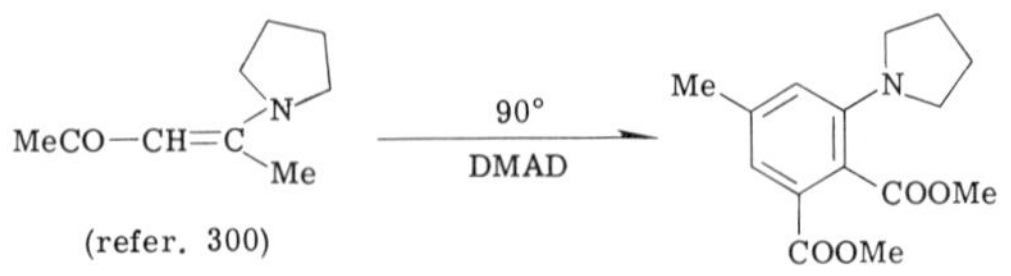

(refer. 297)

$$\tag{5-150}$$

(refer. 300)

a two carbon ring expansion which has seen considerable use to form seven- to ten-numbered rings of both alicyclic and heterocyclic types. The early work was confined largely to alicyclic systems [Eq. (5-151)], though Brannock *et al.*[296] do mention formation of a dihydrooxepin [Eq. (5-152)].

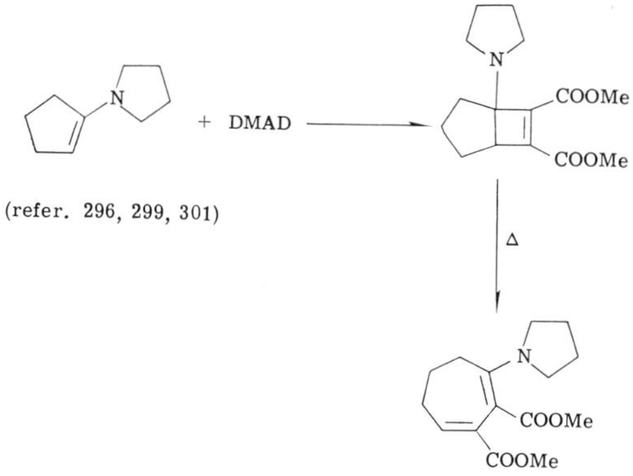

(refer. 296, 299, 301)

$$(5\text{-}151)$$

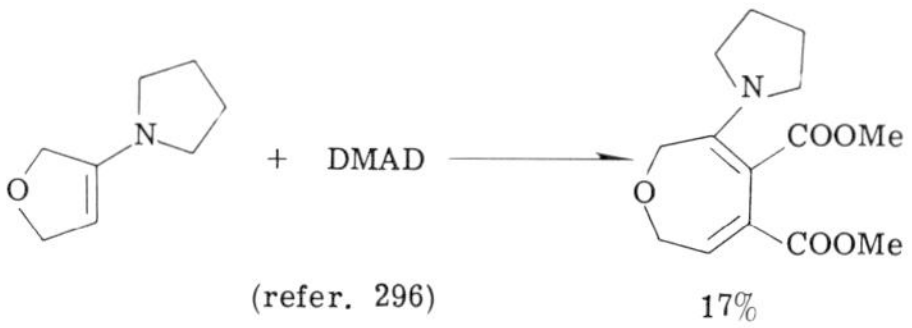

(refer. 296)   17% $$(5\text{-}152)$$

Prior to the Woodward–Hoffmann papers these ring openings would occasion no surprise, but today it is clear that these substituted examples of forbidden reactions occur with unexpected facility. The mechanism of these ring opening reactions has not been delineated and whether the push–pull effect of Epiotis on a concerted retroelectrocyclization or a retroaldol process is involved is uncertain. However, it is clear that the ease of ring opening can be expected to be a very sensitive function of the substitution pattern. A persuasive example has been reported by Ficcini and Durèault[234] in that **118** reacts as shown while the analog missing the 3-ene will not react under the same conditions.

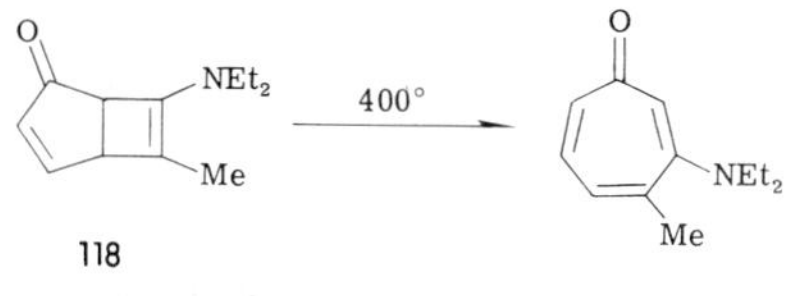

118

$$(5\text{-}153)$$

The method has seen limited use for synthesis of alicyclic compounds, but recent approaches to the synthesis of ophiobolins,[302] and the syntheses of steganicin[303] and steganone illustrate the potential. Clark and Untch[465,466] have suggested its use as a general two carbon ring expansion. Preparation of a hydroazulene derivative using this ring expansion has also been reported,[304] and vellerolactone was prepared via this process.[464] A rather unusual example involving cycloaddition to a benzocyclobutadiene might serve as a route to heavily substituted naphthalenes.[305] The ring expansion has seen use in the syntheses of a number of pleiadienes.[232,233,306] A similar route was used for the synthesis of **119** [Eq. (5-154)].[307] Gassman and

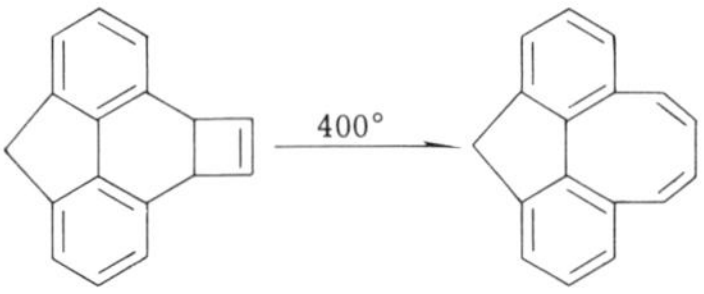

$$\text{(5-154)}$$

119

Creary[308] prepared **120** in high yield via expansion of a trans fused bicyclo-[4.2.0]octene [Eq. (5-155)]. Kitahara prepared 1,4-cyclooctatrienedione[309]

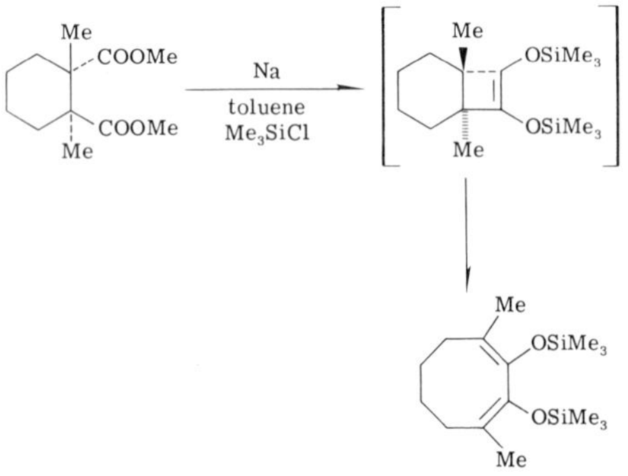

$$\text{(5-155)}$$

120

and a series of alkylidene derivatives from bicyclo[4.2.0]octenes.[310,311] Unsymmetrically substituted cyclobutenes sometimes show unexpected regioselectivity.[467]

Ring expansion has also served for the synthesis of a series of heterocycles with seven or eight-membered rings. Aside from the early reference[296] to the formation of dihydrooxepin, the formation of a benzazepin from a 2-ethoxyindole [Eq. (5-156)] constitutes one of the earliest uses of this

$$(5\text{-}156)$$

method.[312] Liu and Snieckus[313] used the same scheme to prepare benzazep-ines from 3-dialkylamino-*N*-acetylindoles. Eight-membered heterocycles have been prepared in a similar manner [Eq. (5-157)].[314,315] Early attempts

$$(5\text{-}157)$$

to apply the scheme to the preparation of thiepins were unsuccessful, owing to rapid collapse of the thiepins to aromatic compounds via expulsion of sulfur [Eq. (5-158)].[316,317] Dihydrothiepins can be prepared by this ring

$$(5\text{-}158)$$

expansion,[318] and eventually thiepins[319] and benzthiepins[320] were also successfully prepared. An attempt[321] to prepare 1,4-dithiacyclooctatriene failed and benzselenipins extruded selenium too rapidly to permit their isolation.[322]

In some instances [2 + 2]cycloadditions will permit the reaction to be used for the preparation of aliphatic dienes.[468]

Ring opening of benzocyclobutenes was demonstrated by trapping the quinodimethide as a Diels–Alder adduct,[109] but the first important use of this reversible ring opening was developed by Oppolzer in 1971.[323,324] The basic procedure illustrated in Eq. 5-159 was used to prepare *dl*-chelidonine.[324] Stereochemistry of the ring juncture is controlled by the side chain,

$$(5\text{-}159)$$

$$(5\text{-}160)$$

when X = O, Y = H_2, or X = H_2, Y = O the juncture is cis, but with X = Y = H_2 the juncture is mainly trans.[325] Kametani has developed the method for the synthesis a variety of alkaloids, steroids and intermediates for terpene synthesis. Routes were developed to several isoquinoline alkaloids,[326–330] to compounds in the yohimbane group,[331,332] to D-homoestrone,[333] estradiol, estrone,[334] intermediates for terpene synthesis,[335,336] and a variety of other compounds.[337–339] The general approach is illustrated by the reactions of Eq. (5-161). A particularly elegant example of this

$$(5\text{-}161)$$

scheme has been devised by Vollhardt and Funk,[340] who earlier had prepared a tetralin derivative by an intermolecular Diels–Alder reaction of a
silylated benzocylobutene.[341]

(5-162)

Browne and Halton[342] used diphenylbenzocyclobutene as a diene source
in their synthesis of 1,4-diphenyldichloro-2,3-naphthocyclopropene. Paquette utilized the activity of o-xylylene as a diene to prepare a hetero[4.4.3]-
propellane,[343] and dibenzocyclooctadienes have been obtained from benzocyclobutenes.[344] Aside from the synthetic value of benzocyclobutene as a
source of a diene for the Diels–Alder reaction, ring opening of other cyclobutenes has been used to prepare some otherwise difficulty accessible dienes.
Both 2,3-dicyano and dicarbomethoxy-1,3-butadienes have been obtained
in this manner.[345,346,347] Boeckman et al.[348] has used the photo adducts of
α,β-unsaturated ketones with ketene acetals to obtain some interesting dienes
potentially useful for the syntheses of polycyclic molecules. Wilson and
Phillips[349] have suggested the use of 1-methylcyclobutene as an isoprene
synthon.

Bond[98] has shown that pyrolysis of methylenecyclobutenes can provide
a convenient route to vinylallenes. Often the stereospecificity of the ring
opening reaction can be used to good advantage as, for example, in the
synthesis of cis, trans-2,4-hexadiene-1,6-dial.[350] Cyclobutenediones can
serve as very useful synthetic intermediates for the preparation of a variety
of 1,4-diones [Eq. (5-163)].[351,352] Thus, the ring opening of cyclobutenes
shows promise of becoming a very effective synthetic process.

(5-163)

II. CYCLOBUTENES CONTAINING HETEROATOMS

A. Oxetenes

The early literature concerning oxetenes is a consistent series of reports involving oxetenes as possible transient intermediates in reactions leading to unsaturated carbonyl products. Thus, for example, an oxetene was suggested as a possible intermediate in the formation of methyl 2-octenoate from dimethyl ketene acetal and hexanal.[353] More recent studies[354] leave some doubt about the reality of the oxetene as an intermediate in that type of reaction. On the other hand, the suggestion by Büchi et al.[355] that oxetenes were reasonable intermediates in the photochemistry of acetylenes with carbonyl compounds [Eq. (5-164)] has been longer lived. The cycloaddition

Mixture of isomers

(5-164)

step is photochemical but the ring opening of the oxetene is considered to be a thermal process. Following additional examples, i.e., addition of diphenylacetylene to benzoquinones,[356,357] a spate of examples appeared in the literature.[358-365] These included additions of thioalkyl-,[363] alkoxy-,[364,365] and alkyl acetylenes[361] to aldehydes and ketones as well as addition of diphenylacetylene to esters.[362] In all cases the products were in accord with the intermediacy of an oxetene, though neither isolation nor identification of this postulated intermediate was accomplished.

Exactly equivalent results were obtained for these cases where an acetylene will undergo a thermal [2 + 2]cycloaddition to a carbonyl compound.[366] Examples include the reactions of acetylenic ethers with aldehydes, ketones and α,β-unsaturated aldehydes, ketones with ethynyl ethers catalyzed by boron trifluoride,[367] with ynamines,[368,369] of ynamines with imines,[368] of ynamines with ketenes,[370,371] of ynamines with isocyanates,[372,373,376] and of ynamines with carbon dioxide.[374] The reactions of ynamines with carbonyl compounds have been reviewed.[375] Again the products generally are in complete accord with an oxetene intermediate.

$$ \text{(5-165)} $$

$$ \text{(5-166)} $$

The first oxetene to be isolated and identified was derived from the cyclo-addition of ethoxyacetylene and hexafluoroacetone.[377] When the reaction was carried out in the cold the oxetene could be isolated, but at 70° ring opening occurred. Clearly the substitution of an oxygen for a methylene in cyclobutene facilitates the ring opening reaction. Kuehne and Sheeran[370] reported the isolation of oxetenes in the reactions of ynamines with diphenyl-

$$ \text{(5-167)} $$

ketene, but later showed these compounds were in reality allenic amides.[378] Thus, the first non-fluorinated oxetene actually isolated appears to be 8-oxabicyclo[5.2.0]nona-1(9),3,5-triene.[379] No thermal ring opening was reported, though the oxetene was isolated by preparative glc.

Friedrich and Schuster[380] isolated and identified 2,3,4,4-tetramethyl-oxetene from irradiation of 3,4-dimethyl-3 penten-2-one. The oxetene

$$ \text{(5-168)} $$

reverted to the reactant on attempted distillation; a $t_{1/2} \simeq 12$ hrs in refluxing pentane was initially reported. Careful study of the thermolysis in heptane between $40°$ and $60°$ gave $k(41.1°) = 1.15 \times 10^{-5}$ sec^{-1} and $\Delta H^{\ddagger} = 25.1$, $\Delta S^{\ddagger} = -1.4$.[381] This result corresponds to a rate of ca. 1×10^7 times faster than cis-1,2,3,4-tetramethylcyclobutene. When dimethylacetylene and benzaldehyde are irradiated at $-78°$, the nmr spectrum of 4-phenyl-2,3-dimethyloxetene can be observed though the compound is too unstable to be isolated.[382] A half-life of several hours at $-35°$ has been estimated, and this oxetene thus undergoes retroelectrocyclization about 1×10^6 times faster than the tetramethyl derivative.

Oxetene intermediates have been suggested for the reactions of benzynes with α,β-unsaturated aldehydes [Eq. (5-169)].[383,384,385,51] This process

(5-169)

provides a very convenient synthesis for chromenes. It is interesting and perhaps somewhat surprising that little evidence has accumulated showing that o-quinomethides can revert to benzoxetenes. Generally the o-quinomethides trimerize,[386,387] but in some rather special cases [Eq. (5-170)] ring closure does occur.[388,389]

(5-170)

B. Azetines

There are two possible azetines, 1-azetine (1-azacyclobut-1-ene) and 2-azetine (1-azacyclobut-2-ene). Conceptually both could undergo a retroelectrocyclic reaction, but only 1-azetines could be examined for the stereochemistry of the reaction. The reaction has been studied theoretically using the simple HMO method and checked by CNDO and extended Hückel procedures.[390] The results indicate that the presence of the nitrogen (1-azetine) atom does not alter the stereo preference for a conrotatory thermal reaction. Unfortunately no experimental test of this conclusion appears to have been made.

Simple 1-azetines seem to be surprisingly stable thermally. Thus, for example, the first azetines without an alkoxy group at C_2 were prepared by heating dichlorocyclopropyl azides in boiling toluene for several hours.[391]

$$\text{(5-171)}$$

R_1	R_2	R_3
Ph	H	H
Ph	Me	H
Me	Me	H

(yields 50–75%)

The sole thermal reaction reported was cleavage to a nitrile and a dichloro-alkene. Thermal stability seems to characterize those 1-azetines which lack an alkoxy group at C_2. Photochemical cycloaddition of nitriles to electron-rich alkenes (four alkyl substituents, two alkoxy or two alkyl and one alkoxy substituents)[394] gives 1-azetines and 2-azabutadienes.[392,393,394] The 1-azetines generally are stable to sublimation or to gas chromatographic analysis at 170°, and thus it seems reasonable to conclude that the azabutadienes were produced by photochemical ring opening. Recently Wendling and Bergman[395] have invoked the formation of a 1-azetine from a 2-azabutadiene to explain the formation of hydrogen cyanide during the pyrolysis of an azirine [Eq. (5-172)].

$$\text{(5-172)}$$

This thermal stability does not extend to 1-azetines having an alkoxy group at C_2. Thus both *cis*- and *trans*-3,4-diethyl-2-methoxy-1-azetines gave ring opened products upon thermolysis at 600°.[396] Pyrolysis in the gas phase is a necessity, but ring opening occurs at a reasonable rate at 160°–200°.[397]

$$\text{(5-173)}$$

Though the direct ring-opened 2-azabutadiene was not obtained, since it always undergoes a 1,5-hydrogen shift to form an iminoether, the relative rates of a series of methyl substituted azetines were suggestive of a conrotatory reaction. Thus, the 3,3-dimethyl derivative [Eq. (5-174)] reacted slower

$$k(160°) = 0.75 \times 10^{-5} \ sec^{-1}$$

(5-174)

than the 3,3,4-trimethyl analog and the tetramethyl homolog was slowest of the three.[397] The retro-electrocyclization rate is very notably enhanced by addition of another strained ring. The bicyclic lactone (120) reacts at $-30°$

120

(5-175)

with $t_{1/2} \sim 1$ min.[398] Similarly, Dewar pyridine reverts to pyridine at $0°$ with a half-life of 36 min,[399] and 121 goes to the aromatic form with $t_{1/2} = 5$

121

$$R = -CF(CF_3)_2$$

(5-176)

min at $100°$.[400]

Not much is known explicitly about the thermal behavior of 2-azetines. The simplest 2-azetine known appears to be the N-carbomethoxy derivative which is stable to $50°$, but polymerizes readily above that temperature.[401] Whether polymerization involves ring opening is not known. N,2-diphenyl 2-azetinone has been prepared, and it melts with decomposition at $121°$.[402] Ring opened products apparently derived from N-methyl-2-azetinone have been isolated, but it seems quite likely that ring cleavage is photochemical.[403] The 2-azetines serve more often as unobserved intermediates in thermal reactions of ynamines. For example, 122 reacts with sulfonylimines to give sulfonamidines [Eq. (5-177)].[370] Also, 2-azetinones have been implicated

$$\text{PhC}{\equiv}\text{CNEt}_2 \; + \; \text{ArCH}{=}\text{NSO}_2\text{Tol} \quad \longrightarrow \quad \left[\begin{array}{c} \text{Ph} \qquad \text{Ar} \\[6pt] \text{Et}_2\text{N} \qquad \underset{\text{SO}_2\text{Tol}}{\text{N}} \end{array} \right]$$

$$\textbf{122}$$

$$(5\text{-}177)$$

$$\begin{array}{c} \text{Ar} \\ \text{Ph} \qquad \text{H} \\[6pt] \text{Et}_2\text{N} \qquad \text{NSO}_2\text{Tol} \end{array}$$

Ar = Ph,

as intermediates in the reactions of ynamines with isocyanates,[370,373] but again the retroelectrocyclic reaction must be accepted as occurring rapidly under extraordinarily mild conditions. The azine of hexafluoroacetone reacts with an ynamine to give a monoadduct which is converted thermally to a 2,3-diazahexatriene.[470]

Some 1,2-diazabicyclo[3.2.0]-6-heptene-4-ones undergo a very facile apparent retroelectrocyclization.[404] However, the intervention of the unshared pan of electrons on N_2 is indicated by the rate sequence Me > H > Ac [Eq. (5-178)] with $k = 21 \times 10^{-3}$ sec^{-1} for R = Me at 23°.[405] Presence of

$$(5\text{-}178)$$

R = H, Me, Ac

the oxo moiety is necessary for the reaction, since the alcohol does not give a diazepine in boiling butanol. The rate in methanol (R = Me) is ca. 30 × faster than in hexane, and a 2-ene function markedly stabilized the bicyclic molecule.[406]

The interesting Dewar pyridine (**123**) is relatively stable thermally, revert-

$$\xrightarrow{\;175°\;}$$

$$(5\text{-}179)$$

123

ing to the monocyclic form only when heated vigorously.[407] N-Phenyl-benzazetine undergoes ring opening at 200°, and the unstable intermediate can be intercepted by a good dienophile [Eq. (5-180)].[408] Apparently ben-

$$(5\text{-}180)$$

zazetinones are in mobile equilibrium with the iminoketene form at 25°
and the equilibrium can be displaced by another electrocyclization [Eq.
(5-181)].[409] Azetinium ions have also been proposed as intermediates in
cycloaddition reactions of ynamines.[410]

$$(5\text{-}181)$$

$$(5\text{-}182)$$

C. Thietes

Thiete itself is stable to distillation at 30°, but several substituted analogs
such as 7-thiabicyclo[4.2.0]oct-1(8)-ene (124) are much less stable.[411] All of
the thietes are unstable with respect to thermal polymerization, a process
which may well involve initial electrocyclic ring opening. The bicyclic thiete
(124) reacts with dinitrophenylhydrazine to produce the DNF of 1-cyclo-
hexenylthiocarboxaldehyde, which suggests ring opening prior to hydrazone
formation.[411] Thietes have been implicated as transient intermediates in
the reaction of thioketones with ynamines.[412] An orange solid obtained
when (125) was treated with triethyl phosphite has been tentatively assigned

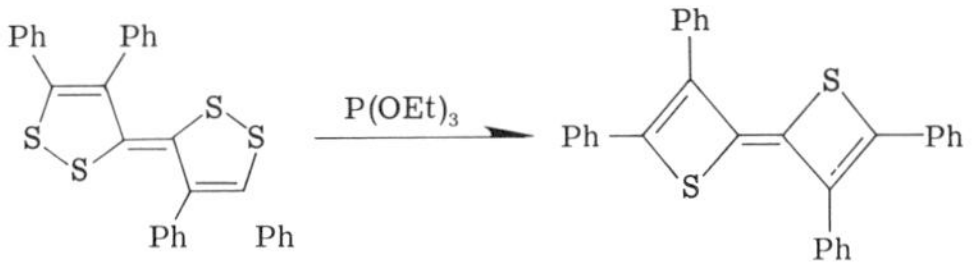

(5-183)

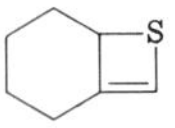

the unusual structure **126**, largely on the basis of its thermal conversion to the condensed dithiophene.[413] A benzthiete has been reported to be stable to mild heating.[414] However, benzthiete is thermally unstable and at 100° it forms dibenzodithiacyclooctadiene, presumably via thiaquinomethide.[471]

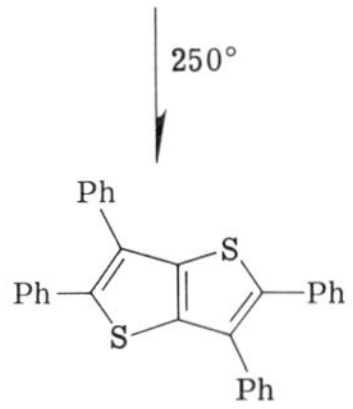

Thiet-1,1-dioxides have been reported to be thermally stable,[415,416] but under appropriate conditions ring opening to unsaturated sulfenes occurs, followed by formation of sulfinates [Eq. (5-184)] and under more drastic conditions other products as well.[417,418]

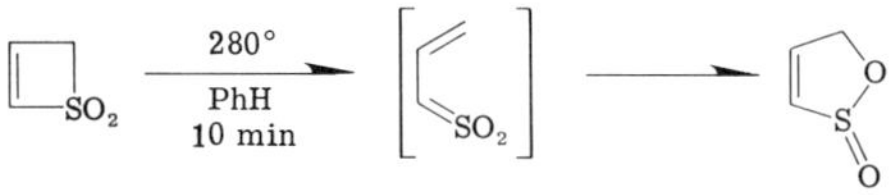

(5-184)

D. Siletenes

Both 1,1,2-triphenylsilacyclobut-2-ene[472] and 1,1-dimethyl-2-phenylsila-cyclobut-2-ene[473] have been prepared but no indication of thermal ring opening has been reported. However, 1,1-dimethyl-2-siletene has been prepared, probably via ring closure of a silabutadiene.[474]

E. Two Heteroatoms

Dioxetenes, potential ephemeral intermediates in addition of oxygen to acetylenes, would be expected to form α-diones via a very exothermic retro-electrocyclic reaction. Hoffmann[419] has indicated that this ring opening is a forbidden electrocyclic reaction if planar symmetry is preserved during the process. Unfortunately oxygen shows no special tendency to add to acetylenes. However, two acetylenes have been converted to α-diones, sensitized oxidation of an ynamine [Eq. (5-185)],[420] and reaction of an acetylene with

$$\text{MeC}\equiv\text{C}-\text{NEt}_2 + \text{O}_2 \xrightarrow[\text{sens.}]{h\nu} \text{Me}-\text{CO}-\text{CO}-\text{NEt}_2 \qquad (5\text{-}185)$$

$$\text{Ph}-\text{C}\equiv\text{CR} + \text{O}_3 + \text{TCNE} \xrightarrow{-78°} \text{Ph}-\text{CO}-\text{CO}-\text{R} + (\text{NC})_2-\text{C}\overset{\displaystyle O}{\frown}\text{C(CN)}_2 \qquad (5\text{-}186)$$

ozone in the presence of TCNE at $-78°$ [Eq. (5-186)].[421] In neither case was any adequate evidence adduced to establish the presence of a transient intermediate such as a dioxetene.

An elegant investigation of the addition of either singlet or triplet oxygen to strained acetylenes has, however, gone a long way toward establishing the presence of a dioxetene intermediate.[422] Addition of triplet oxygen in the dark at elevated temperature, or of singlet oxygen at $-30°$ to **127** gives an α-dione and the solution is chemiluminescent. Reaction with singlet oxygen at $-90°$ gives an intermediate which is stable for days at that temperature and shows no ketone bands in the infrared. The activation energy for decomposition of the intermediate is 18 kcal/mol.

(5-187)

127

Of the three possible isomeric diazetines, the 1,2-diaza form would be expected to undergo preferential loss of nitrogen and this indeed is the sole reaction normally observed [Eq. (5-188)].[423] In one unusual case thermal

(5-188)

cleavage of the 3,4-bond was observed [Eq. (5-189)].[424] This reaction is only

(5-189)

formally related to the retroelectrocyclic reaction. One example of ring opening of a 1,4-diazetine has been uncovered in a rather exotic case [Eq.

(5-190)].[425] At 35° in ethanol the rate is 1.6×10^{-4} sec^{-1}, and reaction proceeds about twice as fast in methylene chloride.

$$ (5\text{-}190) $$

Dithietenes have been the most thoroughly investigated cyclobutenes having two heteroatoms. This unusual situation arises partly because of the relative accessibility of dithietenes[426,427] and partly because of the similar stability of the dithietenes and the α-dithiones.[428] Though an early report of the preparation of benzodithietene[429] was proven incorrect,[430] bis-tri-fluoromethyldithietene proved to be quite simply prepared [Eq. (5-191)].[426]

$$ F_3C-C\equiv C-CF_3 \ + \ S \ (\text{vapor}) \ \xrightarrow{\ 200°\ } \qquad (5\text{-}191) $$

The dithietene appears to undergo ring opening thermally to an unstable intermediate which can dimerize or revert to the reactant.[431] The intermediate was trappable by olefins,[426] or with alkynes.[432] Dicyanodithietene has not been isolated but has been postulated as an intermediate in several reactions, based on trapping results.[433] More precise data were obtained with diaryldithietenes where both valence isomers could be observed.[427] Photochemical generation of dithioketones has been observed, but some questions about its generality have been raised.[434] However, some spectral evidence suggestive of an equilibrium mixture of dithietene and dithioketone

$$ (5\text{-}192) $$

$$ Ar = Me_2N- $$

at 19.8° $k_1 = 3.88 \times 10^{-3}$ sec^{-1} $\Delta H^{\ddagger} = 17.5$ $\Delta S^{\ddagger} = 11.7$
for reverse step $\Delta H^{\ddagger} = 22.4$ $\Delta S^{\ddagger} = 0.5$
at 21.9° $K = 7.9$

has been presented.[435] A benzodithietene has been isolated but little of its chemistry has been studied.[436]

Disiletenes are not heavily represented in the literature, but the interesting reaction of Eq. (5-193) has been reported.[437] No other mention of disiletenes

$$ (5\text{-}193) $$

has appeared. Cyclobutenes having two different heteroatoms present figure heavily as possible transient intermediates in a number of reactions. Specially stabilized examples have been isolated in two cases. The monomeric nitroso compound **128** is stable at 25°, but gives the interesting oxazetine as a stable

$$
\begin{array}{c}
\text{Me}_3\text{C} \\
\diagdown \\
\text{C}=\text{CH}-\text{N}=\text{O} \\
\diagup \\
\text{Me}_3\text{C}
\end{array}
\xrightarrow{220°}
\begin{array}{c}
\text{CMe}_3 \\
\text{Me}_3\text{C} \\
\diagup\!\!\!\square \\
\text{O}-\text{N}
\end{array}
\xrightarrow{240°}
(\text{Me}_3\text{C})_2\text{C}=\text{O} \ + \ \text{HCN} \quad (5\text{-}194)
$$

128

molecule in 54% yield.[438] Oxazetines having less hindered substituents are not isolable. A thiazetine, stabilized by trifluoromethyl groups, has also been

$$
\begin{array}{c}
\text{N}=\!\!\!=\!\!\!\diagup^{\text{Ar}} \\
\text{F}_3\text{C}-\!\!\!\square\!\!\!-\text{S} \\
\text{F}_3\text{C}
\end{array}
\underset{\substack{135° \\ \text{xylene}}}{\overset{\Delta}{\rightleftharpoons}}
\left[
\begin{array}{c}
\text{Ar} \\
\diagdown \\
\text{N}-\text{C} \\
(\text{CF}_3)_2\text{C}\!\!\diagdown\!\!\!_{\text{S}}
\end{array}
\right]
\xrightarrow{\text{RNC}}
\begin{array}{c}
\text{Ar} \\
\text{N}=\!\!\!=\!\!\!\diagup \\
(\text{F}_3\text{C})_2\!\!\!-\!\!\!\square\!\!\!-\text{S} \\
\diagdown \\
\text{N}-\text{R}
\end{array}
\quad (5\text{-}195)
$$

isolated,[439] but undergoes reversible ring scission as indicated by trapping with isocyanides.[440] Oxazetines have often been proposed as intermediates in reactions of nitrites,[441,442] and of unsaturated nitroso compounds formed by scavenging vinyl radicals with nitric oxide.[443,444]

Both of the other isomeric thiazetines have been proposed as reactive intermediates [Eq. (5-196)[445] and (5-197)].[446,447] Winterfeldt[448] has shown that DMSO reacts with DMAD in a reaction completely analogous to that

$$
\begin{array}{c}
\text{Ph} \\
\diagdown \\
\text{C}=\text{S} \ + \ \text{R}'\text{CN} \\
\diagup \\
\text{R}
\end{array}
\xrightarrow{h\nu}
\left[
\begin{array}{c}
\text{S}\!\!\!=\!\!\!\diagup^{\text{R}'} \\
\text{Ph}-\!\!\!\square\!\!\!-\text{N} \\
\text{R}
\end{array}
\right]
\longrightarrow
\begin{array}{c}
\text{S} \\
\| \\
\text{Ph}-\text{C}=\text{N}-\text{C}-\text{R}' \\
| \\
\text{R}
\end{array}
\quad (5\text{-}196)
$$

$$
\text{Me}_2\text{S}=\text{NAr} \ + \ \text{EtOOC}-\text{C}\!\equiv\!\text{C}-\text{COOEt}
\longrightarrow
\left[
\begin{array}{c}
\text{EtOOC}\diagdown\diagup\text{COOEt} \\
\square \\
\text{Me}_2\text{S}-\text{N}\diagdown_{\text{Ar}}
\end{array}
\right]
$$

129

$$(5\text{-}197)$$

$$
\begin{array}{c}
\text{EtOOC}\text{COOEt} \\
\diagup\!\!\!\diagdown \\
\text{EtOOC}\text{N}\text{COOEt} \\
| \\
\text{Ar}
\end{array}
\xleftarrow{\textbf{129}}
\begin{array}{c}
\text{EtOOC}\text{COOEt} \\
\diagup\!\!\!\diagup \\
\text{Me}_2\text{S}\text{N}-\text{Ar} \\
\text{(isolable)}
\end{array}
$$

Ar = p-ClPh, p-NO$_2$Ph

of Eq. (5-197) to give an isolable acylsulfonium ylide which can react further to give substituted furans.

A cyclobutene having three heteroatoms was suggested as an intermediate in the thermal decomposition of **130**.[449]

$$\underset{130}{\text{Ph}-\overset{\overset{\displaystyle O}{\|}}{\text{C}}-\text{N}=\text{S}=\text{O}} \quad \xrightarrow{\Delta} \quad \left[\text{Ph}-\overset{\overset{\displaystyle O-SO}{|}}{\underset{|}{C}}=\text{N}\right] \quad \longrightarrow \quad \text{PhCN} \; + \; \text{SO}_2 \qquad (5\text{-}198)$$

III. FOUR ELECTRON–FIVE ATOM SYSTEMS

The electrocyclic reaction of this section is the conversion of a pentadienyl to a cyclopentenyl cation [Eq. (5-199)]. The reaction should be thermally a

$$\text{(pentadienyl cation)} \quad \xrightarrow{\Delta} \quad \text{(cyclopentenyl cation)} \qquad (5\text{-}199)$$

conrotatory process and it does follow that stereochemical course. A good review of the research in this area has been prepared by Sorensen and Rauk.[475] Recently the reaction has figured prominently in a number of syntheses for cyclopentenones. A divinyldichloromethane served as reactant in one case [Eq. (5-200)],[476] and apparently was formed *in situ* in a second.[477]

$$(5\text{-}200)$$

Electrocyclization may be the basic step in two diverse reactions which lead to cyclopentenones [Eqs. (5-201) and (5-202)].[478,479] Both Cooke *et al.*[480] and Fristad *et al.*[481] have been developing the Nazarov reaction [Eq. (5-203)]

$$(5\text{-}201)$$

$$(5\text{-}202)$$

$$(5\text{-}203)$$

as a method for cyclopentannulation. An unusual tetraza analog has been advanced to rationalize the ring closure of Eq. (5-204).[482] Rho values for

$$(5\text{-}204)$$

the several aryl groups were measured to support the proposed mechanism.

REFERENCES

1. K. B. Wiberg and R. A. Fenoglio, *J. Am. Chem. Soc.* **90**, 3395 (1968).
2. R. Willstätter and W. von Schmädel, *Chem. Ber.* **38**, 1992 (1905).
3. R. Willstätter and J. Bruce, *Chem. Ber.* **40**, 3979 (1907).
4. J. D. Roberts and C. W. Sauer, *J. Am. Chem. Soc.* **71**, 3925 (1949).
5. K. Alder and H. A. Dortmann, *Chem. Ber.* **85**, 556 (1952).
6. K. Alder and O. Ackermann, *Chem. Ber.* **87**, 1567 (1954).
7. R. Criegee, W. Hörauf, and W. O. Schellenberg, *Chem. Ber.* **86**, 126 (1953).
8. See footnote 21 in C. T. Genaux, F. Kern, and W. D. Walters, *J. Am. Chem. Soc.* **75**, 6196 (1953).
9. E. Vogel, *Angew. Chem.* **66**, 640 (1959).
10. W. Cooper and W. D. Walters, *J. Am. Chem. Soc.* **80**, 4220 (1958).
11. G. Filer, *Theor. Chem. Acta* **12**, 412 (1968).
12. W. Th. A. M. von der Lugt and L. J. Oosterhoff, *Chem. Commun.* p. 1235 (1968).
12a. W. Th. A. M. von der Lugt and L. J. Oosterhoff, *J. Am. Chem. Soc.* **91**, 6042 (1969).
13. M. J. S. Dewar and S. Kirschner, *J. Am. Chem. Soc.* **93**, 4290, 4291, 4292 (1971).
14. M. J. S. Dewar, S. Kirschner, and H. W. Kollmar, *J. Am. Chem. Soc.* **96**, 5240 (1974).
15. M. J. S. Dewar and S. Kirschner, *J. Am. Chem. Soc.* **96**, 6809 (1974).
16. K. Hsu, R. J. Buenker, and S. D. Peyerimhoff, *J. Am. Chem. Soc.* **93**, 2117 (1971).
17. K. Hsu, R. J. Buenker, and S. D. Peyerimhoff, *J. Am. Chem. Soc.* **94**, 5639 (1972).
18. R. J. Buenker, S. D. Peyerimhoff, and K. Hsu, *J. Am. Chem. Soc.* **93**, 5005 (1971).
19. J. W. McIver, Jr. and A. Komornicki, *J. Am. Chem. Soc.* **94**, 2625 (1972).
20. A. Rastelli, A. S. Pozzoli, and G. Del Re, *J. Chem. Soc. Perkin Trans. 2* p. 1571 (1972).
21. O. S. Tee, *J. Am. Chem. Soc.* **91**, 7144 (1969).
21a. O. S. Tee and K. Yates, *J. Am. Chem. Soc.* **94**, 3074 (1972).

22. E. Vogel, *Angew. Chem.* **66,** 640 (1954).
22a. E. Vogel, *Justus Liebigs Ann. Chem.* **615,** 14 (1958).
23. R. Criegee, D. Seeback, R. E. Winter, B. Borretzen, and H.-A. Brune, *Chem. Ber.* **98,** 2339 (1965).
24. R. E. K. Winter, *Tetrahedron Lett.* p. 1207 (1965).
25. R. Srinivasan, *J. Am. Chem. Soc.* **91,** 7557 (1969).
26. J. I. Brauman and W. C. Archie, Jr., *Tetrahedron* **27,** 1275 (1971).
27. R. Walsh, D. M. Golden, and S. W. Benson, *J. Am. Chem. Soc.* **88,** 650 (1966).
28. J. I. Brauman and W. C. Archie, Jr., *J. Am. Chem. Soc.* **94,** 4262 (1972).
29. G. F. Emerson, L. Watts, and R. Pettit, *J. Am. Chem. Soc.* **87,** 131 (1965).
30. R. Criegee and K. Noll, *Justus Liebigs Ann. Chem.* **627,** 1 (1959).
31. W. Adam, *Chem. Ber.* **97,** 1811 (1964).
32. G. R. Branton, H. M. Frey, and R. F. Skinner, *Trans. Faraday Soc.* **62,** 1546 (1966).
33. H. M. Freedman, G. A. Doorakian, and V. R. Sandel, *J. Am. Chem. Soc.* **87,** 3019 (1965).
34. R. E. K. Winter and M. L. Hönig, *J. Am. Chem. Soc.* **93,** 4616 (1971).
35. D. L. Dalrymple and W. B. Russo, *J. Org. Chem.* **40,** 492 (1975).
36. G. Maier and M. Wiessler, *Tetrahedron Lett.* p. 4987 (1969).
37. H. M. Frey, *Trans. Faraday Soc.* **60,** 83 (1964).
38. H. M. Frey, D. C. Marshall, and R. F. Skinner, *Trans. Faraday Soc.* **61,** 861 (1965).
39. M. Pomerantz and P. H. Hartmann, *Tetrahedron Lett.* p. 991 (1968).
40. H. M. Frey, J. Metcalf, and B. M. Pope, *Trans. Faraday Soc.* **67,** 750 (1971).
41. R. N. McDonald and E. P. Lyznicki, Jr., *J. Am. Chem. Soc.* **93,** 5920 (1971).
42. C. W. Jefford, A. F. Boschung, and C. G. Rimbault, *Tetrahedron Lett.* p. 3387 (1974).
43. P. Courtot and R. Rumin, *Tetrahedron* **32,** 441 (1976).
44. M. Pomerantz, R. N. Wilke, G. W. Gruber, and U. Roy, *J. Am. Chem. Soc.* **94,** 2752 (1972).
45. H. M. Frey and R. K. Solly, *Trans. Faraday Soc.* **65,** 448 (1969).
46. G. O. Schenck, W. Hartmann, and R. Steinmetz, *Chem. Ber.* **96,** 498 (1963).
47. G. A. Doorakian and H. H. Freedman, *J. Am. Chem. Soc.* **90,** 5310 (1968); see also **90,** 6896 (1968).
48. B. J. Arnold, P. G. Sammes, and T. W. Wallace, *J. Chem. Soc., Perkin Trans. 1* p. 409 (1974).
49. W. Oppolzer, *J. Am. Chem. Soc.* **93,** 3833 (1971).
50. M. R. De Camp, R. H. Levin, and M. Jones, Jr., *Tetrahedron Lett.* p. 3575 (1974).
51. A. T. Bowne and R. H. Levin, *Tetrahedron Lett.* p. 2043 (1974).
52. R. Hug, H.-J. Hansen, and H. Schmid, *Helv. Chem. Acta* **55,** 10 (1972).
53. L. Carpino, *J. Am. Chem. Soc.* **84,** 2196 (1962).
54. R. Huisgen and H. Seidel, *Tetrahedron Lett.* p. 3381 (1964).
55. G. Quinkert, K. Opitz, W. W. Wiersdorf, and M. Finke, *Tetrahedron Lett.* p. 3009 (1965).
56. G. Quinkert, W. W. Wiersdorf, M. Finke, K. Opitz, and F.-G. vonder Haar, *Chem. Ber.* **101,** 2302 (1968).
57. B. J. Arnold, P. G. Sammes, and T. W. Wallace, *J. Chem. Soc. Perkin Trans. 1* p. 415 (1974).
58. B. J. Arnold, S. M. Mellows, P. G. Sammes, and T. W. Wallace, *J. Chem. Soc. Perkin Trans. 1* p. 401 (1974).
59. P. H. J. Ooms, J. W. Scheeren, and R. J. F. Nivard, *Synthesis* p. 263 (1975).
60. R. K. Boekman, Jr., M. H. Delton, T. Nagasaka, and T. Watanabe, *J. Org. Chem.* **42,** 2946 (1977).
61. T. Kametani, M. Tsubuki, Y. Shiratori, V. Kato, H. Nemoto, M. Ihara, K. Fukumoto, F. Satoh, and H. Inoue, *J. Org. Chem.* **42,** 2672 (1977).

62. R. W. Carr, Jr. and W. D. Walters, *J. Phys. Chem.* **69,** 1073 (1965).
63. H. M. Frey, *Trans. Faraday Soc.* **58,** 957 (1962).
64. H. M. Frey and R. F. Skinner, *Trans. Faraday Soc.* **61,** 1918 (1965).
65. D. Dickens, H. M. Frey, and R. F. Skinner, *Trans. Faraday Soc.* **65,** 453 (1969).
66. D. Dickens, H. M. Frey, and J. Metcalf, *Trans. Faraday Soc.* **67,** 2328 (1971).
67. S. F. Sarner, D. M. Gale, H. K. Hall, Jr., and A. B. Richmond, *J. Phys. Chem.* **76,** 2817 (1972).
68. H. M. Frey, *Trans. Faraday Soc.* **59,** 1619 (1963).
69. H. M. Frey, D. C. Montague, and I. D. R. Stevens, *Trans Faraday Soc.* **63,** 372 (1967).
70. M. A. Battiste and M. E. Burns, *Tetrahedron Lett.* p. 523 (1966).
71. H. M. Frey, B. M. Pope, and R. F. Skinner, *Trans. Faraday Soc.* **63,** 1160 (1967).
72. G. L. Doorakian and H. M. Freedman, *J. Am. Chem. Soc.* **90,** 3582 (1968).
73. H. M. Frey, R. G. Hopkins, and I. C. Vinall, *J. Chem. Soc. Faraday I,* **68,** 1874 (1972).
74. J. P. Chesick, *J. Am. Chem. Soc.* **88,** 4800 (1966).
75. E. W. Schlag and W. B. Peatman, *J. Am. Chem. Soc.* **86,** 1676 (1964).
76. H. M. Frey and B. M. Pope, *Trans. Faraday Soc.* **65,** 441 (1969).
77. R. Huisgen and H. Seidel, *Tetrahedron Lett.* p. 3381 (1964).
78. A. T. Cocks and H. M. Frey, *J. Chem. Soc. B* p. 952 (1970).
79. D. H. Aue and R. N. Reynolds, *J. Am. Chem. Soc.* **95,** 2027 (1973).
80. W. A. Dolbier, Jr., D. Lomas, and P. Tarrant, *J. Am. Chem. Soc.* **90,** 3594 (1968).
81. E. L. Eliel, N. L. Allinger, S. J. Angyal, and G. A. Morrison, "Conformational Analysis," pp. 449–456, Wiley, New York, 1965.
82. E. L. Eliel, "Stereochemistry of Carbon Compounds," p. 331, McGraw-Hill, New York, 1962.
83. K. B. Wiberg and R. A. Fenoglio, *J. Am. Chem. Soc.* **90,** 3395 (1968).
84. J. G. Aston, G. Szasz, W. W. Wooley, and F. G. Brickwedde, *J. Chem. Phys.* **14,** 67 (1946).
85. A. Danti, *J. Chem. Phys.* **27,** 1227 (1957).
86. M. Prober and W. T. Miller, Jr., *J. Am. Chem. Soc.* **71,** 598 (1949).
87. R. N. Haszeldine and J. E. Osborne, *J. Chem. Soc.* p. 3880 (1955).
88. C. R. Brundle and M. B. Robin, *J. Am. Chem. Soc.* **92,** 5550 (1970).
89. J. L. Anderson, R. E. Putnam, and W. H. Sharkey, *J. Am. Chem. Soc.* **83,** 389 (1961).
90. R. E. Putnam and J. E. Castle, *J. Am. Chem. Soc.* **83,** 389 (1961).
91. H. A. Brune and W. Schwab, *Tetrahedron* **25,** 4375 (1969).
92. L. Skattebøl and S. Solomon, *J. Am. Chem. Soc.* **87,** 4506 (1965).
93. H. Hopf, *Angew Chem.* **82,** 203 (1970).
93a. H. Hopf, *Chem. Ber.* **104,** 1499 (1971).
94. K. Kleveland and L. Sakttebøl, *Acta Chem. Scand.* **29,** 191 (1975).
95. W. D. Huntsman and H. J. Wristers, *J. Am. Chem. Soc.* **85,** 3308 (1963).
96. W. D. Huntsman, J. A. De Boer, and M. H. Woosley, *J. Am. Chem. Soc.* **88,** 5846 (1964).
97. W. D. Huntsman and H. J. Wristers, *J. Am. Chem. Soc.* **89,** 342 (1967).
98. F. T. Bond, *J. Org. Chem.* **31,** 3057 (1966).
99. E. F. Jenney and J. D. Roberts, *J. Am. Chem. Soc.* **78,** 2005 (1956).
100. E. F. Jenney and J. Druey, *J. Am. Chem. Soc.* **82,** 3111 (1960).
101. F. B. Mallory and J. D. Roberts, *J. Am. Chem. Soc.* **83,** 393 (1961).
102. E. F. Silversmith, Y. Kitihara, M. C. Caserio, and J. D. Roberts, *J. Am. Chem. Soc.* **80,** 5840 (1958).
103. E. F. Silversmith, Y. Kitihara, and J. D. Roberts, *J. Am. Chem. Soc.* **80,** 4088 (1958).
104. H. Mayr, *Angew. Chem. Internl. Edit. Engl.* **14,** 500 (1975).
105. P. Schiess and P. Radimerski, *Angew. Chem. Int. Edit. Engl.* **11,** 288 (1972).
105a. P. Schiess and P. Radimerski, *Helv. Chem. Acta* **57,** 2583 (1974).

106. A. T. Blomquist and E. A. Lalancette, *J. Am. Chem. Soc.* **83,** 1387 (1961).
107. L. A. Errede, *J. Am. Chem. Soc.* **83,** 949 (1961).
108. M. P. Cava and A. A. Deana, *J. Am. Chem. Soc.* **81,** 4266 (1959).
109. F. R. Jensen, W. E. Coleman, and A. J. Berlin, *Tetrahedron Lett.* p. 15 (1962).
110. M. P. Cava, R. L. Shirley, and B. W. Erickson, *J. Org. Chem.* **27,** 755 (1962).
111. M. P. Cava and R. L. Shirley, *J. Am. Chem. Soc.* **82,** 654 (1960).
112. M. P. Cava, A. A. Deana, and K. Muth, *J. Am. Chem. Soc.* **81,** 6458 (1959).
113. M. P. Cava and M. J. Mitchell, *J. Am. Chem. Soc.* **81,** 5409 (1959).
114. W. Baker, J. F. McOmie, and D. R. Preston, *J. Chem. Soc.* 2971 (1961).
115. L. Carpino, *J. Am. Chem. Soc.* **84,** 2196 (1962).
116. L. Carpino, *Chem. Commun.* p. 494 (1966).
117. H. Hart, J. A. Hartlage, R. W. Fish, and R. R. Rafos, *J. Org. Chem.* **31,** 2244 (1966).
118. D. S. Weiss, *Tetrahedron Lett.* p. 4001 (1975).
119. L. Friedman and R. J. Osiewicz as noted in D. R. Arnold, E. Hedaya, V. Y. Merritt, L. A. Karnischky, and M. E. Kent, *Tetrahedron Lett.* p. 3917 (1972).
120. K. M. Shumate, P. N. Neuman, and G. J. Fonken, *J. Am. Chem. Soc.* **87,** 3996 (1965).
122. R. S. H. Liu, *J. Am. Chem. Soc.* **89,** 112 (1967).
121. L. L. Barber, O. L. Chapman, and J. D. Lassila, *J. Am. Chem. Soc.* **91,** 531 (1969).
123. D. I. Schuster, B. R. Sckolnick, and F.-T. H. Lee, *J. Am. Chem. Soc.* **90,** 1300 (1968).
124. R. Gompper, E. Kutter, and H. Kast, *Angew. Chem. Int. Edit. Engl.* **6,** 171 (1968).
125. B. S. Rabinovitch and K.-W. Michel, *J. Am. Chem. Soc.* **81,** 5065 (1959).
126. W. von E. Doering and G. W. Beasley, *Tetrahedron* **29,** 2231 (1973).
127. W. R. Roth, G. Rüf, and P. W. Ford, *Chem. Ber.* **107,** 48 (1974).
128. H. M. Frey, A. M. Lamont, and R. Walsh, *J. Chem. Soc. Chem. Commun.* p. 1583 (1970).
129. L. M. Stephenson, R. V. Gemmer, and J. I. Brauman, *J. Am. Chem. Soc.* **94,** 8620 (1972).
130. E. N. Marvell, J. Seubert, G. Vogt, G. Zimmer, G. Moy, and J. R. Siegmann, *Tetrahedron* **34,** 1323 (1978).
131. J. I. Brauman, L. E. Ellis, and E. E. van Tamelen, *J. Am. Chem. Soc.* **88,** 846 (1966).
132. J. I. Brauman and D. M. Golden, *J. Am. Chem. Soc.* **90,** 1920 (1968).
133. R. B. Turner, P. Goebel, B. J. Mallon, W. von E. Doering, J. F. Coburn, Jr., znd M. Pomerantz, *J. Am. Chem. Soc.* **90,** 4315 (1968).
134. J. E. Baldwin, R. K. Pinschmidt, Jr., and A. H. Andrist, *J. Am. Chem. Soc.* **92,** 5249 (1970).
135. J. E. Baldwin and A. H. Andrist, *J. Chem. Soc. Chem. Commun.* p. 1561 (1970).
136. W. R. Roth, *Tetrahedron Lett.* p. 1009 (1964).
137. S. McLean and D. M. Findlay, *Can. J. Chem.* **48,** 3107 (1970).
138. S. McLean, D. M. Findlay, and G. I. Dimitrienko, *J. Am. Chem. Soc.* **94,** 1380 (1972).
139. J. E. Baldwin and G. D. Andrews, *J. Am. Chem. Soc.* **94,** 1775 (1972).
140. M. C. Flowers and H. M. Frey, *J. Am. Chem. Soc.* **94,** 8636 (1972).
141. J. I. Brauman, W. E. Farneth, and M. B. D'Amore, *J. Am. Chem. Soc.* **95,** 5043 (1973).
142. G. D. Andrews, M. Davalt, and J. E. Baldwin, *J. Am. Chem. Soc.* **95,** 5045 (1973).
143. W. E. Farneth, M. B. D'Amore, and J. I. Brauman, *J. Am. Chem. Soc.* **98,** 5546 (1976).
144. G. D. Andrews and J. E. Baldwin, *J. Am. Chem. Soc.* **99,** 4853 (1977).
145. M. J. S. Dewar and S. Kirschner, *J. Chem. Soc. Chem. Commun.* p. 461 (1975).
146. M. Franck-Neumann, *Tetrahedron Lett.* p. 341 (1966).
147. R. N. McDonald and C. E. Reinecke, *J. Org. Chem.* **32,** 1878 (1967).
148. Cf. footnote 34 in W. L. Jorgensen, *J. Am. Chem. Soc.* **97,** 3082 (1975).
149. D. Seebach, *Chem. Ber.* **97,** 2953 (1964).
150. F. van Rantwijk and H. van Bekkum, *Tetrahedron Lett.* p. 3341 (1976).
151. P. Reeves, J. Henery, and R. Pettit, *J. Am. Chem. Soc.* **91,** 5888 (1969).

152. P. Reeves, T. Devon, and R. Pettit, *J. Am. Chem. Soc.* **91,** 5890 (1969).
153. E. K. G. Schmidt, *Chem. Ber.* **108,** 1609 (1975).
154. E. N. Cain and R. K. Solly, *J. Am. Chem. Soc.* **95,** 4791 (1973).
155. N. L. Bauld, F. R. Farr, and S.-C. Chang, *Tetrahedron Lett.* p. 2443 (1972).
156. R. Criegee, H. Kristinsson, D. Seebach, and F. Zanker, *Chem. Ber.* **98,** 2331 (1965).
157. H. Ona, H. Yamaguchi, and S. Masamune, *J. Am. Chem. Soc.* **92,** 7495 (1970).
158. W. G. Dauben and G. J. Fonken, *J. Am. Chem. Soc.* **81,** 4060 (1959).
159. A. M. Bloothood-Kruisbeck and J. Lugtenberg, *Rec. Trav. Chem. Pays-Bas* **91,** 1369 (1972).
160. W. G. Dauben and R. M. Coates, *J. Am. Chem. Soc.* **86,** 2490 (1964).
161. W. H. Schuller, R. N. Moore, J. E. Hawkins, and R. V. Lawrence, *J. Org. Chem.* **27,** 1178 (1962).
162. J. Fincini, J. Besseyre, and C. Barbara, *Tetrahedron Lett.* p. 3151 (1975).
163. M. Avram, I. G. Dinulescu, E. Marica, G. Meteescu, E. Eliam, and C. D. Nenitzescu, *Chem. Ber.* **97,** 382 (1969).
164. H. M. Frey, M.-D. Martin, and M. Hickman, *J. Chem. Soc. Chem. Commun.* p. 204 (1975).
165. R. Gleiter, E. Heilbronner, M. Hickman, and M.-D. Martin, *Chem. Ber.* **106,** 28 (1973).
166. R. S. Case, M. J. S. Dewar, S. Kirschner, R. Pettit, and W. Slegeir, *J. Am. Chem. Soc.* **96,** 7581 (1974).
167. L. A. Paquette and M. J. Carmody, *J. Am. Chem. Soc.* **98,** 8175 (1976).
167a. L. A. Paquette and M. J. Carmody, *J. Am. Chem. Soc.* **99,** 6152 (1977).
168. E. E. van Tamelen and D. Carty, *J. Am. Chem. Soc.* **93,** 6102 (1971).
169. J. R. Peyser and T. W. Flechtner, *J. Org. Chem.* **41,** 2028 (1976).
170. R. Srinivasan and K. A. Hill, *J. Am. Chem. Soc.* **87,** 4653 (1965).
171. R. Criegee, W. Eberius, and H.-A. Brune, *Chem. Ber.* **101,** 94 (1968).
172. R. Criegee, W.-D. Wirth, W. Engel, and H.-A. Brune, *Chem. Ber.* **96,** 2230 (1963).
173. R. Criegee, G. Schröder, G. Maier, and H.-G. Fischer, *Chem. Ber.* **93,** 1553 (1960).
174. R. Criegee and G. Schröder, *Justus Liebigs Ann. Chem.* **623,** 1 (1959).
175. R. Criegee and G. Louis, *Chem. Ber.* **90,** 417 (1957). For a review of the series of studies see R. Criegee, *Angew. Chem. Int. Edt. Engl.* **1,** 519 (1962).
176. R. Criegee and R. Huber, *Chem. Ber.* **103,** 1862 (1970).
177. G. Wittig and U. Mayer, *Chem. Ber.* **96,** 342 (1963).
178. E. Müller and H. Kessler, *Tetrahedron Lett.* p. 3037 (1968).
179. G. J. Fonken and W. Moran, *Chem. and Ind.* (*London*) p. 1891 (1963).
180. G. Maier, *Chem. Ber.* **96,** 2238 (1963).
181. E. H. White and H. C. Dunathan, *J. Am. Chem. Soc.* **86,** 453 (1964).
182. K. Nagarajan, M. C. Caserio, and J. D. Roberts, *J. Am. Chem. Soc.* **86,** 449 (1964).
183. B. Sket and M. Zupan, *J. Chem. Soc. Chem. Commun.* p. 365 (1977).
184. M. P. Cava and D. P. Napier, *J. Am. Chem. Soc.* **79,** 1701 (1957).
185. M. Avram, D. Dinulescu, G. Mateescu, and C. D. Nenitzescu, *Chem. Ber.* **93,** 1789 (1960).
186. M. P. Cava and J. F. Stucker, *J. Am. Chem. Soc.* **79,** 1706 (1957).
187. M. P. Cava and K. Muth, *J. Org. Chem.* **27,** 757 (1962).
188. M. Avram, D. Constantinescu, I. G. Dinulescu, and C. D. Nenitzescu, *Tetrahedron Lett.* p. 5215 (1969).
189. H. Straub and J. Hambrecht, *Chem. Ber.* **110,** 3221 (1977).
190. A. Roedig, G. Bonse, and R. Helm, *Chem. Ber.* **106,** 2825 (1973).
191. C. M. Sharts and J. D. Roberts, *J. Am. Chem. Soc.* **83,** 871 (1961).
192. C. E. Berkoff, R. C. Cookson, J. Hudec, and R. D. Williams, *Proc. Chem. Soc.* p. 312 (1961).

193. E. E. van Tamelen and S. P. Pappäs, *J. Am. Chem. Soc.* **84,** 3789 (1962).

194. E. E. van Tamelen and S. P. Pappäs, *J. Am. Chem. Soc.* **85,** 3297 (1963).

195. R. Criegee and F. Zanker, *Angew. Chem.* **76,** 716 (1964).

196. H. C. Volger and M. Hogeveen, *Rec. Trav. Chem. Pays-Bas* **86,** 830 (1967).

197. J. F. M. Oth, *Rec. Trav. Chem. Pays-Bas* **87,** 1185 (1968).

197a. J. F. M. Oth, *Angew Chem. Int. Edit. Engl.* **7,** 646 (1968).

198. W. Adams and J. C. Chang, *Int. J. Chem. Kinet.* **1,** 487 (1969).

199. I. Haller, *J. Phys. Chem.* **72,** 2882 (1968).

200. E. Ratajczak and A. F. Trotman-Dickenson, *J. Chem. Soc. A* p. 509 (1968).

201. R. Breslow, J. Napierski, and A. H. Schmidt, *J. Am. Chem. Soc.* **94,** 5906 (1972).

202. D. M. Lemal and L. H. Dunlap, Jr., *J. Am. Chem. Soc.* **94,** 6562 (1972).

203. R. Criegee and R. Huber, *Chem. Ber.* **103,** 1855 (1970).

204. S. L. Kammula, L. D. Isoff, M. Jones, Jr., J. W. van Straten, W. H. de Wolf, and F. Bickelhaupt, *J. Am. Chem. Soc.* **99,** 5815 (1977).

205. J. W. van Straten, I. J. Landheer, W. H. de Wolf, and F. Bickelhaupt, *Tetrahedron Lett.* p. 4499 (1975).

206. G. P. Semeluk and R. D. S. Stevens, *Chem. Commun.* p. 1720 (1970).

207. G. Camaggi and F. Gozzo, *J. Chem. Soc. C* p. 489 (1969).

208. A.-M. Dabbagh, W. T. Flowers, R. N. Hazeldine, and P. J. Robinson, *Chem. Commun.* p. 323 (1975).

209. P. Lechtken, R. Breslow, A. H. Schmidt, and N. J. Turro, *J. Am. Chem. Soc.* **95,** 3025 (1973).

210. N. J. Turo and A. Devaquet, *J. Am. Chem. Soc.* **97,** 3859 (1975).

211. N. J. Turro, G. B. Schuster, R. G. Bergman, K. J. Shea, and J. H. Davis, *J. Am. Chem. Soc.* **97,** 4758 (1975).

212. M. J. S. Dewar and S. Kirschner, *Chem. Commun.* p. 463 (1975).

213. M. J. S. Dewar and S. Kirschner, *J. Am. Chem. Soc.* **96,** 7579 (1974).

214. M. D. Newton, J. M. Schulman, and M. M. Manus, *J. Am. Chem. Soc.* **96,** 17 (1974).

215. J. H. Dopper, B. Greijdanus, and H. Wynberg, *J. Am. Chem. Soc.* **97,** 216 (1975).

216. O. L. Chapman and D. J. Pasto, *J. Am. Chem. Soc.* **82,** 3642 (1960).

217. W. G. Dauben and R. L. Cargill, *Tetrahedron* **12,** 186 (1961).

218. O. L. Chapman, D. J. Pasto, G. W. Borden, and A. A. Griswold, *J. Am. Chem. Soc.* **84,** 1220 (1962).

219. W. G. Dauben, K. Koch, S. L. Smith, and O. L. Chapman, *J. Am. Chem. Soc.* **85,** 2616 (1963).

220. R. Criegee and H. Farrer, *Chem. Ber.* **97,** 2949 (1964).

221. R. Askani, *Chem. Ber.* **98,** 2323 (1965).

222. R. Criegee, U. Zirngible, H. Furrer, D. Seeback, and G. Freund, *Chem. Ber.* **97,** 2942 (1964).

223. G. R. Branton, H. M. Frey, D. C. Montague, and I. D. R. Stevens, *Trans. Faraday Soc.* **62,** 659 (1966).

224. M. R. Willcott and E. Goerland, *Tetrahedron Lett.* p. 6391 (1966).

225. T. Miyashi, M. Nitta, and T. Mukai, *J. Am. Chem. Soc.* **93,** 3441 (1971).

226. N. J. Turro, V. Ramamurthy, R. M. Pagne, and J. A. Butcher, Jr., *J. Org. Chem.* **43,** 92 (1977).

227. H. M. Frey, J. Metcalf, and J. M. Brown, *J. Chem. Soc. B* p. 1586 (1970).

228. T. Nozoe, T. Hirai, and T. Kobayashi, *Tetrahedron Lett.* p. 3501 (1970).

229. R. Criegee, H. Hofmeister, and G. Bolz, *Chem. Ber.* **98,** 2327 (1965).

230. T. Mukai and T. Miyashi, *Tetrahedron* **23,** 1613 (1967).

231. L. Cannell, *Tetrahedron Lett.* p. 5967 (1966).

232. J. E. Shields, D. Gavrilovic, J. Kopecky, W. Hartmann, and H.-G. Heine, *J. Org. Chem.* **39,** 515 (1974).

233. J. Meinwald, G. E. Samuelson, and M. Ikeda, *J. Am. Chem. Soc.* **92,** 7604 (1970).

234. J. Ficini and A. Durèault, *Tetrahedron Lett.* p. 809 (1977).

235. A. C. Day and M. A. Ledlie, *Chem. Commun.* p. 1265 (1970).

236. M. Kimura and T. Mukai, *Tetrahedron Lett.* p. 4207 (1970).

237. L. A. Paquette and J. H. Barrett, *J. Am. Chem. Soc.* **88,** 1718 (1966).

238. D. N. Reinhoudt and C. G. Leliveld, *Tetrahedron Lett.* p. 3119 (1972).

239. D. N. Reinhoudt and C. G. Kouwenhoven, *Chem. Commun.* p. 1233 (1972).

240. T. Q. Mink, L. Christiaens, P. Grandclaudon, and A. Lablanche-Combier, *Tetrahedron* **33,** 2225 (1977).

241. T. Mukai and T Shishido, *J. Org. Chem.* **32,** 2744 (1967).

242. T. Miyashi, M. Nitta, and T. Mukai, *Tetrahedron Lett.* p. 3433 (1967).

243. J. E. Baldwin and M. S. Kaplan, *Chem. Commun.* p. 1354 (1969).

244. J. E. Baldwin and M. S. Kaplan, *J. Am. Chem. Soc.* **93,** 3969 (1971).

245. J. E. Baldwin and M. S. Kaplan, *J. Am. Chem. Soc.* **94,** 668 (1972).

246. R. Crigee, G. Bolz, and R. Askani, *Chem. Ber.* **102,** 275 (1969).

247. R. Criegee, *Angew. Chem. Int. Edit. Engl.* **7,** 559 (1968).

248. R. S. H. Liu, *J. Am. Chem. Soc.* **89,** 112 (1967).

249. J. S. McConaghy, Jr. and J. J. Bloomfield, *Tetrahedron Lett.* p. 3719 (1969).

250. J. J. Bloomfield, J. S. McConaghy, Jr., and A. G. Hartmann, *Tetrahedron Lett.* p. 2723 (1969).

251. A. Padwa, W. Koehn, J. Masaracchia, C. L. Osborn, and D. J. Trecker, *J. Am. Chem. Soc.* **93,** 3633 (1971).

252. W. R. Roth and B. Peltzer, *Angew. Chem. Int. Edit. Engl.* **3,** 440 (1964).

253. J. E. Baldwin and M. S. Kaplan, *J. Am. Chem. Soc.* **94,** 4696 (1970).

254. L. T. Scott and M. Jones, Jr., *Chem. Rev.* **72,** 181 (1972).

255. M. Avram, E. Sliam, and C. D. Nenitzescu, *Justus Liebigs Ann. Chem.* **636,** 184 (1960).

256. C. D. Nenitzescu, M. Avram, I. I. Pogany, G. D. Mate, and M. Farcasiu, *Acad. Rep. Pop. Rom. Stud. Cercet. Chim.* **11,** 7 (1963).

257. W. von E. Doering and J. W. Rosenthal, *J. Am. Chem. Soc.* **88,** 2078 (1966).

258. E. Vedejs, *Tetrahedron Lett.* p. 4963 (1970).

259. R. C. Cookson, J. Hudec, and J. Marsden, *Chem. Ind. (London)* 21 (1961).

260. E. E. von Tamelen and B. C. T. Pappas, *J. Am. Chem. Soc.* **93,** 6111 (1971).

261. M. Avram, G. Mateescu, and C. D. Nenitzescu, *Justus Liebigs Ann. Chem.* **636,** 174 (1960).

262. G. Maier, *Chem. Ber.* **102,** 3310 (1969).

263. L. A. Paquette and J. C. Stowell, *Tetrahedron Lett.* p. 2259 (1970).

264. E. Vedejs and E. S. C. Wu, *J. Am. Chem. Soc.* **97,** 4706 (1975).

265. L. A. Paquette and J. C. Stowell, *J. Am. Chem. Soc.* **93,** 5735 (1971).

266. P. J. Collins and W. H. F. Sasse, *Tetrahedron Lett.* p. 1689 (1968).

266a. P. J. Collin and W. H. F. Sasse, *Austr. J. Chem.* **24,** 2325 (1971).

267. Y. Kayama, M. Oda, and Y. Kitihara, *Tetrahedron Lett.* p. 3293 (1974).

268. M. Oda, Y. Kayama, H. Miyazaki, and Y. Kitihara, *Angew. Chem. Int. Edit. Engl.* **14,** 418 (1975).

269. Y. Kitihara, M. Oda, and Y. Kayama, *Angew. Chem. Int. Edit. Engl.* **15,** 487 (1976).

270. Y. Kitihara, M. Oda, and H. Miyazaki, *Angew. Chem. Int. Edit. Engl.* **15,** 487 (1976).

271. P. Radlick and W. Fenical, *Tetrahedron Lett.* p. 4901 (1967).

272. K. M. Schumate, Ph.D. Thesis, Univ. of Texas, 1966. As noted in refer. 250.

273. M. Jones, Jr., S. D. Reich, L. T. Scott, and L. E. Sullivan, *Angew. Chem. Int. Edit. Engl.* **7,** 644 (1968).

274. R. Srinivasan, *Tetrahedron Lett.* p. 4029 (1973).

275. A. H. Braun, *J. Org. Chem.* **35,** 1208 (1969).

276. M. Kato, T. Sawa, and T. Miwa, *Chem. Commun.* p. 1235 (1971).

277. S. Masamune, C. G. Chen, K. Hojo, and R. T. Seidner, *J. Am. Chem. Soc.* **89,** 4804 (1967).

278. S. Masamune, G. G. Chen, H. Zenda, K. Hojo, and R. T. Seidner, *Angew. Chem. Int. Edit. Engl.* **7,** 645 (1968).

279. S. Masamune and R. T. Seidner, *Chem. Commun.* p. 542 (1969).

280. G. Schröder and Th. Martini, *Angew. Chem. Int. Edit. Engl.* **6,** 806 (1967).

281. L. A. Paquette, M. R. Short, and J. F. Kelly, *J. Am. Chem. Soc.* **93,** 7179 (1971).

282. K. J. Crowley, *Proc. Chem. Soc.* p. 334 (1962).

283. K. J. Crowley, *Tetrahedron* **21,** 1001 (1964).

284. R. S. H. Liu, *J. Am. Chem. Soc.* **89,** 112 (1967).

285. W. G. Dauben, R. L. Cargill, R. M. Coates, and J. Saltiel, *J. Am. Chem. Soc.* **88,** 2742 (1966).

286. W. J. Ball and S. P. Landor, *J. Chem. Soc.* p. 2298 (1962).

287. L. Skattebøl and S. Solomon, *J. Am. Chem. Soc.* **87,** 4506 (1965).

288. K. Untch and D. J. Martin, *J. Am. Chem. Soc.* **87,** 4501 (1965).

289. R. Criegee and H. G. Reinhardt, *Chem. Ber.* **101,** 102 (1968).

290. M. B. D'Amore and R. G. Bingman, *J. Am. Chem. Soc.* **91,** 5694 (1969).

291. J. I. Brauman and D. I. Golden, *J. Am. Chem. Soc.* **90,** 1920 (1968).

292. F. C. Lupton, Jr., *Tetrahedron Lett.* p. 4209 (1968).

293. G. A. Doorakian and H. H. Friedman, *J. Am. Chem. Soc.* **90,** 5310, 6896 (1968).

294. B. S. Rabinovitch, E. W. Schlag, and K. B. Wiberg, *J. Chem. Phys.* **28,** 504 (1958).

295. M. C. Flowers and H. M. Frey, *Proc. Royal Soc.* (*London*) **A257,** 121 (1960).

296. K. C. Brannock, R. D. Burpitt, V. W. Goodlett, and J. G. Thweat, *J. Org. Chem.* **28,** 1464 (1963).

297. K. C. Brannock, R. D. Burpitt, and J. G. Thweat, *J. Org. Chem.* **28,** 1697 (1963).

298. K. C. Brannock, R. O. Burpitt, V. W. Goodlett, and J. G. Thweat, *J. Org. Chem.* **29,** 813, 818 (1964).

299. G. A. Berchtold and G. F. Uhlig, *J. Org. Chem.* **28,** 1459 (1963).

300. C. F. Huebner and E. Donoghue, *J. Org. Chem.* **28,** 1732 (1963).

301. C. F. Huebner, L. Dorfman, M. M. Robinson, E. Donoghue, W. G. Pierson, and P. Strachan, *J. Org. Chem.* **28,** 3134 (1963).

302. W. G. Dauben and D. J. Hart, *J. Org. Chem.* **42,** 922 (1977).

303. D. Becker, L. R. Hughes, and R. A. Raphael, *J. Chem. Soc. Perkin Trans. 1* p. 1674 (1977).

304. T. Fex, J. Froborg, G. Magnusson, and S. Thoren, *J. Org. Chem.* **41,** 3518 (1976).

305. A. Huth, H. Straub, and E. Müller, *Justus Liebigs Ann. Chem.* p. 1893 (1973).

306. S. F. Nelsen and J. P. Gillespie, *J. Am. Chem. Soc.* **95,** 1874 (1973).

307. I. Willner and M. Rabinovitz, *Tetrahedron Lett.* p. 1223 (1976).

308. P. Gassman and X. Creary, *Chem. Commun.* p. 1214 (1972).

309. M. Oda, Y. Kayama, H. Miyazaki, and Y. Kitahara, *Angew. Chem. Int. Edit. Engl.* **14,** 418 (1975).

310. Y. Kitahara, M. Oda, and Y. Kayama, *Angew. Chem. Int. Edit. Engl.* **15,** 487 (1976).

311. Y. Kitahara, M. Oda, and H. Miyazaki, *Angew. Chem. Int. Edit. Engl.* **15,** 487 (1976).

312. H. Plieninger and D. Wild, *Chem. Ber.* **99,** 3070 (1966).

313. M.-S. Liu and V. Snieckus, *J. Org. Chem.* **36,** 645 (1971).

314. P. S. Mariano, M. E. Osborn, D. Dunaway-Mariano, B. C. Gunn, and R. C. Pettersen, *J. Org. Chem.* **42,** 2903 (1977).

315. D. J. Haywood and S. T. Reid, *J. Chem. Soc. Perkin Trans. 1* p. 2457 (1977).

316. D. C. Neckers, J. H. Dopper, and H. Wynberg, *Tetrahedron Lett.* p. 2913 (1969).

317. J. H. Dopper and D. C. Neckers, *J. Org. Chem.* **36,** 3755 (1971).

318. D. N. Reinhoudt and C. G. Leliveld, *Tetrahedron Lett.* p. 3119 (1972).
319. D. N. Reinhoudt and C. G. Kouwenhoven, *Chem. Commun.* p. 1233 (1972).
320. D. N. Reinhoudt and C. G. Kouwenhoven, *Tetrahedron* **30,** 2431 (1974).
321. D. L. Coffen, Y. C. Poon, and M. L. Lee, *J. Am. Chem. Soc.* **93,** 4627 (1971).
322. T. Q. Mink, L. Christiaens, P. Grandclaudon, and A. Lablache-Combier, *Tetrahedron* **33,** 2225 (1977).
323. W. Oppolzer, *J. Am. Chem. Soc.* **93,** 3833, 3834 (1971).
324. W. Oppolzer and K. Keller, *J. Am. Chem. Soc.* **93,** 3836 (1971).
325. W. Oppolzer, *Tetrahedron Lett.* p. 1001 (1974).
326. T. Kametani, T. Takahashi, and K. Ogasawara, *J. Chem. Soc. Perkin Trans. 1* p. 1464 (1973).
327. T. Kametani, K. Ogasawara, and T. Takahashi, *Tetrahedron* **29,** 73 (1973).
328. T. Kametani, M. Takemura, K. Ogasawara, and K. Fukumoto, *J. Heterocycl. Chem.* **11,** 179 (1974).
329. T. Kametani, T. Takahashi, T. Honda, K. Ogasawara, and K. Fukumoto, *J. Org. Chem.* **39,** 447 (1974).
330. T. Kametani, Y. Katoh, and K. Fukumoto, *J. Chem. Soc. Perkin Trans. 1* p. 1712 (1974).
331. T. Kametani, M. Kajiwara, and K. Fukumoto, *Tetrahedron* **30,** 1053 (1974).
332. T. Kametani, M. Kajiwara, T. Takahashi, and K. Fukomoto, *J. Chem. Soc. Perkin Trans. 1* p. 737 (1975).
333. Kametani, H. Nemoto, H. Ishikawa, K. Shiroyama, and K. Fukumoto, *J. Am. Chem. Soc.* **98,** 3378 (1976).
334. T. Kametani, H. Nemoto, H. Ishikawa, K. Shiroyama, H. Matsumoto, and K. Fukumoto, *J. Am. Chem. Soc.* **99,** 3461 (1977).
335. T. Kametani, Y. Hirai, F. Sotoh, and K. Fukumoto, *Chem. Commun.* p. 16 (1977).
336. T. Kametani, Y. Kato, F. Satoh, and K. Fukumoto, *J. Org. Chem.* **42,** 1177 (1977).
337. T. Kametani, Y. Kato, T. Honda, and K. Fukumoto, *J. Am. Chem. Soc.* **98,** 8185 (1976).
338. T. Kametani, T. Takahashi, K. Ogasawara, and K. Fukumoto, *Tetrahedron* **30,** 1047 (1974).
339. T. Kametani, T. Kato, T. Honda, and K. Fukumoto, *J. Chem. Soc. Perkin Trans. 1* p. 2001 (1975).
340. R. L. Funk and K. P. C. Vollhardt, *J. Am. Chem. Soc.* **99,** 5483 (1977).
341. W. G. L. Aalbersberg, A. J. Barkovich, R. L. Funk, R. L. Hilliard, and K. P. C. Vollhardt, *J. Am. Chem. Soc.* **97,** 5600 (1975).
342. A. R. Browne and B. Halton, *Chem. Commun.* p. 1341 (1972).
343. L. A. Paquette, T. G. Wallis, T. Kempe, G. G. Christoph, J. P. Springer, and J. Clardy, *J. Am. Chem. Soc.* **99,** 6948 (1977).
344. F. H. Marquardt, *Tetrahedron Lett.* p. 4989 (1967).
345. D. Bellus and C. D. Weiss, *Tetrahedron Lett.* p. 999 (1973).
346. D. Bellus, K. von Bredow, H. Sauter, and C. O. Weiss, *Helv. Chim. Acta* **56,** 3004 (1973).
347. P. Dowd and K. Kang, *Synth. Commun.* **4,** 151 (1974).
348. R. K. Boeckman, Jr., M. H. Delton, T. Nagasaka, and T. Wantanabe, *J. Org. Chem.* **42,** 2946 (1977).
349. S. R. Wilson and L. R. Phillips, *Tetrahedron Lett.* p. 3047 (1975).
350. J. C. Hinshaw, *J. Org. Chem.* **29,** 3951 (1974).
351. J. Hambrecht, H. Straub, and E. Müller, *Chem. Ber.* **107,** 3962 (1974).
352. J. Hambrecht and H. Straub, *Tetrahedron Lett.* p. 1079 (1976).
353. S. M. McElvain, E. R. Degginger, and J. D. Bekun, *J. Am. Chem. Soc.* **76,** 5736 (1954).
354. H. W. Scheeren, R. W. M. Aben, P. H. J. Ooms, and R. J. F. Nivard, *J. Org. Chem.* **42,** 3128 (1977).

355. G. Büchi, J. T. Kofron, E. Koller, and D. Rosenthal, *J. Am. Chem. Soc.* **78,** 876 (1956).

356. H. E. Zimmerman and L. Craft, *Tetrahedron Lett.* p. 2131 (1964).

357. D. Bryce-Smith, G. I. Fray, and A. Gilbert, *Tetrahedron Lett.* p. 2137 (1964).

358. J. A. Barltrop and B. Hesp, *J. Chem. Soc. C* p. 1625 (1967).

359. S. P. Pappas and N. A. Portnoy, *J. Org. Chem.* **33,** 2200 (1968).

360. D. Bryce-Smith, A. Gilbert, and M. G. Johnson, *Tetrahedron Lett.* p. 2863 (1968).

361. S. Farid, W. Kothe, and G. Pfundt, *Tetrahedron Lett.* p. 4147 (1968).

362. T. Miyamoto, Y. Shigemitsu, and Y. Odaira, *Chem. Commun.* p. 1410 (1969).

363. H. J. T. Bos and J. Boleij, *Rec. Trav. Chim. Pays-Bas* **88,** 465 (1969).

364. H. J. T. Bos, H. T. Van Der Bend, J. S. M. Boleij, C. J. A. Everaars, and H. Polman, *Rec. Trav. Chim. Pays-Bas* **91,** 65 (1972).

365. H. Polman, J. S. M. Boleij, and H. J. T. Bos, *Rec. Trav. Chim. Pays-Bas* **91,** 1088 (1972).

366. H. G. Viehe, "Chemistry of Acetylenic Compounds," Marcel Dekker, New York, 1969.

367. H. Vieregge, H. M. Schmidt, J. Renema, H. J. T. Bos, and J. F. Arens, *Rec. Trav. Chim. Pays-Bas* **85,** 929 (1966).

368. R. Fuks, R. Buijle, and H. G. Viehe, *Angew. Chem. Int. Edit. Engl.* **5,** 585 (1966).

369. F. Ficini and A. Krief, *Tetrahedron Lett.* p. 2497 (1967).

370. M. E. Kuehne and P. J. Sheeran, *J. Org. Chem.* **33,** 4406 (1968).

371. M. Delanois and L. Ghosez, *Angew. Chem. Int. Edit. Engl.* **8,** 72 (1969).

372. J. Ficini and J. Pouliquen, *Tetrahedron Lett.* p. 1139 (1972).

373. M. E. Kuehne and H. Linde, *J. Org. Chem.* **37,** 1846 (1972).

374. J. Ficini and J. Pouliquen, *J. Am. Chem. Soc.* **93,** 3295 (1971).

375. J. Ficini, *Tetrahedron* **32,** 1449 (1976).

376. J. U. Piper, M. Allard, M. Faye, L. Hamel, and V. Chow, *J. Org. Chem.* **42,** 461 (1977).

377. W. J. Middleton, *J. Org. Chem.* **30,** 1307 (1965).

378. M. E. Kuehne and P. J. Sheeran, *J. Org. Chem.* **34,** 3715 (1969).

379. J. M. Holovka, P. D. Gardner, C. B. Strow, M. L. Hill and T. V. Van Anken, *J. Am. Chem. Soc.* **90,** 5041 (1968).

380. L. E. Friedrich and G. B. Schuster, *J. Am. Chem. Soc.* **91,** 7204 (1969).

381. L. E. Friedrich and G. B. Schuster, *J. Am. Chem. Soc.* **93,** 4602 (1971).

382. L. E. Friedrich and J. D. Bower, *J. Am. Chem. Soc.* **95,** 6869 (1973).

383. H. Heaney and J. M. Jablonski, *Chem. Commun.* p. 1139 (1968).

384. H. Heaney and C. T. McCarty, *Chem. Commun.* p. 123 (1970).

385. H. Heaney, J. M. Jablonski, and C. T. McCarty, *J. Chem. Soc. Perkin Trans. 1* p. 2903 (1972).

386. S. B. Cavitt, H. Sarrafizadeh, and P. O. Gardner, *J. Org. Chem.* **27,** 1211 (1962).

387. C. L. McIntosh and O. L. Chapman, *Chem. Commun.* p. 771 (1971).

388. E. Müller, R. Mayer, B. Narr, A. Riecker, and K. Scheffer, *Justus Liebigs Ann. Chem.* **645,** 25 (1961).

389. H.-D. Becker and K. Gustafsson, *J. Org. Chem.* **42,** 2966 (1977).

390. Z. Neiman, *J. Chem. Soc. Perkin Trans. 2* p. 1746 (1972).

391. A. B. Levy and A. Hassner, *J. Am. Chem. Soc.* **93,** 2051 (1971).

392. T. S. Cantrell, *J. Am. Chem. Soc.* **94,** 5929 (1972).

393. N. C. Yang, B. Kim, W. Chiang, and T. Hamada, *Chem. Commun.* p. 729 (1976).

394. T. S. Cantrell, *J. Org. Chem.* **42,** 4238 (1977).

395. L. A. Wendling and R. G. Bergman, *J. Org. Chem.* **41,** 831 (1976).

396. L. A. Paquette, M. J. Wyvratt, and G. R. Allen, Jr., *J. Am. Chem. Soc.* **92,** 1763 (1970).

397. D. H. Aue and D. Thomas, *J. Org. Chem.* **40,** 1349 (1975).

398. G. Maier and U. Schäfer, *Tetrahedron Lett.* p. 1053 (1977).

399. K. E. Wilzbach and D. J. Rausch, *J. Am. Chem. Soc.* **92,** 2178 (1970).

400. R. D. Chambers, J. R. Maslakiewicz, and K. C. Srivastava, *J. Chem. Boc. Perkin Trans. 1* p. 1130 (1975).
401. R. N. Warrener, G. Kretschmer, and M. N. Paddon-Row, *Chem. Commun.* p. 806 (1977).
402. K. R. Henery-Logan and J. V. Rodericks, *J. Am. Chem. Soc.* **85**, 3524 (1963).
403. G. Kretschmer and R. N. Warrèner, *Tetrahedron Lett.* p. 1335 (1975).
404. W. J. Theuer and J. A. Moore, *Chem. Commun.* p. 468 (1965).
405. J.-L. Derocque, W. J. Theuer, and J. A. Moore, *J. Org. Chem.* **33**, 4381 (1968).
406. E. J. Volker, M. G. Pleiss, and J. A. Moore, *J. Org. Chem.* **35**, 3615 (1970).
407. R. D. Chambers and R. Middleton, *J. Chem. Soc. Perkin Trans. 1* p. 1500 (1977).
408. E. M. Burgess and L. McCullagh, *J. Am. Chem. Soc.* **88**, 1580 (1966).
409. E. M. Burgess and G. Milne, *Tetrahedron Lett.* p. 93 (1966).
410. R. Fuks, G. S. D. King, and H. G. Viehe, *Angew. Chem. Int. Edit. Engl.* **8**, 675 (1969).
411. D. C. Dittmer, P. L.-F. Chang, F. A. Davis, H. Iwanami, J. K. Stamos, and K. Takahashi, *J. Org. Chem.* **37**, 1111 (1972).
412. A. C. Brouwer and H. J. T. Bos, *Tetrahedron Lett.* p. 209 (1976).
413. H. Behringer and E. Meinetsberger, *Tetrahedron Lett.* p. 3473 (1975).
414. L. A. Paquette, *J. Org. Chem.* **30**, 629 (1965).
415. D. C. Dittmer and N. Takashima, *Tetrahedron Lett.* p. 3809 (1964).
416. W. R. Truce, R. H. Bavry, and P. S. Bailey, Jr., *Tetrahedron Lett.* p. 5651 (1968).
417. J. F. King, K. Piers, D. J. H. Smith, C. L. McIntosh, and P. deMayo, *Chem. Commun.* p. 31 (1969).
418. C. L. McIntosh and P. deMayo, *Chem. Commun.* 32 (1969).
419. R. Hoffmann, *Helv. Chim. Acta* **53**, 2331 (1970).
420. C. S. Foote and J. W.-P. Liu, *Tetrahedron Lett.* p. 3267 (1968).
421. N. C. Yang and J. Libman, *J. Org. Chem.* **39**, 1782 (1974).
422. N. J. Turro, V. Ramamurthy, K.-C. Liu, A. Krebs, and R. Kemper, *J. Am. Chem. Soc.* **98**, 6758 (1976).
423. N. Rieber, J. Albert, J. Lipsky, and D. Lemal, *J. Am. Chem. Soc.* **91**, 5668 (1969).
424. D. Applequist, M. Lintner, and R. Searle, *J. Org. Chem.* **33**, 254 (1968).
425. G. A. Closs, H. A. Böll, M. Heyn, and V. Dev, *J. Am. Chem. Soc.* **90**, 173 (1968).
426. C. G. Crespan, B. C. McCusick, and T. L. Cairns, *J. Am. Chem. Soc.* **82**, 1515 (1960).
427. W. Kusters and P. de Mayo, *J. Am. Chem. Soc.* **96**, 3502 (1974).
428. H. E. Simmons, D. C. Blomstrom, and R. D. Vest, *J. Am. Chem. Soc.* **84**, 4782 (1962).
429. P. C. Guha and M. N. Chakladar, *Quart. J. Indian Chem. Soc.* **2**, 318 (1925).
430. L. Field, W. D. Stevens, and E. L. Lyspert, Jr., *J. Org. Chem.* **26**, 4782 (1961).
431. C. G. Crespan, *J. Am. Chem. Soc.* **83**, 3434 (1961).
432. C. G. Crespan and B. C. McCusick, *J. Am. Chem. Soc.* **83**, 3438 (1961).
433. H. E. Simmons, D. C. Blomstrom, and R. D. Vest, *J. Am. Chem. Soc.* **84**, 4772 (1962).
434. W. Schroth, H. Bahn, and R. Zschernitz, *Z. Chem.* **13**, 424 (1973).
435. S. Wawzonek and S. M. Heilmann, *J. Org. Chem.* **39**, 511 (1974).
436. R. B. Boar, D. W. Hawkins, J. F. McGhie, and D. H. R. Barton, *J. Chem. Soc. Perkin Trans. 1* p. 515 (1977).
437. T. J. Barton and J. A. Kilgour, *J. Am. Chem. Soc.* **96**, 7150 (1974).
438. K. Wieser and A. Berndt, *Angew. Chem. Int. Edit. Engl.* **14**, 70 (1975).
439. K. Burger, J. Albanbauer, and M. Eggersdorfer, *Angew. Chem. Int. Edit. Engl.* **14**, 766 (1975).
440. K. Burger, J. Albanbauer, and W. Foag, *Angew. Chem. Int. Edit. Engl.* **14**, 767 (1975).
441. J.-M. Surzur, C. Dupuy, M. P. Bertrand, and R. Nougier, *J. Org. Chem.* **37**, 2782 (1972).
442. A. S. Monahan, J. D. Freilich, J.-J. Fong, and D. Kronenthal, *J. Org. Chem.* **43**, 232 (1978).

443. A. G. Sherwood and H. E. Gunning, *J. Am. Chem. Soc.* **85**, 3506 (1963).

444. A. G. Sherwood and H. E. Gunning, *J. Phys. Chem.* **69**, 1732 (1965).

445. D. S. L. Blackwell, P. de Mayo, and R. Suan, *Tetrahedron Lett.* p. 91 (1974).

446. P. Barraclough, M. Edwards, T. L. Gilchrist, and C. J. Harris, *J. Chem. Soc. Perkin Trans. 1* p. 716 (1976).

447. Y. Hayashi, T. Iwagami, A. Kadoi, T. Shono, and D. Swern, *Tetrahedron Lett.* p. 1071 (1974).

448. E. Winterfeldt, *Chem. Ber.* **98**, 1581 (1965).

449. E. S. Levchenko and E. M. Dorokhova, *J. Org. Chem.* (*USSR*) **8**, 2573 (1972).

450. C. W. Wilcox, Jr., B. K. Carpenter, and W. R. Dolbier, Jr., *Tetrahedron* **35**, 707 (1979).

451. R. Mündnick, H. Plieninger, and H. Vogler, *Tetrahedron* **33**, 2661 (1977).

452. R. Mündnick and H. Plieninger, *Tetrahedron* **34**, 887 (1978).

453. M. F. Semmelhack and R. J. De Franco, *J. Am. Chem. Soc.* **94**, 2116 (1972).

454. J. E. Baldwin and M. C. McDaniel, *J. Am. Chem. Soc.* **89**, 1537 (1967).

455. J. E. Baldwin and M. C. McDaniel, *J. Am. Chem. Soc.* **90**, 6118 (1968).

456. S. M. Krueger, J. A. Kapecki, J. E. Baldwin, and I. A. Paul, *J. Chem. Soc B* p. 796 (1969).

457. H. W. Moore, L. Hernandez, Jr., and R. Chambers, *J. Am. Chem. Soc.* **100**, 2245 (1978).

458. M. J. S. Dewar, G. P. Ford, and M. S. Rzepa, *Chem. Commun.* p. 728 (1977).

459. J. J. Bloomfield, *Tetrahedron Lett.* p. 587 (1968).

460. W. R. Dolbier, Jr., K. Matsui, H. J. Dewey, D. V. Horak, and J. Michl, *J. Am. Chem. Soc.* **101**, 2136 (1979).

461. R. P. Steiner and J. Michl, *J. Am. Chem. Soc.* **100**, 6413 (1978).

462. R. P. Steiner, R. D. Miller, H. J. Dewey, and J. Michl, *J. Am. Chem. Soc.* **101**, 1820 (1979).

463. G. R. Krow, K. M. Damodaran, E. Michenor, R. Wolf, and J. Guare, *J. Org. Chem.* **43**, 3950 (1978).

464. J. Froborg and G. Magnusson, *J. Am. Chem. Soc.* **100**, 6728 (1978).

465. R. D. Clark and K. G. Untch, *J. Org. Chem.* **44**, 248 (1979).

466. R. D. Clark and K. G. Untch, *J. Org. Chem.* **44**, 253 (1979).

467. H. Fienemann and M. M. R. Woffmann, *J. Org. Chem.* **44**, 2802 (1979).

468. G. Desimoni, P. Righetti, and G. Tacconi, *J. Chem. Soc. Perkin Trans. 1* p. 856 (1979).

469. T. Kametani, H. Matsumoto, H. Nemoto, and K. Fukumoto, *J. Am. Chem. Soc.* **100**, 6218 (1978).

470. K. Burger, H. Schickaneder, and A. Meffert, *Z. Naturforsch. B: Anorg. Chem., Org. Chem. B* **30**, 622 (1975).

471. W. J. M. van Tilborg and R. Plomp, *Chem. Commun.* p. 130 (1977).

472. H. Gilman and W. H. Atwell, *J. Am. Chem. Soc.* **87**, 2678 (1965).

473. P. B. Valkovich and W. P. Weber, *Tetrahedron Lett.* p. 2153 (1975).

474. E. Block and L. K. Revelle, *J. Am. Chem. Soc.* **100**, 1630 (1978).

475. T. S. Sorenson and A. Rauk, "Pericyclic Reactions" (A. P. Marchand and R. E. Lehr, eds.), Vol. II, pp. 21–29, Academic Press, New York, 1977.

476. T. Hiyama, M. Shinoda, and H. Nozaki, *Tetrahedron Lett.* p. 771 (1978).

477. Y. Gaoni, *Tetrahedron Lett.* p. 3277 (1978).

478. T. Hiyama, M. Shinoda, and W. Nozaki, *J. Am. Chem. Soc.* **101**, 1599 (1979).

479. G. Piancatelli, A. Scettri, G. David, and M. D'Auria, *Tetrahedron* **34**, 2775 (1978).

480. F. Cooke, J. Schwindeman, and P. Magnus, *Tetrahedron Lett.* p. 1995 (1979).

481. W. E. Fristad, D. S. Dime, T. R. Bailey, and L. Paquette, *Tetrahedron Lett.* p. 1999 (1979).

482. A. F. Hegarty, J. H. Coy, and F. L. Scott, *J. Chem. Soc. Perkin Trans. 2* p. 104 (1975).

CHAPTER 6
SIX ELECTRON–FIVE ATOM SYSTEMS

Only two all carbon types come under this classification, pentadienylidenes and pentadienylanions. Conversely, there are many heteroatom systems which belong here, though careful mechanistic studies are relatively sparse.

I. PENTADIENYLIDENES

A carbene conjugated to a *cis*-butadienyl moiety could undergo an electrocyclic closure to form a cyclopentadiene [Eq. (6-1)]. The stereochemistry

$$(6-1)$$

of such a reaction is not open to experimental study, and the mechanism is a matter of conjecture since two additional routes can be devised [Eqs. (6-2) and (6-3)]. The facile thermal 1,5-H shifts in cyclopentadienes complicate

$$(6-2)$$

$$(6-3)$$

any experimental tests of mechanism. The best known example is the formation of fluorene in the thermolysis of diphenylcarbene.[1-5] Fluorene formation does not proceed directly, but via a series of rearrangements leading to *o*-phenylphenylcarbene (see Chapter 4, Section III). Since the direct route would necessitate formation of a five membered ring allene and two consecutive 1,3-hydrogen shifts, reaction via an alternative path is not sur-

$$(6-4)$$

prising. Use of substituted diarylcarbenes has shown that the route to fluorenes is generally that of Eq. (6-4).[5,6,7] However, di-α-naphthyldiazomethane has been reported to cyclize without rearrangement.[8] Thus, the reaction does occur, but no evidence which clearly delineates the mechanism has as yet been reported.

In Chapter 4, Section III it was noted that pyrolysis of cyclopropenes can lead to vinylcarbenes, and when these have an appropriately placed aromatic ring indenes may be formed. Thus, for example, tetraphenylcyclopropene gives 1,2,3-triphenylindene when heated to 200°.[53] Other examples have been reported more recently.[54,273]

Reactions which appear to belong in this category have been observed with one nitrogen atom in the pentadienylidene skeleton. Thus, when N-phenylbenztriazole is heated carbazole is formed [Eq. (6-5)].[9] A related

$$(6-5)$$

reaction leading to the formation of an oxazole has been reported [Eq. (6-6)].[10] This case is particularly persuasive of the electrocyclic nature of the ring closure.

$$(6-6)$$

II. BUTADIENYLNITRENES

In strong contrast to pentadienylidenes, conjugated nitrenes have attracted considerable attention. Since these dienylnitrenes can be only transient intermediates, studies have been initiated with azidodienes, aryl nitro compounds, and azirines or their precursors. Most of the work has involved azidodienes as reactants and in many instances the presence of a nitrene as a discrete intermediate is doubtful. The stereochemical consequences are not measurable, so mechanistic work has leaned heavily on kinetic studies and trapping experiments.

The thermal decomposition of azidodienes to give pyrroles and nitrogen offers numerous mechanistic possibilities. Thus, for example, the initial step may be internal addition of azide to a double bond, concerted elimination of nitrogen with cyclization, or direct loss of nitrogen to form either a 2H-pyrrole or a vinylazirine [Eq. (6-7)]. If the nitrene does form, it may give

$$(6\text{-}7)$$

the pyrrole by direct C-H insertion, addition to the double bond and cleavage of the bicyclic ring, or electrocyclic closure followed by a 1,5-hydrogen shift. Little information exists which would permit a choice among these paths, and there appears to have been no consideration of the electrocyclic closure as a mechanistic possibility.

$$(6\text{-}8)$$

The early work on this reaction category involved formation of carbazoles from 2-azidobiphenyls.[11-15] The work has been reviewed.[16] Smith and Hall[15] concluded that direct loss of nitrogen to form a nitrene occurred in the initial step and that there was little or no participation by the second aromatic ring. Evidence has also been presented that the nitrene is in the singlet state.[17] Smolinsky[14] found that thermolysis of **131** gave an N-Me carbazole [Eq. (6-9)]. This suggested that the second step was an electrophilic

$$(6\text{-}9)$$

substitution, but further studies[18] raised some doubts about that route. Smith[16] has called attention to the higher yields obtained where an aromatic ring rather than a saturated C-H participated in the second step. With the possible exception of the result of Eq. (6-10) [19] all of the experimental in-

$$(6\text{-}10)$$

formation can be readily accommodated by the mechanism of Eq. (6-11).

$$(6\text{-}11)$$

Since the source of the N-H [Eq. (6-10)] was not determined that result could indicate migration via the protonated benzene ring.

Pyrolysis of o-azidostyrenes[20] or of β-azidostyrenes[44,45,46] leads to the formation of substituted indoles in good yield [Eq. (6-12)].[20] Both of these

$$(6\text{-}12)$$

Y = Pr, Ph, COCOOMe, COPh

might have been expected to proceed via a mechanism analogous to that of Eq. (6-11). However, neither reaction appears to follow such a mechanism. Evidence indicating participation by the vinyl group during nitrogen evolution from o-azidostyrenes has been presented.[24] Azirines are clearly intermediates in the pyrolysis of β-azidostyrenes [Eq. (6-13)] since they are isolable

$$(6\text{-}13)$$

under less vigorous conditions, but react further to give indoles at higher temperatures.[22,25,26,28,28a] Nishiwaki[27] has indicated that vinyl nitrenes and azirines are in equilibrium at temperatures slightly above ambient. Presumably the final step in Eq. (6-13) occurs by the same mechanism as the ring closure for carbazole formation. A more complex mechanism, but still utilizing β-phenylazidoalkenes, has been postulated to account for the formation of indoles via pyrolysis of 1,2,3-triazoles.[29]

Vinylazirines form butadienyl nitrenes during thermolysis, and these intermediates can be trapped.[30,31] Where undiverted the nitrenes close to give pyrroles in good yield.[30,55] The clearest indication that ring closure is an electrocyclic process is shown by Eq. (6-14).[31] Here migration of the

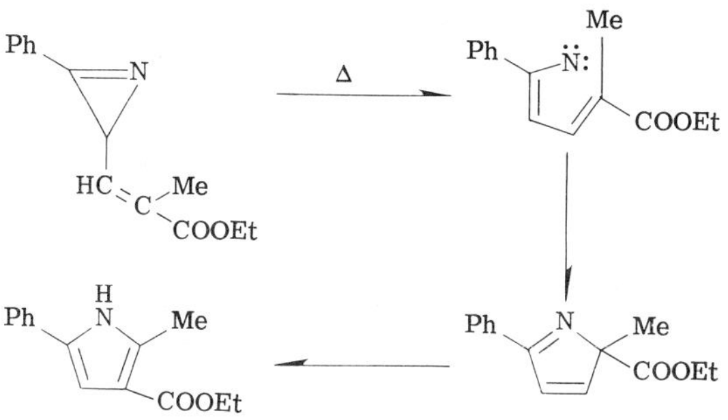

$$ \tag{6-14} $$

carbethoxy group is most reasonably explicable via the ring closure route shown.

The literature of heterocyclic ring formation via pyrolysis of compounds of the general type 132 is quite extensive,[16,19,32,33,34] but in these reactions

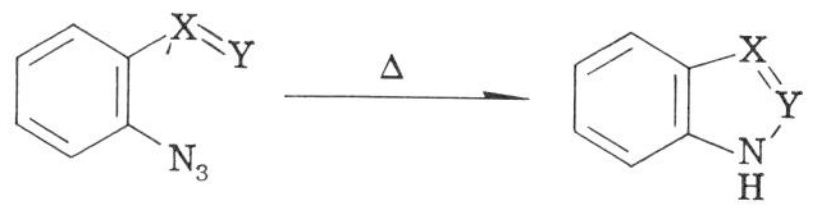

$$ \tag{6-15} $$

132

X, Y = C, N, O

generally the unsaturated moiety participates in the elimination of nitrogen.[33] Azidovinylketones form isoxazoles when heated [Eq. (6-16)], but

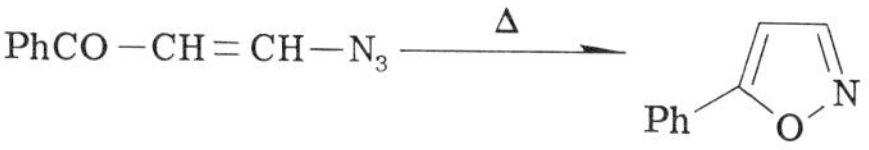

$$ PhCO-CH=CH-N_3 \xrightarrow{\Delta} \tag{6-16} $$

little evidence indicative of mechanism has been accumulated.[35,36,37] Though nitrenes may not be implicated in these reactions, nitrenes can be formed by thermolysis of substituted azirines, and they do undergo an electrocyclic ring closure.[30,38,39] Catalysis of the reaction by $Mo(CO)_6$ has been reported.[40]

$$Y = NPh, O$$
$$R = H, Ph$$

While nitrenes may not play a role in isoxazole formation from azido precursors, nitrenes can be formed in the pyrolysis of isoxazoles.[41–46] Some of this work has been reviewed by Nishiwaki.[47] These results show that the electrocyclic reaction [Eq. (6-18)] is reversible. Where nitrenes do not revert

to isoxazoles, they may form azirines,[45] but other reactions may also ensue.[41,42] Nitrenes formed by ring opening of isoxazoles or by ring fission of tetrazoles[48,49,50] or of oxadiazolones[51] cyclize to new five-membered ring heterocycles.[48–52] The cyclization–migration of Eq. (6-19) is particularly interesting.[50,274]

III. PENTADIENYL ANIONS

A thorough review of the electrocyclic reactions of anions has been published recently.[56] Therefore, the coverage here will be brief. Concerted electrocyclization should be allowed in disrotatory fashion, but no thorough theoretical study of the process appears to have been made. Under circumstances which clearly make the pentadienyl anion the more stable entity thermodynamically, retroelectrocyclization is observed. Thus, for example, 6,6-diphenyl-2-bicyclo[3.1.0]hexene produces monocyclic dienes when treated with sodium *tert*-amylate.[57] Other examples of a similar nature have been observed.[58] A number of heterocyclic ring cleavage reactions, for example formation of crotonaldehyde enolate from dihydrofuran, have been reported by Kloosterziel and his coworkers.[59,60,61]

Pentadienyl anion ring closures are apparently difficult to bring about. A number of pentadienyl anions have been prepared and their conformations studied by nmr.[62,63,64,69] None cyclize thermally. Cycloheptadienide ion does not cyclize at 100°.[65] Apparently the first cyclization observed was the formation of bicyclo[3.3.0]octene from 1,3- or 1,4-cyclooctadiene under rather drastic conditions, phenylpotassium at 175°,[66] or potassium hydride at 190°.[67] Bates and McCombs[68] studied the cyclooctadienide ion, presenting nmr evidence for the U anion, and showed that the cyclization product was *cis*-bicyclo[3.3.0]octenide ion (**133**). A $\Delta G^{\ddagger} = 18.4$ kcal/mol was ob-

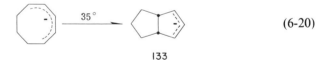

$$\text{(6-20)}$$

133

tained for this disrotatory reaction, with $t_{1/2} = 80$ min at 35°.

Shoppee and Henderson were unable to induce 1,5-diphenylpentadienide ions to cyclize, even at 190°.[71] However, cyclized products were obtained when the semicarbazone of 1,5-diphenyl-1,4-pentadien-3-one was heated with *tert*-butoxide ion at 225°,[70,71] or the tosylhydrazone was treated with lithium aluminum hydride in ether at 35°.[72] The authors suggested that 1,5-diphenylpentadienide ion does form under the Wolff-Kischner and Caglioti conditions and cyclizes in disrotatory fashion. Just why the ion formed *in situ* should behave so differently from the ion formed by removal of a proton was not discussed. Hunter *et al.*[64] were unable to induce cyclization of 1,3,5-triphenylpentadienide ion at 130°.

Heteroatoms in the pentadienide skeleton alter the results dramatically. Hunter and Sim[73,73a] examined the mechanism of what is probably the oldest electrocyclic reaction known. Thermal ring closure of hydrobenzamide to amarine was reported in 1844.[74] Reaction was found[73,73a] to occur at −70° when hydrobenzamide was treated with phenyllithium in THF. Later the reaction was found to proceed rapidly at $-130°$, but no evidence for the intermediacy of an open chain ion was obtained. Amarine, the less stable

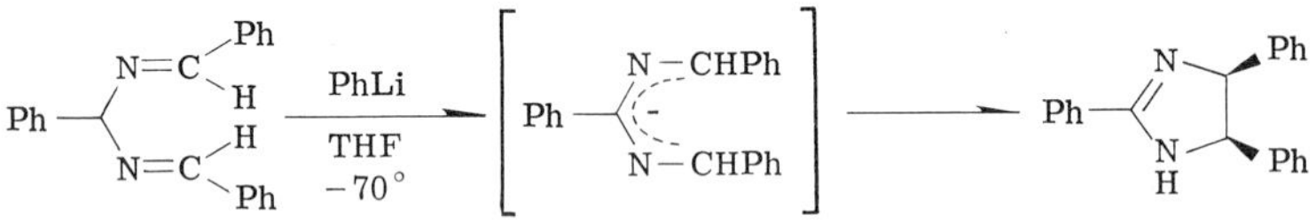

(6-21)

cis isomer, is found in at least 99.7% purity. If one of the nitrogens of hydrobenzamide is replaced by a CH group, the rate of closure is retarded by about six powers of ten.[75] Furthermore, the ring closure gives both cis and trans products in nearly equal amounts (21% cis, 26% trans), but the stereochemistry appears to be kinetically controlled. Arguments in favor of competing allowed and forbidden reactions have been presented.[64]

Schmidt and co-workers have studied the behavior of antiaromatic cyclic anions, which generally undergo ring closure to form bicyclic allyl anions via what appears to be an electrocyclic process. For example, **134** gives a deep blue anion which undergoes ring closure even at $-120°$ [Eq. (6-21)].[76,]

(6-21a)

[77,78] Analogs with sulfur in place of oxygen lose sulfur and give pyrroles.[79] Cyclopentadienes are formed in low yield when γ-pyrans are treated in similar fashion.[80] The subject has been reviewed by Schmidt.[78]

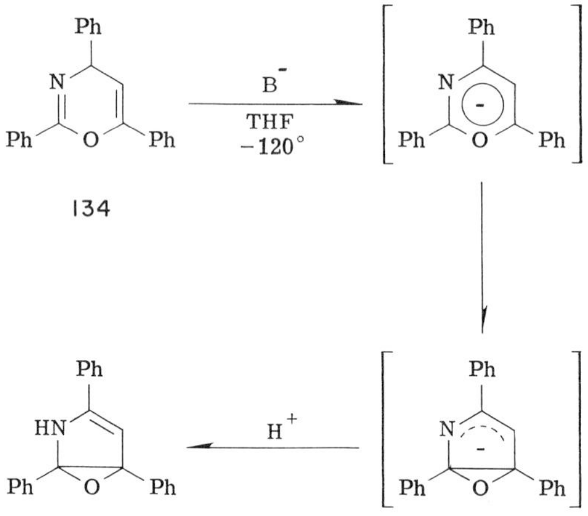

IV. HETEROATOM ANALOGS OF PENTADIENYL ANIONS

Heteroatoms having an unshared pair of electrons should be able to replace the anionic carbon atom, but by supplying an electron pair the heteroatom must accept a positive charge [Eq. (6-22)]. Elguero has suggested that incor-

$$(6\text{-}22)$$

poration of a positive nitrogen into the acyclic moiety should enhance the possibility for the electrocyclic reaction [Eq. (6-23)],[81] and he has surveyed

$$(6\text{-}23)$$

a group of cyclizations which may fit this pattern. The reaction may be considered a vinylogous nucleophilic addition to an immonium ion as well as an electrocyclic process, but the latter designation is useful since it would require a disrotatory process and a specific stereochemical outcome.

Coispeau and Elguero[82] have surveyed the literature on the reactions of hydrazines to form pyrazoles or their reduced forms, and they have suggested the ring closure fits the above pattern. Similar mechanistic suggestions were made earlier,[83,84] based on kinetics, Hammett ρ values ($Ar_1 = -1.44$ (ρ^+), $Ar_2 = -2.54$),[89] and the steric effects of substituents [Eq. (6-24)]. The stereo-

$$(6\text{-}24)$$

chemistry of the ring closure could not be verified because the final proton shift was too rapid.[85] However, acid catalyzed formation of pyrazolines from azines was found to lead under kinetic control preferentially to the cis

$$(6\text{-}25)$$

product, though acid catalyzed equilibration occurred to give the thermodynamic product under normal synthesis conditions.[86] Electrocyclization appears to be the logical explanation. Other examples, generally catalyzed by acids, have been reported.[275-277] A related cyclization has been postulated as one step in the deep seated rearrangement of Eq. (6-26).[87] The

$$(6\text{-}26)$$

special relationship between the nucleophilic addition and an electrocyclic mechanism is nicely illustrated by consideration of the two resonance contributors. Elguero[81] had postulated that the electrocyclic mechanism applies to formation of thiadiazoles via treatment of hydrazones with carbon disulfide.[88] A similar mechanism could also be applied to the reaction of azomethine imines with carbon disulfide.[89] Many of Elguero's proposed electrocyclic mechanisms involve reaction of less stable tautomeric forms

$$(6\text{-}27)$$

of esters, amides, etc., where the more stable form could equally readily give the same product via a conventional carbonyl addition mechanism. While the electrocyclic mechanism may be applicable in some other cases, the literature pertaining to these ring closures is so voluminous that a complete survey was not attempted.

In at least one case it appears that the direct cyclization of a divinylamine did occur though the anionic moiety of the dipolar cyclic entity was stabilized by a carbethoxy group.[278]

V. DIPOLAR SYSTEMS

Dipolar cycloaddition could really be called Huisgen's world, so thoroughly has he dominated this area of chemistry.[90,91,92] As the Diels–Alder cycloaddition is the intermolecular analog of the electrocyclic reaction, so dipolar cycloadditions have their conjugated intramolecular counterparts which are the subject of this section. The process was discussed by Huisgen in 1960,[93] and was more explicitly presented as a general electrocyclic reaction by Riemlinger in 1970.[94] Riemlinger noted that the general process represented by Eq. (6-28) encompasses potentially 98 variants. As Table 6-1

TABLE 6-1

Potential Variations of the Dipolar Electrocyclization

One heteroatom

$$\underset{C=Y-C}{+ \quad -} \overset{}{\diagup\hspace{-0.3em}=} \quad \longrightarrow \quad \text{(ring with Y)} \qquad Y=O,S,NR$$

$$-C\equiv\overset{+}{N}-\overset{-}{C}\diagup\hspace{-0.3em}= \quad \longrightarrow \quad \text{(ring with N)}$$

Two heteroatoms

$$\underset{C=Y-C}{+ \quad -}\overset{}{C=Z} \quad \longrightarrow \quad \text{(ring with Y, Z)} \qquad \begin{array}{l} Y=O,S,NR \\ Z=O,S,NR \end{array}$$

$$\underset{C=Y-C}{+ \quad -}\overset{}{N=C} \quad \longrightarrow \quad \text{(ring with Y, N)} \qquad Y=O,S,NR$$

$$-\overset{+}{C}\equiv N-\overset{-}{C}\overset{}{C=Z} \quad \longrightarrow \quad \text{(ring with N, Z)} \qquad Z=O,S,NR$$

$$-C\equiv\overset{+}{N}-\overset{-}{C}\overset{}{N=C} \quad \longrightarrow \quad \text{(ring with N, N)}$$

$$\underset{C=X-Y}{+ \quad -} \quad \longrightarrow \quad \text{(ring with X, Y)} \qquad \begin{array}{l} X=O,S,NR \\ Y=O,S,NR \end{array}$$

$$\underset{C=X-N}{+ \quad -} \quad \longrightarrow \quad \text{(ring with X, N)} \qquad X=O,S,NR$$

$$\underset{C-N\equiv N}{- \quad +} \quad \longrightarrow \quad \text{(ring with N, N)}$$

$$-C\equiv\overset{+}{N}-\overset{-}{N} \quad \longrightarrow \quad \text{(ring with N, N)}$$

(continued)

TABLE 6-1 (*Continued*)

Three Heteroatoms[a]

Y=O,S,NR
X=N
Z=O,NR

X=N
Z=O,NR

X=O,S,NR
Y=O,S,NR
Z=O,S,NR

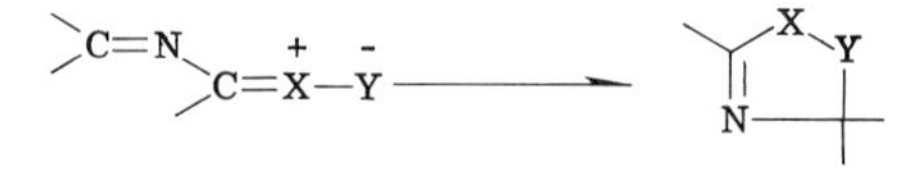

X=O,S,NR
Y=O,S,NR

X=O,S,NR
Z=O,S,NR

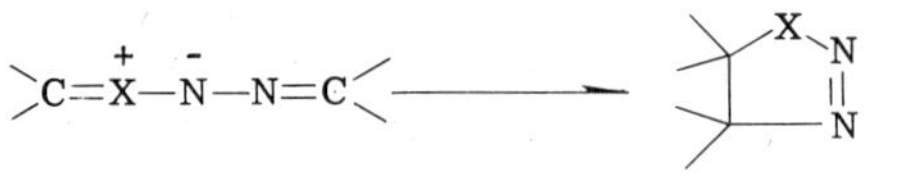

X=O,S,NR

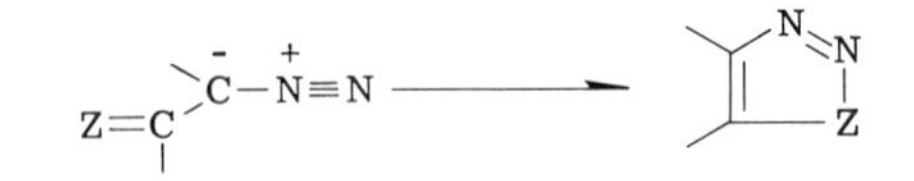

Z=O,S,NR

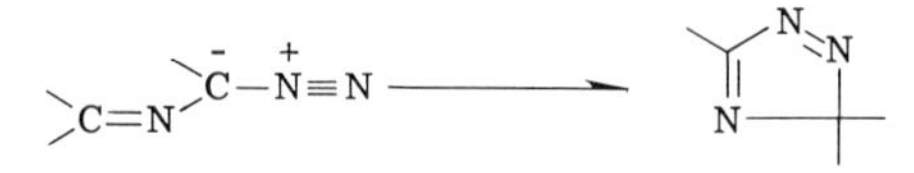

Y=O,S,NR

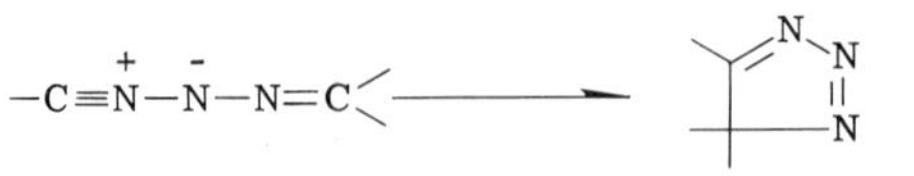

X=O,S,NR
Y=O,S,NR

TABLE 6-1 (Continued)

Four Heteroatoms*

$$A{=}B\diagdown \atop C{=}\overset{+}{X}{-}\overset{-}{Y}$$ ⟶ (cyclic: X–Y / B–A, C bridging) A=O,S,NR / B=N / X=O,S,NR / Y=O,S,NR

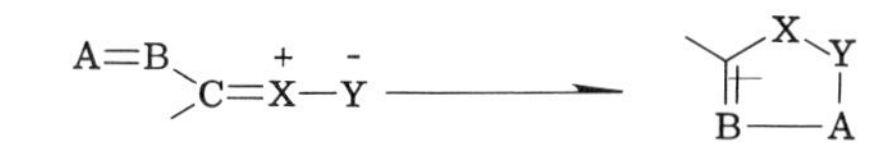

$$\diagup C{=}\overset{+}{X}{-}\overset{-}{N}{-}A{=}B$$ ⟶ (cyclic: X–N / B–A) X=O,S,NR / A=N / B=O,S,NR

$$A{=}B\diagdown \atop C{-}\overset{-}{N}{\equiv}N$$ ⟶ (cyclic: N–N / B–A) A=O,S,NR / B=N

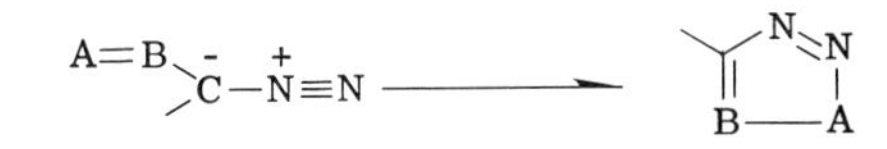

$$-C{\equiv}\overset{+}{N}{-}\overset{-}{N}{-}B{=}A$$ ⟶ (cyclic: N–N / A–B) A=O,S,NR / B=N

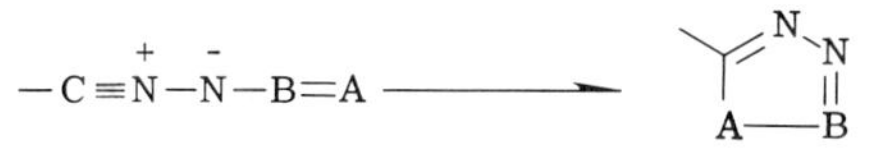

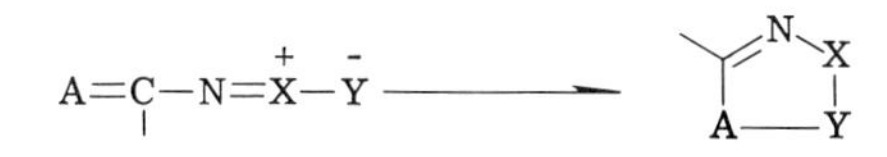

$$X{=}\underset{|}{C}{-}\overset{-}{N}{-}\overset{+}{N}{\equiv}N$$ ⟶ (cyclic: N–N / X–N) X=O,S,NR

$$\diagup C{=}N{-}\overset{-}{N}{-}\overset{+}{N}{\equiv}N$$ ⟶ (cyclic: N–N / N=N)

$$A{=}\underset{|}{C}{-}\overset{+}{N}{=}\overset{-}{X}{-}\overset{-}{Y}$$ ⟶ (cyclic: N–X / A–Y) A=O,S,NR / X=O,S,NR / Y=O,S,NR

$$\diagup C{=}N{-}\overset{+}{N}{=}\overset{-}{X}{-}\overset{-}{Y}$$ ⟶ (cyclic: N–N / Y–X) X=O,S,NR / Y=O,S,NR

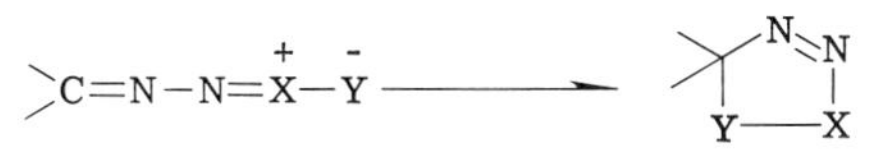

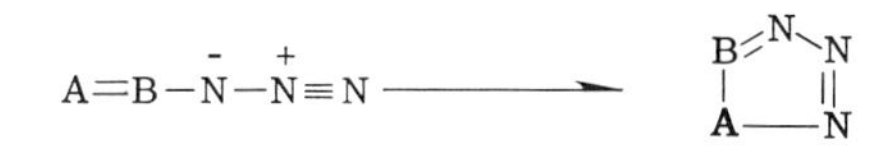

Five Heteroatoms*

$$A{=}B{-}\overset{-}{N}{-}\overset{+}{N}{\equiv}N$$ ⟶ (cyclic: B–N–N / A–N) A=O,S,NR / B=N

$$A{=}B{-}\overset{+}{N}{=}\overset{-}{X}{-}\overset{-}{Y}$$ ⟶ (cyclic: B–N–X / A–Y) A=O,S,NR / B=N / X=O,S,NR / Y=O,S,NR

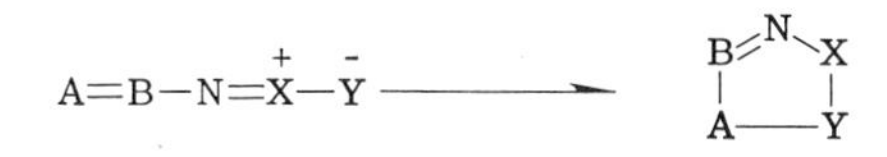

a Clearly a number of combinations which include three or more adjacent oxygen atoms or sulfur atoms are not likely to exist and should be discounted.

$$a \overset{+}{=} b - \overset{-}{c} - d = e \longrightarrow \underset{d=c}{\overset{a}{\underset{e \quad b:}{\bigcirc}}} \tag{6-28}$$

illustrates there are more than 100 possible cases if sulfur can be expected to replace oxygen in most positions.

A. One Heteroatom

The most carefully investigated example in this class has oxygen as the central atom and the dipolar intermediate results from heterolysis of an epoxide [Eq. (6-29)]. Generally the formation of the dipolar intermediate is

$$\text{(6-29)}$$

favored by groups which will help stabilize the charges (Chapter 4, Section II,2). The ring closure step was first disclosed by Pommelet *et al.* (1970),[95] who showed that *trans*-divinylethylene oxide gave 5-vinyl-4,5-dihydrofuran. The stereochemical result [Eq. (6-30)] indicated that ring closure was dis-

$$\text{(6-30)}$$

(cis only)

rotatory, and the overall reaction scheme was the relatively complex one shown. This route has received strong experimental support from the elegant studies of Crawford *et al.* on 3,4-epoxy-1-butene.[96] The optically active epoxide racemizes forty times faster than the *cis*-9-deuterio form isomerizes to the trans isomer, and 4,4-dideuterio labelled material gives only 5,5-dideuterio-4,5-dihydrofuran. Similar results were obtained with active *trans*-2,3-divinyloxirane[97] and diisopropenyloxirane.[98]

Study of **135** showed that isomerization of the double bond could also occur.[99] Thus a small amount of *trans*-substituted dihydrofuran was ob-

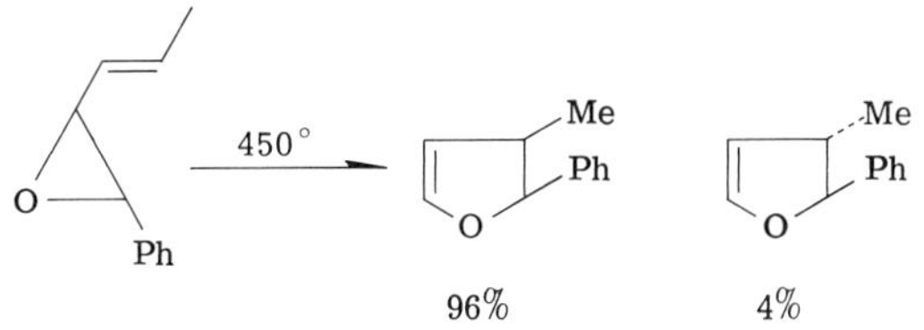

$$(6\text{-}31)$$

96%   4%

135

tained, but a forbidden conrotatory ring closure was excluded. Ring opening was facilitated by a 2-cyano group, and conrotatory ring opening of the oxirane was confirmed by trapping with maleic anhydride.[100] Manisse and Chuche[101] found that *trans*-2-vinyl-3-ethynyloxirane gave 5-ethynyl-4,5-dihydrofuran as one of several thermolysis products. A dienyloxirane (**136**)

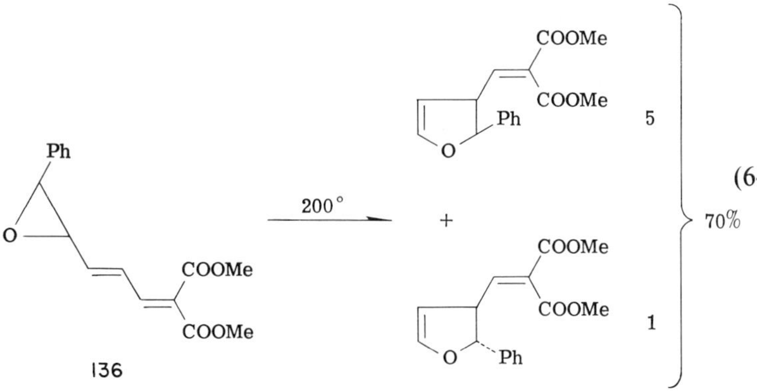

$$(6\text{-}32)$$

136

gave an 83:17 ratio of *cis*- and *trans*-dihydrofurans, and the initial conrotatory opening of the oxirane was confirmed by trapping.[102]

Azomethine ylides with a properly situated double bond undergo a facile electrocyclization even when the double bond is part of an aromatic ring. Benzindolizines have been prepared via a reaction which appears to involve such a ring closure [Eq. (6-33)].[103,104] The method has been extended to

$$+ \quad MeCOCH_2 - COOEt \quad + \qquad\qquad (6\text{-}33)$$

other dichlorides.[105] Augenstein and Kröhnke[106] developed a similar syn-

$$(6\text{-}34)$$

thesis for substituted indolizines starting from a pyridinium ylid [Eq. (6-35)].

$$(6\text{-}35)$$

More recently a number of examples of this indolizine synthesis using non-aromatic double bonds have appeared.[107-111,267,268,278] Typical examples are shown in Eq. (6-36). Much of the work has been reviewed.[112] In

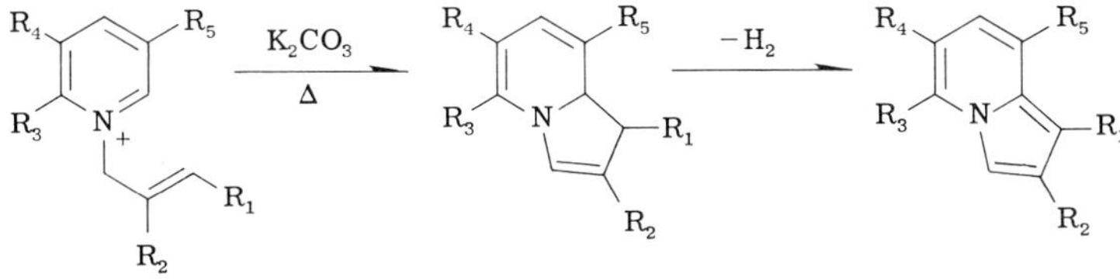

$$
\text{(6-36)}
$$

R_1	R_2	R_3	R_4	R_5
Ph	H	H	H	H
PhCO	Ph	H	H	H
COOMe	Me	Me	H	H
COOMe	COOMe	H	Me	H
COOMe	COOMe	Me	H	Me

some cases the presence of an alkyl group at C_2 of the pyridinium ring leads to an alternate reaction path.[113] Vinylaziridines form pyrrolidines undoubtedly via the ring closure of an intermediate azomethine ylide.[279]

The 1,5-dipolar electrocyclization of thiocarbonyl ylides appears to be represented by one general example only. Smutny[114] has prepared dihydro-

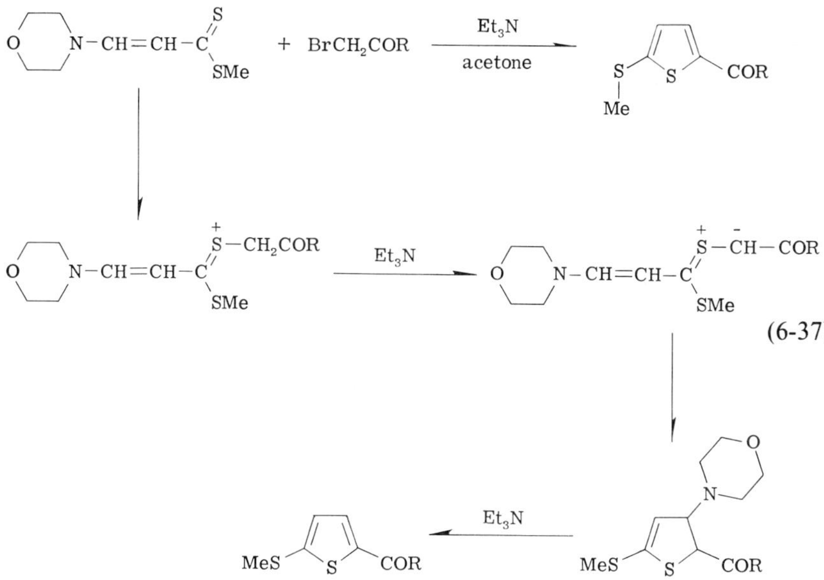

$$
\text{(6-37)}
$$

thiophenes, and eventually thiophenes, by reaction of α,β-unsaturated dithioesters with α-halocarbonyl compounds in the presence of a base [Eq. (6-37)]. The stereochemistry of the ring closure step was not observed.

Vinyl substituted nitrile ylides cyclize to form, after appropriate hydrogen shifts, arylpyrroles. Nitrile ylides are prepared by photolyses of arylazirines and are trappable by electron deficient olefins.[115] Addition of a vinyl substituent leads to 1,5-electrocyclization in a thermal step.[30] The intermediate can be trapped by an external electron-deficient olefin.

$$(6\text{-}38)$$

Y = COOMe, CN, COPh, CHO, Ph

B. Two Heteroatoms

Only a few of the approximately thirty possible variants of two heteroatom 1,5-dipolar electrocyclizations have been realized. In most cases this seems to be an investigative lack rather than a non-occurrence.

Azomethine ylides are formed by thermal ring opening of aziridines (see Chapter 4, Section VI,A), and in certain cases a carbonyl group can act as an internal trap.[116,117] For example, N-tert-butyl-2-benzoyl-3-phenyl-aziridine gives 2,5-diphenyloxazole when heated at 220°. The reaction undoubtedly follows the route of Eq. (6-39) but isobutane is lost in a final step.[117]

$$(6\text{-}39)$$

R_1	H	Ph	Ph	Ph
R_2	MeS	t-Bu	$PhCH_2$	C_6H_{11}
X	COOMe	Ph	Ph	p-Tol
Y	COOMe	H	H	H

There are known a number of interconversions of isoxazoles to oxazoles,[38] 4-acyloxazoles to isomeric 4-acyloxazoles,[93] and acylazirines to both oxazoles and isoxazoles.[30] For all of these nitrile ylides appear to play a central role. The interesting 4-acyloxazole rearrangement was first observed by Conforth,[118,118a] and is often called the Conforth rearrangement [Eq. (6-40)]. The intermediate formation of a nitrile ylide is supported by ^{14}C

$$ (6\text{-}40) $$

labeling studies.[119] Dewar et al.[120] has suggested that the ylide might be better represented as a carbenoid structure, but his conclusions, based on the solvent influence on rate, seem tenuous. Recently a decarboxylative example has been noted.[122]

Isoxazoles are isomerized into oxazoles both thermally and photo-chemically.[38,121] The intermediate formation of an acylazirine in both processes has been proposed, but this raises some interesting questions. Padwa et al.[30] have shown that acylazirines undergo different reactions thermally and photochemically [Eqs. (6-41) and (6-42)]. A number of exam-

$$ (6\text{-}41) $$

$$ (6\text{-}42) $$

ples of thermal isoxazole to oxazole isomerizations have been found.[123-125] The intermediate formation of an azirine is bolstered by ^{13}C labeling studies,[124] but the process apparently requires the mechanism of Eq. (6-43).

$$ (6\text{-}43) $$

The thermal ring opening to the nitrene is facile but reversible, and apparently cleavage to the nitrile ylide does occur thermally as well as photochemically, albeit less readily than to the nitrene.

Iminoazirines undergo photochemical ring scission to an imino substituted

$$R = Ph^-, CH_2{=}CH{-}CH_2{-}$$

nitrile ylide, which closes to an imidazole.[30,31]

Tamura et al.[109,126,127] and Sasaki et al.[128,129] and their co-workers have provided a number of examples of pyrazoline formation from azomethine imines. Often the final product is a pyrazole. All of the examples reported have been initiated with N-vinyliminopyridinium ylides.[269-272] In one case

the dihydro product was isolated and was reported to be the cis isomer [Eq. (6-46)], thus suggesting a disrotatory ring closure. However, the double

bond geometry of the reactant was not established. The yields are often rather modest and some ylides will not undergo the 1,5-electrocyclization.[128]

Vinyldiazomethanes are known to form pyrazoles, undoubtedly via pyrazolenines and an electrocyclic reaction.[130,131] Vinyldiazomethane itself cyclizes on standing in ether at 25°.[131] The kinetics of the reaction have been studied and $\Delta H^{\ddagger} = 22.0$ kcal/mol, $\Delta S^{\ddagger} = 3.6$ eu.[132,133] Further mechanistic studies were carried out by Brewbaker and Hart[134] who noted that 1-alkyl substitution notably enhances the rate of the reaction, and 3-aryl derivatives showed a modest rate acceleration by electron donors, $\rho = -0.40$. Obvious questions about mechanistic details arise because of the linear diazoalkyl grouping. These are not readily investigated experimentally, but can be approached via theoretical calculations. While no studies of this system have been made the isoelectronic vinyl azide has been treated (see Section V,C).

The reaction constitutes a convenient and useful synthesis of pyrazoles. Dornow and Bartsch[135] found that tosylhydrazones of unsaturated aldehydes and ketones form pyrazoles in good yield. [Eq. (6-48)]. The same reac-

$$PhCH{=}CH{-}CH{=}N{-}NHTos \xrightarrow{\quad MeO^- \quad} PhCH{=}CH{-}CH{=}N\bar{N}Tos$$

(6-48)

$$PhCH{=}CH{-}CH{=}N_2$$

tion was noted by Meerwein in 1950.[136] Closs developed the process and showed that the reaction fails when two alkyl groups are on the β-carbon.[137,138] Sharp has shown that pyrazolenine formation is rapid and reversible.[139,140] Use of protic solvents does not interfere with pyrazole formation. Pyrazolenines are formed in modest yield in certain cases.[139,140,141] Since the ring closure step is reversible other reactions may occur with

(6-49)

$R_1, R_2 = (CH_2)_4, R_3 = R_4 = Me$
$R_1 = Me, R_2 = H, R_3 = R_4 = Ph$
$R_1 = R_2 = H, R_3 = R_4 = Ph$

the diazo compound.[137,139,139a,142,143] The intermediate diazo compound has been trapped.[139,139a]

N-arylnitrilimines formed thermally from oxadiazolones [Eq. (6-50)] under vigorous conditions will cyclize to form indazoles.[144] The same reaction has been observed with diphenyltetrazoles.[145]

$$\text{(6-50)}$$

R = Ph, Me

C. Three Heteroatoms

The versatility of the 1,5-dipolar electrocyclization for heterocycle synthesis is clearly illustrated by the diverse examples covered in this section. The general utility is really limited only by the accessibility of the conjugated 1,3-dipolar species.[280]

Cyclizations of such conjugated dipolar moieties as azomethine imines, and thiocarbonyl imines (thione S-imides) have been observed, but examples are rare and have generally been uncovered serendipitously. Cyclizations of thiocarbonyl imines with conjugation at either the carbon or nitrogen end of the dipole have appeared. Research aimed at preparing stable thiocarbonyl imines showed that N-carbonyl stabilized examples will cyclize at temperatures above −30° [Eq. (6-51)].[146] Cyclization of a carbon con-

$$\text{(6-51)}$$

jugated thiocarbonyl imine was encountered by chance [Eq. (6-52)].[147] One step in the rationale of the complex reaction of Eq. (6-53) is the ring closure of an N-benzoylazomethine imine.[148]

$$\text{(6-52)}$$

(6-53)

Ring closure of a α-diazoimines and α-diazothiocarbonyl compounds to 1,2,3-triazoles and thiadiazoles has been known since 1902,[149] when Wolff noted that treatment of α-diazo-β-dicarbonyl compounds with primary amines or hydrogen sulfide led to heterocycles. Initially it was presumed that α-diazocarbonyl compounds also form oxadiazoles, but Dimroth showed that this does not occur.[150] Numerous studies of this ring-chain

(6-54)

tautomerism have been reported, and these show the equilibrium is generally very sensitive to structural and solvent effects. After Curtius and Thompson[154] had shown that α-diazoacetamide cyclizes to 5-hydroxy-1,2,3-triazole, Dimroth studied the equilibrium between triazoles and diazoamides in several cases.[151,152] Regitz[155] has shown that electron attractors favor the diazoimine form and electron donors the triazole form where N-(p-substituted phenyl) imines are used. With N-cyanoimines polar solvents favor the diazoimine,[156] while the cyclic form predominates with 1-arylsulfonyl-4-methyl-5-diethylaminotriazoles.[157]

Dimroth[151,152] determined the rates of ring opening of 4-substituted-5-hydroxy-1,2,3-triazoles. Brown and Hammick[153] reinterpreted Dimroth's data and suggested that the rate determining step was a bimolecular reaction between two oppositely charged ions [Eq. (6-55)], i.e., not an electrocyclic

(6-55)

reaction. However, in acetonitrile and dimethylformamide the reaction follows first order kinetics, independent of added acid.[158] When R = X—⟨⟩—
[Eq. (6-56)] the rate is increased by electron attracting substituents and is

$$(6\text{-}56)$$

decreased by added salts such as lithium chloride. The mechanism thus is adequately represented as a simple retroelectrocyclization converting a dipolar reactant to a less polar product.[158] The bicyclic triazole (137) was shown to be in equilibrium with α-diazomethylpyridine [Eq. (6-57)] since

137

$$(6\text{-}57)$$

the diazo compound could be trapped, and the rate of decomposition of the triazole was accelerated by the addition of the dinitrile.[159] Thus, the electrocyclic reaction accounts for the tautomerism.

That the diazoimine–triazole tautomerism permits interconversion of two 5-aminotriazoles was discovered by Dimroth,[160] who provided arguments for the intermediacy of a diazoimine species.[161] When $R_1 = X-\!\!\left\langle\;\right\rangle\!\!-$, $R_2 = H$, $R_3 = COOEt$ both K (as shown in Eq. 6-58) and the rate of the

$$(6\text{-}58)$$

forward reaction increase with electron attracting substituents X.[162] A low concentration of acid increases the rate, but at higher concentrations the rate is first order in triazole and independent of added acid, thus suggesting the proton transfer step involves an amidinium ion.[162] In the case $R_1 = X—\langle\ \rangle—$, $R_2 = H$, $R_3 = Ph$ the $\Delta H^{\ddagger}$ varies from ca 20 kcal/mol (p-NO$_2$) to 29.4 (p-OMe) while $\Delta S^{\ddagger}$ shows an inverse trend going from -18 eu (p-NO$_2$) to -3.6 (p-OMe).[163] Again the K values vary as before.[164] Tennant and Ververs[165] followed the degenerate reaction [Eq. (6-59)] by nmr.

$$\text{(6-59)}$$

A similar rearrangement interconverts 1,2,3-triazol-5-thiols and 5-amino-thiadiazoles.[166] In basic solution the triazole is favored,[167] while in acid solution the thiadiazole is the dominant form.[168] Electron attracting substituents [X, Eq. (6-60)] favor the thiadiazole.[169,170] α-Diazothioamides are readily converted to thiadiazoles.[171]

$$\text{(6-60)}$$

$$R = H, PhCH_2, \quad X—\langle\ \rangle—$$

Conjugated nitrilimines have been widely employed for 1,5-dipolar electrocyclizations.[93] Studies of the process were initiated by Huisgen and his students in the late 1950s. Initially this work showed that conjugation with a carbonyl group gave 1,3,4-oxadiazoles,[172] and with an imino group gave

$$\text{(6-61)}$$

$$X = O, NPh$$

1,3,4-triazoles.[173] Tetrazoles serve as the source of the nitrilimine in all cases. Investigations[174,175] of the reaction between nitriles and azidocarboxylates showed that oxadiazoles were obtained in good yield. Formation of the nitrilimine followed by a 1,5-dipolar cyclization was suggested as the mechanism, but this was rejected by Huisgen and Blaschke.[176] The tetrazole process was extended to include formation of 1,3,4-thiadiazoles.[177] The

$$\text{Ph} - \underset{\underset{N=N}{\overset{N}{\diagdown}}}{\overset{N}{\diagup}} N - \overset{\overset{S}{\parallel}}{C} - \text{NHPh} \quad \xrightarrow{\Delta} \quad \text{Ph} - \underset{S}{\overset{N-N}{\diagup \diagdown}} - \text{NHPh} \qquad (6\text{-}62)$$

limits and uses of these heterocyclic syntheses were explored in a series of papers by Huisgen and his students,[178–182] and the work has been reviewed.[93,183,184]

Gibson[185,186] found that N-acylhydrazones react with bromine and sodium acetate in acetic acid giving oxadiazoles, presumably via the nitrilimine. The formation of N-acylnitrilimines from tetrazoles has been supported by ^{15}N labeling.[187] An interesting case involves competition between a C-vinyl and an N-acyl group on the nitrilimine [Eq. (6-63)].[188] An o-nitro-

$$\text{(6-63)}$$

phenyl at either end of the nitrilimine leads to different results but in the presence of either an N-acyl or N-imine substituent only the oxadiazole or triazole forms.[189] Riemlinger has shown that 1-chloroisoquinoline can function as an iminochloride in the tetrazole–triazole rearrangement,[190] and Könnecke and Lippman[191] have uncovered a further example.

Vinyl azides do not form triazoles, but lose nitrogen to form azirines (see Section II). However it has not been possible to exclude thoroughly the possible route to azirines via an intermediate v-triazole [Eq. (6-64)]. The

$$\text{CH}_2{=}\text{CH} \diagdown \overset{+}{\text{N}}{=}\text{N}{=}\overset{-}{\text{N}} \quad \longrightarrow \quad \underset{\underset{N \diagdown N^{\diagup}N}{}}{\overset{\text{H}_2\text{C}-\text{CH}}{}} \quad \xrightarrow{-\text{N}_2} \quad \underset{\text{N}}{\overset{\text{H}_2\text{C}-\text{CH}}{\diagdown \diagup}} \qquad (6\text{-}64)$$

cyclization step has been investigated theoretically, and it was concluded that ring formation could occur via a 6-electron electrocyclic reaction with

E_a approximately 30–40 kcal/mol.[192] Electronic reorganization during ring closure is pictured as converting initially the orthogonal π bond between the terminal nitrogens into the unshared pair on the penultimate nitrogen, followed by monorotatory closure of what amounts to a dienyl nitrene. The reaction is isoelectronic with the known cyclization of 3-diazopropenes (see Section V,B). The non-aromatic *v*-triazole should be stabilized by a hydrogen shift to give the aromatic 1,2,3-triazole [Eq. (6-65)]. It is not surprising then

(6-65)

that when the blocking proton is removed from a vinyl azide, cyclization occurs readily [Eq. (6-66)].[193,193a] Similar cyclizations occur when azide ion

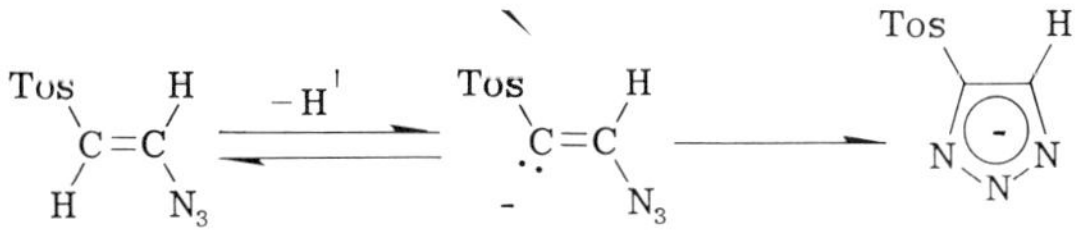

(6-66)

is added to phenyl vinyl ketone in dimethylformamide,[194] and to methyl propiolate or phenylacetylene.[195]

An unusual example of the type C=C—N=X—Y⁻ has been uncovered where X=Y=S.[196] Thus, **138** is in equilibrium with a bicyclic dithiazole, the bicyclic form being favored at equilibrium.

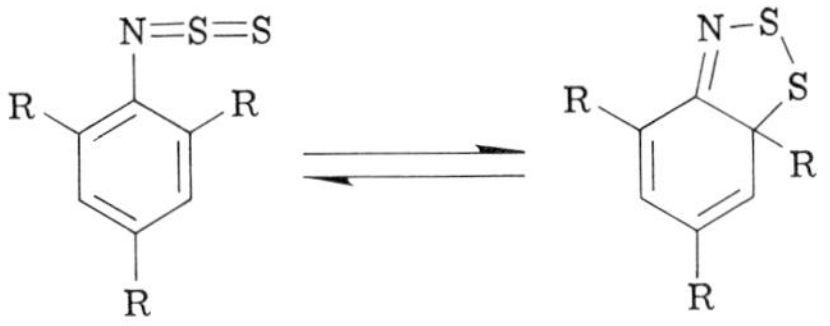

138

(6-67)

D. Four Heteroatoms

The restrictions which come into play when four contiguous heteroatoms must be present in a five-membered heterocycle have limited experimental examples here. Dimroth[197] and Curtius and Thompson[198] had noted that 4-phenylazo-5-hydroxytriazoles rearrange thermally to a colorless product. The structure of the product and the nature of the reaction was clarified by Pedersen,[199] who showed that a carboxamidotetrazole was obtained [Eq. (6-68)]. This is a special case of the general ring closure of conjugated diazo

$$\text{PhN=N} \quad \text{OH} \qquad \xrightarrow[\text{HOAC}]{\Delta} \qquad \text{PhN=N}\diagdown_{\underset{N_2}{C}}\diagup^{\overset{OH}{|}}_{\underset{}{C}}\diagdown_{NR} \qquad \longrightarrow \qquad \text{Ph—N} \quad \diagdown_{CONHR} \qquad (6\text{-}68)$$

compounds. Further examples lacking the hydroxyl group are extant.[281]

A broad range of thiatriazoles and tetrazoles show ring-chain tautomerism with thiocarbonyl azides or imino azides [Eq. (6-69)]. An excellent review of

$$\underset{R_1}{\overset{Y\diagdown N}{\diagdown}}\diagdown_N \quad \rightleftharpoons \quad \underset{R_1}{\overset{Y}{\|}}\diagdown_{N_3} \qquad (6\text{-}69)$$

$$Y = S, NR_2$$

research published prior to 1970 is available,[200] and no attempt will be made here to cover that work in detail. Theoretical studies of both thiatriazole[201] and tetrazole[202] formation from the azide isomers have appeared, and these prove of considerable interest in view of the questions aroused by the linear form of the azido moiety. Conceivably cyclization could occur by a mechanism similar to that of the vinyl azide (see Section V,C), but the calculations indicate that an unshared pair on the imine nitrogen (or the sulfur) becomes the σ bonding electron pair and the imine π bond pair end up as the unshared pair on that nitrogen. Thus, while this electrocyclization could be a six electron type, eight electrons appear to participate. However, the reaction fits best here with the obvious analogs.

The general pattern of this ring-chain tautomerism can be summarized as follows. Normally the ring form is predominant at equilibrium, particularly with the thiatriazoles. The imino azide–tetrazole system is energetically better balanced, and it provides numerous examples where either isomer can be dominant, substituents and solvents playing important roles. Electron attractors on carbon or the imino nitrogen generally favor the azide isomer. Fusion of the imino moiety into another aromatic ring, namely, pyridine, pyrimidine, pyrazole, etc. has a major influence on the equilibrium. Cases of this type have been thoroughly reviewed.[203,204] As would be expected from results with three heteroatoms, the rearrangements of tetrazole-5-thiols to aminothiatriazoles and 5-amino-1-substituted tetrazoles to 5-substituted aminotetrazoles have been observed in a number of cases. Original references to this literature can be found in the three review articles noted above.[200,203,204]

Several examples have been found where both the azide and the tetrazole form can be isolated in solid form.[205,206,207] The influence of hydrogen bonding on the equilibrium between azides and tetrazoles has been studied.[208] The exceptional case of the stable azide, $PhCOCH_2C(N_3) = NR$, has been solved.[209] Hydrogen bonding in the enol form holds the imino nitrogen

$$
\text{(6-70)}
$$

with its lone pair away from the azide group. Since this lone pair was implicated in the electron reorganization, its unavailability was considered reason enough to prevent cyclization. Further evidence for this suggestion was adduced from the failure of 1-azidobenzaloximes to undergo ring closure,

$$
\text{(6-71)}
$$

and the finding that in the presence of acyl chlorides ring closure does occur.[210,211] Eloy[212] has reported that azidoximes can be heated in boiling alcohol and in some cases to 200° without forming tetrazoles. This result is most interesting since the ring closure of diazopropenes, which are isoelectronic with the *cis*-azidoximes, is a very facile process. Azidoazines also must undergo a rate determining syn-anti interconversion prior to cyclization.[282]

Further studies of the azido–tetrazole interplay with fused ring imines have appeared.[213–216] Of these the most interesting indicates that in neutral solution 3-(or 5-) azidopyrazoles exist totally in the azido form, but in basic solution they form tetrazoles via a clean first order reaction.[213,215] Conversion of thiocarbonylazides, which cyclize readily, to the S-oxides prevents ring closure completely.[201]

An iminoazimine ring closure and reopening in another direction has been postulated as the rationale for conversion of benzocinnoline-*N*-(arylbenzimido) imides into benzocinnoline *N*-imide and benzonitrile.[283]

E. Five Heteroatoms

Five atom heterocycles with five heteroatoms are exceedingly rare, and with heteroatoms confined to N, O, or S, only pentazoles appear to be known. Pentazoles are unstable and only arylpentazoles are stable enough to have been identified.[217–219] These decompose in solution at or below room temperature to give aryl azides and nitrogen.[220] These pentazoles are formed by reaction of aryl diazonium ions with azide ion in a rapid second order reaction.[221] Two routes to formation and decomposition of the pentazoles have been considered. One of these involves a diazoazide as an intermediate in both reactions [Eq. (6-72)] while the other treats the formation and decom-

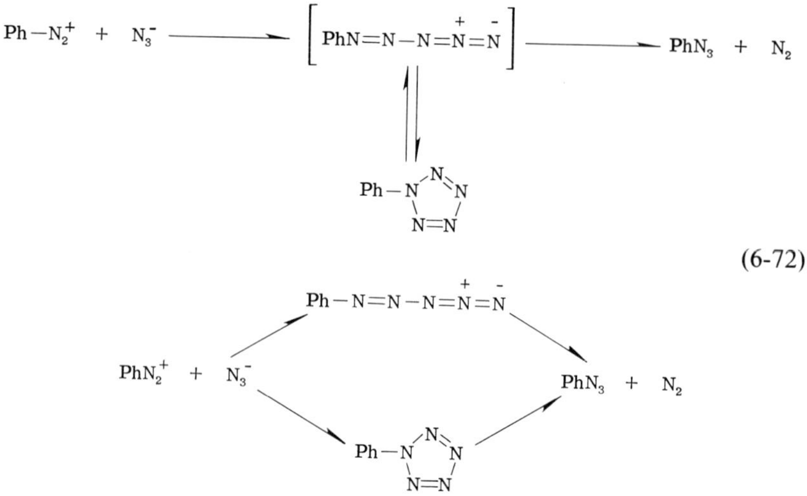

$$Ph-N_2^+ + N_3^- \longrightarrow \left[PhN{=}N{-}\overset{+}{N}{=}N{=}\overset{-}{N} \right] \longrightarrow PhN_3 + N_2$$

$$Ph-N\overset{N}{\underset{N=N}{\diagdown}}N$$

(6-72)

$$Ph-N{=}N{-}\overset{+}{N}{=}N{=}\overset{-}{N}$$

$$PhN_2^+ + N_3^- \qquad\qquad PhN_3 + N_2$$

$$Ph-N\overset{N}{\underset{N=N}{\diagdown}}N$$

position of the pentazole and diazoazide as independent processes.[222] Ugi[223] has presented some evidence in favor of the independent reaction system,[223] but the final resolution of the issue has not yet been accomplished.

VI. HETEROATOM BEARING TRIENES

The cyclization of *cis*-hexatrienes to cyclohexadienes is a classic example of the electrocyclic process. The presence of various heteroatoms (O,N) in the triene chain has a very strong influence on the course of the cyclization. In some cases the normal six atom cyclization proceeds as expected (cf. Chapter 7, Section II,F), but in a number of other examples a five atom cyclization occurs leading to a dipolar product. No complete analysis of the factors controlling the mode of cyclization has been made, but a number of these factors can be identified in certain cases.

Thermolysis of certain *cis*-2-en-1,4-diones has been found to produce γ-lactones.[224-227] The reaction was apparently first observed by Zinin[224]

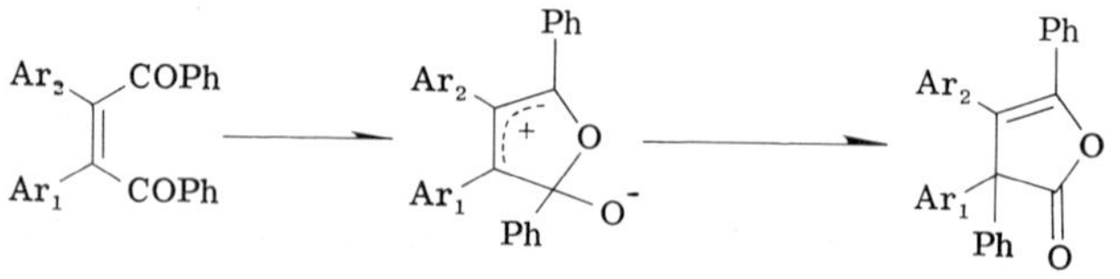

(6-73)

139   140

though he was not aware of the structures of his "oxylepidens." Japp and Klingemann[225] showed that Zinin's compounds corresponded to **139** and **140** (Ar$_1$ = Ar$_2$ = Ph), and they showed that the same reaction proceeds with the triphenyl compound (Ar$_1$ = Ph, Ar$_2$ = H). The latter reaction has been reinvestigated by Blatt[226] and also by Lahiri *et al.*[227] In nitrobenzene a 96% yield of the lactone was obtained.[227] Maier and Wiessler[228] have observed what appears to be a double example of the same type of ring closure [Eq. (6-74)]. In these cases the electrocyclic reactions would not be expected to

$$(6\text{-}74)$$

lead to six-membered rings because neither the 1,2-dioxins nor the α-pyrans would be thermally stable (see Chapter 7, Section II,A). Thus, if the diones equilibrate with both five- and six-atom electrocyclic products, the stabilization of the dipolar isomers by rearrangement or further ring closure would decide the isolable products.

Replacement of one of the terminal oxygens by a nitrogen can, under appropriate conditions, permit closure to a five-membered heterocycle. When 1-phenyl-1,2-dibenzoylethylene is treated with hydroxylamine an unstable monoxime can be isolated by careful acidification of an alkaline solution. In neutral or acid solution this monoxine is converted to a cyclic

$$(6\text{-}75)$$

nitrone.[229] Again, cyclization to a six-membered ring is a possible competitor but 2H-1,2-oxazines are apparently unknown. Presumably if formed, the 2H-1,2-oxazines would revert to the more stable monoxime, and the nitrone represents the product of thermodynamic control. That this might be a correct appraisal is indicated by the thermal behavior of the monoxime of *o*-benzenedicarboxaldehyde [Eq. (6-76)].[230] In this case a 1,2(4H)-oxazine

$$\text{(6-76)}$$

is found reversibly along with amine oxide, but the oxazine is not the product of an electrocyclic process. Either cyclic product may be the major product from this capricious reaction, but it appears reasonable to assume that the oxime geometry would not permit oxazine formation directly and that the isoindole oxide must be the initial product.

Azines of α,β-unsaturated aldehydes undergo ring closure when heated above 175°.[231] The reaction is prevented if no hydrogen is on the β-carbon

$$PhCH{=}CH{-}CH{=}N{-}N{=}CH{-}CH{=}CH{-}Ph$$

$$\text{(6-77)}$$

(see, however, refer. 284), but is otherwise apparently a general process. Ring closure is an intramolecular first order reaction with k_H/k_D for the azine of β-deuteriocinnamaldehyde $= 2.7$.[232] The proposed mechanism is illustrated in Eq. (6-78) with the 1,5-hydrogen shift step being rate determining. If the

$$\text{(6-78)}$$

β-hydrogen is replaced by a halogen, the halogen can be lost rather than transferred.[233,234] This type of ring closure is involved in the so-called criss-cross addition of acetylenes with azines [Eq. (6-79)],[235–238,240] the bicyclic product not being generally isolable.[240] The bicyclic adduct of acetylene (or phenylacetylene) with the azine of hexafluoroacetone is stable enough

$$RCH{=}N{-}N{=}CHR \quad + \quad R'C{\equiv}CR' \longrightarrow$$

(structure of bicyclic product with R', H, R, N, N, R, H, R')

$$(6\text{-}79)$$

(structure) $\longrightarrow$ $(RCH{=}C{-}C{=}N)_2$ with R' R' substituents

to be isolated, but reverts to the 4,5-diazatetraene at 100°.[239] Reaction of hexafluoroacetone azine with an ynamine leads, via an azacyclobutene, to a 2,3-diazatriene which can be cyclized at 80°–90°.[238] Once again the stable product has a five-membered ring.

$$MeC{\equiv}C{-}NEt_2 \quad + \quad [(F_3C)_2C{=}N{-}]_2 \xrightarrow{\ 25°\ }$$

(four-membered ring structure with Me, NEt_2, $(CF_3)_2$, N, $N{=}C(CF_3)_2$)

$$(CF_3)_2CH{-}\underset{\underset{CH_2}{\|}}{\overset{NEt_2}{C}}{-}C{=}N{-}N{=}C(CF_3)_2 \xrightarrow[Et_3N]{\ 80°\ } (CF_3)_2C{=}\underset{Me}{\overset{NEt_2}{C}}{-}C{=}N{-}N{=}C(CF_3)_2$$

$$(6\text{-}80)$$

(five-membered ring structure with $(CF_3)_2CH$, NEt_2, N, $-C(CF_3)_2$) $\longrightarrow$ (five-membered ring structure with $(CF_3)_2CH$, NEt_2, N, $CH(CF_3)_2$)

Yoneda and co-workers[241,242] have shown that pyrazolopyrimidines can be prepared by this ring closure. The intermediate diazatriene can be isolated

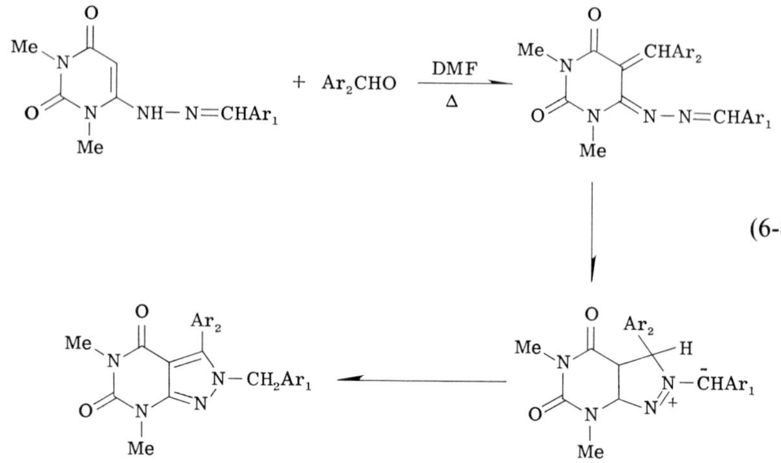

(6-81)

and cyclized in a separate step. All of the cyclizations of 2,3-diazatrienes lead to pyrazoles rather than to 5,6-dihydropyridazines. Under the relatively mild conditions required for the cyclization, the all trans and mono-cis isomers are in equilibrium. Though 5,6-dihydropyridazines are unknown it appears likely that if formed they would be sufficiently stable to persist long enough to be oxidized to a pyridazine. This leads to the assumption that the five atom electrocyclization is more facile in this case.

Electrocyclizations of both 1,2,5-triaza- and 1,2,4,5-tetrazatrienes with a cis-3,4 double bond lead exclusively to five-membered heterocycles. Heating 5-arylazo-6-methyleniminopyrimidinediones brings about cyclization [Eq. (6-82)].[241] The dipolar product is stabilized by loss of an arylnitrene

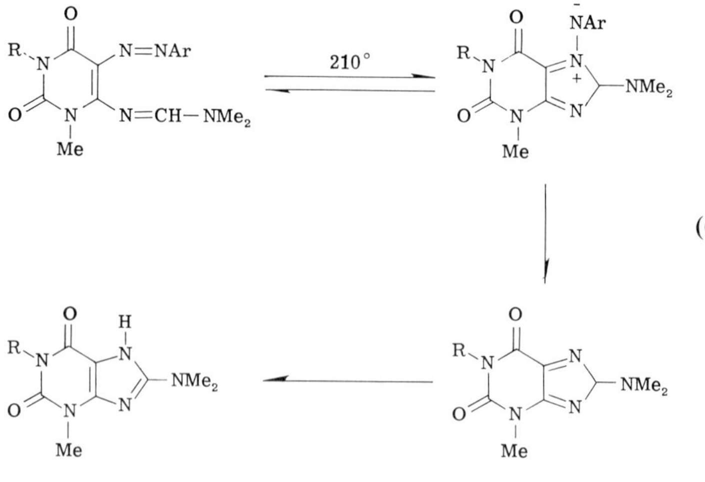

(6-82)

moiety. Studies of the oxidation products of bisphenylhydrazones of α-diketones resulted in an early proposal that dihydro-1,2,3,4-tetrazines were formed,[243] but later work showed that the products were actually bisazoethylenes.[244-246] However, it appears that the bisazoethylene and a triazolium derivative are in mobile equilibrium since the cyclic azomethine imine

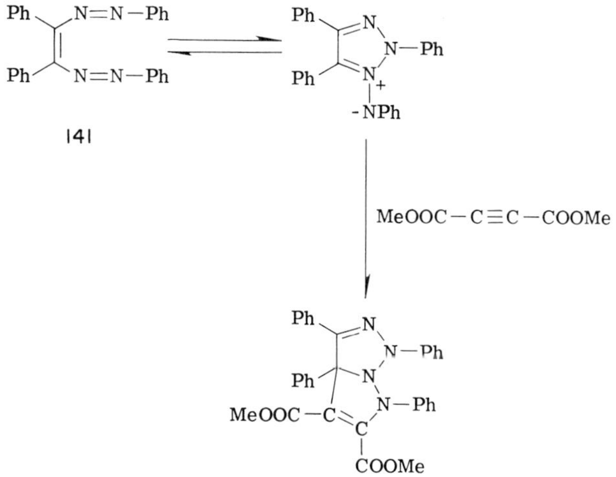

Ph⧵   N=N—Ph   Ph⧵   N
      ⟍   ⟍⟍ N—Ph
Ph⧸   N=N—Ph   ⇌   Ph⧸   N⧸
                                    |+
      141   - NPh

$$\text{MeOOC}-\text{C}\equiv\text{C}-\text{COOMe} \qquad (6\text{-}83)$$

Ph⧵   N
      ⟍⟍ N—Ph
Ph⧸   ⟍N⧸
MeOOC—C⩰   N—Ph
         C
         |
       COOMe

can be trapped by conventional dipolarophiles.[247,247a] When **141** is heated to 170°, it forms 2,4,5-triphenyl-1,2,3-triazol.[247,247a,248] Some substituted bisazoethylenes exist predominantly in the cyclic dipolar form.[246] Oxidation of bisbenzoylhydrazones produces an enol benzoate [Eq. (6-84)].[249] Some

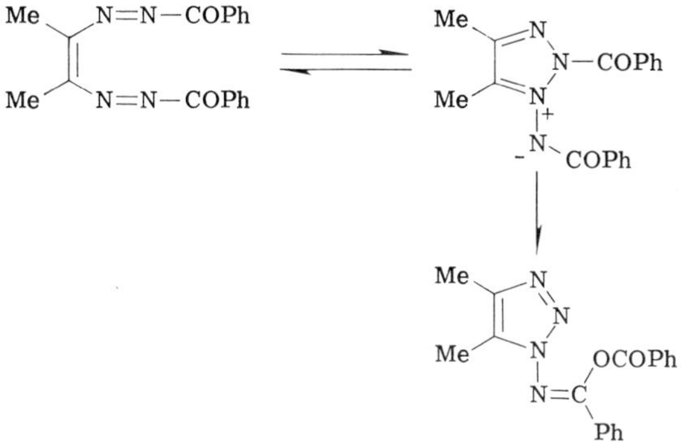

Me⧵   N=N—COPh   Me⧵   N
      ⟍   ⟍⟍ N—COPh
Me⧸   N=N—COPh   ⇌   Me⧸   N⧸
                                  |+
                                  - N⧵COPh

$$(6\text{-}84)$$

Me⧵   N
      ⟍⟍ N
Me⧸   ⟍N⧸   OCOPh
        N=C⧸
            ⧹Ph

arguments[250] about the proper structural assignment for this enol benzoate were resolved by an X-ray analysis.[251] Any questions about the possible role of a dihydro-1,2,3,4-tetrazine in these valence isomerizations were re-

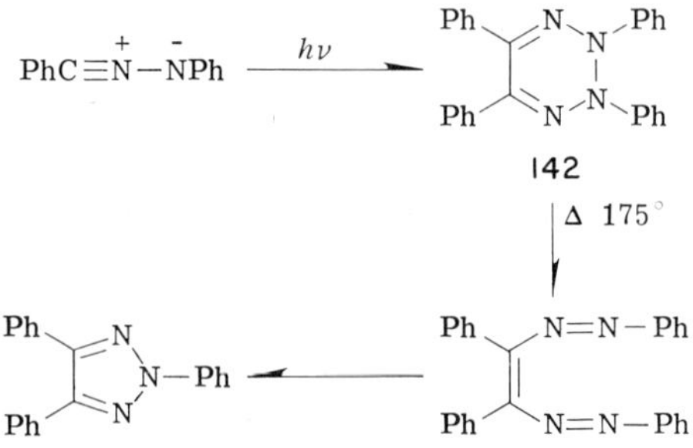

$$
\text{PhC}\equiv\overset{+}{\text{N}}-\overset{-}{\text{NPh}} \quad \xrightarrow{h\nu} \quad \text{142}
$$

(6-85)

$$
\Delta \ 175°
$$

solved when it was shown that the dihydrotetrazine (142) can be thermally converted to triphenyltriazol [Eq. (6-85)].[252]

Trienes composed of a nitroso group and certain unsaturated entities situated cis-1,2 on an ethylene group cyclize spontaneously to five-membered

$$
\begin{array}{c} \text{X=Y} \\ | \\ \text{N=O} \end{array} \quad \longrightarrow \quad \begin{array}{c} \text{X} \\ \text{Y} \\ \text{N} \\ | \\ \text{O}^- \end{array}
$$

(6-86)

$$
\text{X=Y, N=N, N=C, N=O}
$$

heterocycles.[253] Benzimidazole N-oxides are prepared by treating o-nitroso-

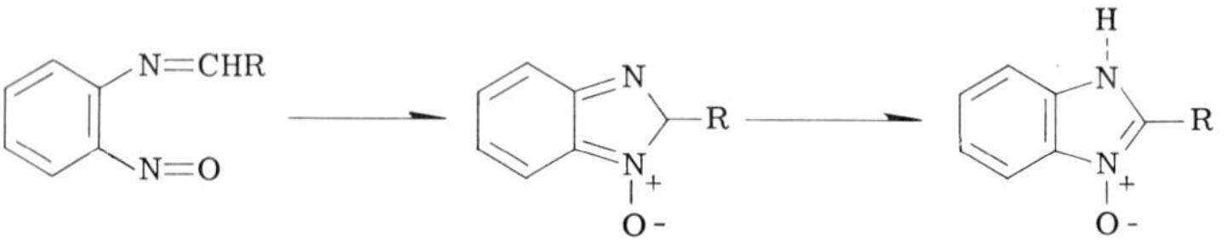

$$
\begin{array}{c} \text{N=CHR} \\ \text{N=O} \end{array} \quad \longrightarrow \quad \longrightarrow
$$

(6-87)

anilines with aldehydes.[254,255] The same reaction can be used to prepare purine oxides.[256] Triazole oxides result when an azo group interacts with the nitroso group. A number of examples of this reaction are listed by Katritzky and Lagowski.[253] The most thoroughly investigated case of these cyclizations is the formation of furoxans from o-dinitroso compounds.[253] Though the dinitroso form has been suggested as the stable one, nmr studies

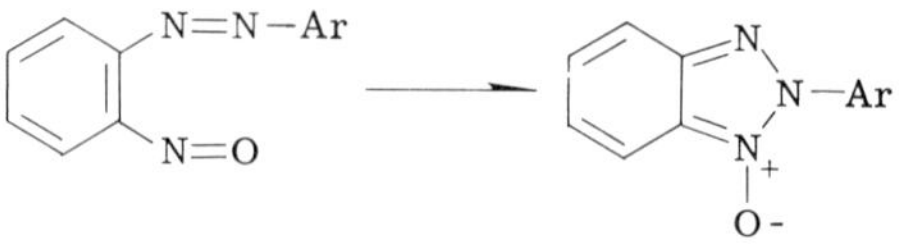

$$
\begin{array}{c} \text{N=N-Ar} \\ \text{N=O} \end{array} \quad \longrightarrow
$$

(6-88)

have indicated unequivocally that the dominant structure is unsymmetrical.[257–259] Despite this, unsymmetrically substituted benzofuroxans do not

$$(6\text{-}89)$$

exist as two isolable isomers, the two forms being very rapidly interconvertible.[258-264] The activation energy for the interconversion is of the order of

$$(6\text{-}90)$$

15–16 kcal/mol, varying slightly with substitution (cf. Table 6-2). Generally, substitution with an electron donating substituent [Eq. (6-90)] favors the isomer with X in the 5-position, and with an electron attracting group the favored isomer has X in the 6-position.[265] Generally ΔG° is of the order of a few hundred calories per mole. Arguments based on comparative energetics of possible intermediates in the interconversion process have implicated an o-dinitrosobenzene as the most likely intermediate, although this isomer has never been identified experimentally.[261]

$$(6\text{-}91)$$

TABLE 6-2

Activation Parameters for the Interconversion of Benzofuroxans

Compound	$\Delta H^\ddagger$	$\Delta S^\ddagger$	$\Delta G^\ddagger$	Reference
Benzofuroxan	15.5	5.1		263
4,7-Dichloro	15.9	4.0		261
4,7-Dibromo	15.7	5.0		261
5,6-Dichloro	14.4	6.0		261
5-Chloro			13.9	265
5-Methoxy			14.6	265
5-Acetoxy			13.8	265
6-Carboxy			14.0	265
6-Carbethoxy			14.0	265
5-Methyl-6-nitro			12.1	265
5-Acetamido			14.1	265

For non-annelated furoxans the interconversion process is slow and both isomers of an unsymmetrically substituted example are isolable. The kinetics of the reactions of Eq. (6-92) have been determined and the $\Delta H^{\ddagger}$ of ca. 34 kcal/mol indicates that interconversion will occur readily only at tempera-

$$\text{(6-92)}$$

R	$\Delta H^{\ddagger}$	$\Delta S^{\ddagger}$
Ph	33.4	+4
Et	34.5	+7

tures above 100°.[266] Again, arguments favoring the intermediacy of a 1,2-dinitroso compound have been advanced.[266]

REFERENCES

1. Carbene reactions including the present process have been quite thoroughly reviewed. See for example C. Wentrup, *Topics Curr. Chem.* **62**, 175 (1976); W. M. Jones and U. H. Brinker, *in* "Pericyclic Reactions" (A. P. Marchand and R. E. Lehr, eds.), Vol. I, pp. 110–198, Academic Press, New York, 1977.
2. H. Staudinger and R. Engle, *Chem. Ber.* **46**, 1437 (1913).
3. F. O. Rice and J. D. Michaelsen, *J. Phys. Chem.* **66**, 1535 (1962).
4. H. D. Harrison and F. P. Lossing, *J. Am. Chem. Soc.* **82**, 1052 (1960).
5. W. M. Jones, R. C. Joines, J. A. Myers, T. Mitsuhashi, K. E. Krajca, E. E. Waali, T. H. Davis, and A. B. Turner, *J. Am. Chem. Soc.* **95**, 826 (1973).
6. C. Wentrup, *Helv. Chim. Acta* **53**, 1459 (1970).
7. E. Hedaya and M. E. Kent, *J. Am. Chem. Soc.* **93**, 3283 (1971).
8. V. Franzen and H.-I. Joschek, *Justus Liebigs Ann. Chem.* **633**, 7 (1960).
9. C. Graebe and F. Ullman, *Justus Liebigs Ann. Chem.* **291**, 16 (1896).
10. R. Huisgen and M. Seidel, *Chem. Ber.* **94**, 2509 (1961).
11. P. A. S. Smith and B. B. Brown, *J. Am. Chem. Soc.* **73**, 2435 (1951).
12. P. A. S. Smith and J. H. Boyer, *J. Am. Chem. Soc.* **73**, 2626 (1951).
13. P. A. S. Smith, J. M. Clegg, and J. H. Hall, *J. Org. Chem.* **23**, 524 (1958).
14. G. Smolinsky, *J. Am. Chem. Soc.* **82**, 4717 (1960).
15. P. A. S. Smith and J. H. Hall, *J. Am. Chem. Soc.* **84**, 480 (1962).
16. P A. S. Smith, *in* "Nitrenes" (W. Lwowski, ed.), pp. 99–162. Wiley (Interscience), New York, 1970.
17. J. M. Lindley, I. M. McRobbie, O. Meth-Cohn, and H. Suschitzky, *Tetrahedron Lett.* p. 4513 (1976).
18. G. Smolinsky, *J. Am. Chem. Soc.* **83**, 2489 (1961).
19. J. H. Hall, F. E. Behr, and R. L. Reed, *J. Am. Chem. Soc.* **94**, 4952 (1972).
20. R. J. Sundberg, H. F. Russell, W. V. Ligon, Jr., and L.-S. Lin, *J. Org. Chem.* **37**, 719 (1972).

21. G. Smolinsky and C. A. Pryde, *J. Org. Chem.* **33,** 2411 (1968).

22. K. Isomura, S. Kobayashi, and H. Taniguchi, *Tetrahedron Lett.* p. 3499 (1968).

23. H. Hemetsberger, D. Knittel, and H. Weidmann, *Monatsh.* **101,** 161 (1970).

24. P. Gemeraad and H. W. Moore, *J. Org. Chem.* **39,** 774 (1974).

25. K. Isomura, K. Uto, and H. Taniguchi, *Chem. Commun.* p. 664 (1977).

26. A. Padwa and P. H. J. Carlsen, *Tetrahedron Lett.* p. 433 (1978).

27. T. Nishiwaki, *Chem. Commun.* p. 565 (1972).

28. J. H. Bowie and B. Nussey, *Chem. Commun.* p. 1565 (1970).

28a. J. H. Bowie and B. Nussey, *J. Chem. Soc. Perkin Trans. 1* p. 1693 (1973).

29. T. L. Gilchrist, G. E. Gymer, and C. W. Rees, *J. Chem. Soc. Perkin Trans. 1* p. 555 (1973).

30. A. Padwa, J. Smolanoff, and A. Tremper, *Tetrahedron Lett.* p. 29 (1974); *J. Am. Chem. Soc.* **97,** 4682 (1975).

31. A. Padwa, J. Smolanoff, and A. Tremper, *J. Org. Chem.* **41,** 543 (1976).

32. J. H. Hall and D. R. Kamm, *J. Org. Chem.* **30,** 2092 (1965).

33. L. K. Dyall and T. E. Kemp, *J. Chem. Soc. B* p. 976 (1968).

34. L. Krbechek and H. Takimoto, *J. Org. Chem.* **29,** 1150 (1969).

35. E. W. Fowler, A. Hassner, and L. A. Levy, *J. Am. Chem. Soc.* **89,** 2077 (1967).

36. S. Sato, *Bull. Chem. Soc. Japan* **41,** 2524 (1968).

37. D. J. Anderson and A. Hassner, *J. Org. Chem.* **38,** 2565 (1973).

38. B. Singh and E. F. Ullman, *J. Am. Chem. Soc.* **89,** 6911 (1967).

39. M. Maeda and M. Kojima, *Chem. Commun.* p. 539 (1973).

40. H. Alper, J. E. Prickett, and S. Wollowitz, *J. Am. Chem. Soc.* **99,** 4330 (1977).

41. P. L. Coe, A. E. Jukes, and J. C. Tatlow, *J. Chem. Soc. C* p. 2020 (1966).

42. R. Kwok and P. Pranc, *J. Org. Chem.* **33,** 2880 (1968).

43. T. Nishiwaki, *Tetrahedron Lett.* p. 2049 (1969).

44. T. Nishiwaki, T. Kitamura, and A. Nakano, *Tetrahedron* **26,** 453 (1970).

45. T. Nishiwaki, A. Nakano, and H. Matsuoka, *J. Chem. Soc. C* p. 1825 (1970).

46. G. Adembri, A. Camparini, F. Ponticelli, and P. Tedeschi, *J. Chem. Soc. Perkin Trans. 1* p. 971 (1977).

47. T. Nishiwaki, *Synthesis* p. 20 (1975).

48. J. Vaugh and P. A. S. Smith, *J. Org. Chem.* **23,** 1909 (1958).

49. P. A. S. Smith and E. Leon, *J. Am. Chem. Soc.* **80,** 4647 (1958).

50. T. L. Gilchrist, C. J. Moody, and C. W. Rees, *Chem. Commun.* p. 414 (1976).

51. T. Bachetti and A. Alemagna, *Atti. Acad. Lincei Rend. Ch. Sci. Fis Mat. Natur.* **22,** 637 (1957); **28,** 824 (1960).

52. G. Wittig, F. Bangert, and H. Kleiner, *Chem. Ber.* **66,** 1140 (1928).

53. M. A. Battiste, B. Halton, and R. H. Grubbs, *Chem. Commun.* p. 907 (1967).

54. A. Padwa, T. J. Blacklock, O. Getman, N. Hatanaka, and R. Loza, *J. Org. Chem.* **43,** 1481 (1978).

55. A. Padwa and P. H. J. Carlsen, *J. Org. Chem.* **43,** 2027 (1978).

56. S. W. Staley, *in* "Pericyclic Reactions" (Lehr and Marchand, eds.), Vol. 1, pp. 199–264, Academic Press, New York, 1977.

57. D. J. Atkinson, M. J. Perkins, and P. Ward, *Chem. Commun.* p. 1390 (1969).

58. M. J. Perkins and P. Ward, *J. Chem. Soc. Perkin Trans. 1* p. 667 (1974).

59. H. Kloosterziel, J. A. A. van Drunen, and P. Golama, *Chem. Commun.* p. 885 (1969).

60. H. Kloosterziel and J. A. A. van Drunen, *Tetrahedron Lett.* p. 1023 (1973).

61. A. E. M. Beyer and H. Kloosterziel, *Rec. Trav. Chim. Pays-Bas* **96,** 178 (1977).

62. R. B. Bates, D. W. Grosselink, and J. A. Kaczynski, *Tetrahedron Lett.* p. 205 (1967).

63. H. Kloosterziel and J. A. A. van Drunen, *Rec. Trav. Chim. Pays-Bas* **89,** 270 (1970).

64. D. H. Hunter, S. K. Sim, and R. P. Steiner, *Can. J. Chem.* **55,** 1229 (1977).

65. R. B. Bates, W. H. Deines, D. A. McCombs, and D. E. Potter, *J. Am. Chem. Soc.* **91,** 4608 (1969).
66. P. R. Stapp and R. F. Kleinschmidt, *J. Org. Chem.* **30,** 3006 (1965).
67. L. H. Slaugh, *J. Org. Chem.* **32,** 108 (1967).
68. R. B. Bates and D. A. McCombs, *Tetrahedron Lett.* p. 977 (1969).
69. S. Brenner and J. Klein, *Israel J. Chem.* **7,** 735 (1969).
70. C. W. Shoppee and G. N. Henderson, *Chem. Commun.* p. 561 (1974).
71. C. W. Shoppee and G. N. Henderson, *J. Chem. Soc. Perkin Trans. 1* p. 765 (1975).
72. C. W. Shoppee and G. N. Henderson, *J. Chem. Soc. Perkin Trans. 1* p. 1028 (1977).
73. D. H. Hunter and S. K. Sim, *J. Am. Chem. Soc.* **91,** 6202 (1969).
73a. D. H. Hunter and S. K. Sim, *Can. J. Chem.* **50,** 678 (1972).
74. M. A. Laurent, *C. R. Acad. Sci. Paris* **19,** 353 (1844).
75. D. H. Hunter and R. P. Steiner, *Can. J. Chem.* **53,** 355 (1975).
76. R. R. Schmidt, *Angew Chem. Int. Edit. Engl.* **10,** 572 (1971).
77. R. R. Schmidt, W. J. W. Mayer, and H. U. Wagner, *Justus Leibigs Ann. Chem.* p. 2010 (1973).
78. R. R. Schmidt, *Angew. Chem. Int. Edit. Engl.* **14,** 581 (1975).
79. R. R. Schmidt and M. Dimmler, *Chem. Ber.* **108,** 6 (1975).
80. R. R. Schmidt, U. Burkert, and R. Prevo, *Tetrahedron Lett.* p. 3477 (1975).
81. J. Elguero, *Bull. Soc. Chim. Fr.* p. 1925 (1971).
82. G. Coispeau and J. Elguero, *Bull. Soc. Chim. Fr.* p. 2717 (1970).
83. G. Coispeau, J. Elguero, P. Jacquier, and D. Tizane, *Bull. Soc. Chim. Fr.* p. 1581 (1970).
84. H. Ferres and W. R. Jackson, *Chem. Commun.* p. 261 (1969).
85. J.-P. Chapell, J. Elguero, R. Jacquier, and G. Tarrago, *Bull. Soc. Chim. Fr.* p. 3147 (1970).
86. J. Elguero, R. Jacquier, and C. Marzin, *Bull. Soc. Chim. Fr.* p. 4119 (1970).
87. J. P. Freeman, D. L. Surbey, and J. E. Kassner, *Tetrahedron Lett.* p. 3797 (1970).
88. F. C. Heugebarth and J. F. Willems, *Tetrahedron* **22,** 913 (1966).
89. R. Grashey, R. Huisgen, K. K. Sun, and R. M. Moriarty, *J. Org. Chem.* **30,** 74 (1965).
90. R. Huisgen, *Proc. Chem. Soc.* p. 357 (1961).
91. R. Huisgen, *Angew. Chem. Int. Edit. Engl.* **2,** 565, 633 (1963).
92. R. Huisgen, R. Grashey, and J. Sauer, "Chemistry of Alkenes" (S. Patai, ed.) pp. 806–878, Wiley (Interscience), New York, 1964.
93. R. Huisgen, *Angew Chem.* **72,** 359 (1960).
94. H. Riemlinger, *Chem. Ber.* **103,** 1900 (1970).
95. J. C. Pommelet, N. Manisse, and J. Chuche, *C. R. Acad. Sci. Paris, Ser. C.* **270,** 1894 (1970).
95a. J. C. Pommelet, N. Manisse, and J. Chuche, *Tetrahedron* **28,** 3929 (1972).
96. R. J. Crawford, S. B. Lutener, and R. D. Cockroft, *Can. J. Chem.* **54,** 3364 (1976).
97. R. J. Crawford, V. Vukov, and H. Tokunaga, *Can. J. Chem.* **51,** 3718 (1973).
98. J. C. Palladini and R. J. Crawford, *Can. J. Chem.* **52,** 2098 (1974).
99. J. C. Palladini and J. Chuche, *Tetrahedron Lett.* p. 4383 (1971).
99a. J. C. Palladini and J. Chuche, *Bull. Soc. Chim. Fr.* p. 197 (1974).
100. M. S. Medinagh and J. Chuche, *Tetrahedron Lett.* p. 793 (1977).
101. N. Manisse and J. Chuche, *Tetrahedron* **33,** 2399 (1977).
102. W. Eberbach and U. Trostmann, *Tetrahedron Lett.* p. 3569 (1977).
103. E. F. Pratt, R. W. Luckenbaugh, and R. L. Erickson, *J. Org. Chem.* **19,** 176 (1954).
104. E. F. Pratt, R. G. Rice, and R. W. Luckenbaugh, *J. Am. Chem. Soc.* **79,** 1212 (1957).
105. E. F. Pratt and J. C. Keresztesy, *J. Org. Chem.* **32,** 49 (1967).
106. W. Augenstein and F. Kröhnke, *Justus Liebigs Ann. Chem.* **697,** 158 (1966).
107. E. Pohjala, *Tetrahedron Lett.* p. 2585 (1972).

108. T. Sasaki, K. Kanematsu, A. Kakehi, and G. Ito, *Tetrahedron* **28,** 4947 (1972).
109. Y. Tamura, N. Tsujimoto, Y. Sumida, and M. Ikeda, *Tetrahedron* **28,** 21 (1972).
110. T. Sasaki, K. Kanematsu, A. Kakehi, and G. Ito, *J. Chem. Soc. Perkin Trans. 1* p. 2089 (1973).
111. Y. Tamura, Y. Sumeda, and M. Ikeda, *J. Chem. Soc. Perkin Trans. 1* p. 2091 (1973).
112. T. Uchida and K. Matsumoto, *Synthesis* p. 209 (1976).
113. Y. Tamura, Y. Sumida, S. Haruki, and M. Ikeda, *J. Chem. Soc. Perkin Trans. 1* p. 575 (1975).
114. E. J. Smutny, *J. Am. Chem. Soc.* **91,** 208 (1969).
115. A. Padwa, M. Dharan, J. Smolanoff, and S. I. Wetmore, *J. Am. Chem. Soc.* **94,** 1395 (1972); **95,** 1945, 1954 (1973).
116. J. E. Baldwin, R. G. Pudussery, A. K. Gureshi, and B. Sklarz, *J. Am. Chem. Soc.* **90,** 5325 (1968).
117. A. Padwa and W. Eisenhardt, *Chem. Commun.* p. 380 (1968).
118. J. W. Conforth, *in* "The Chemistry of Penicillin," p. 700, Princeton Univ. Press, Princeton, New Jersey, 1949.
118a. J. W. Conforth and E. Cookson, *J. Chem. Soc.* p. 1086 (1952).
119. C. G. Stuckwisch and D. D. Powers, *J. Org. Chem.* **25,** 1819 (1960).
120. M. J. S. Dewar, P. A. Spanninger, and I. J. Turchi, *Chem. Commun.* p. 925 (1973).
121. For a brief review see H. C. van der Plas, "Ring Transformations of Heterocycles," Vol. I p. 298, Academic Press, New York, 1973.
122. G. Höfle and W. Steglich, *Chem. Ber.* **104,** 1408 (1971).
123. A. Padwa and E. Chen, *J. Org. Chem.* **39,** 1976 (1974).
124. G. L. Aldous, J. H. Bowie, and M. J. Thompson, *J. Chem. Soc. Perkin Trans. 1* p. 16 (1976).
125. K. L. Davies, R. C. Storr, and P. J. Whittle, *Chem. Commun.* p. 9 (1978).
126. Y. Tamura, N. Tsujimoto, and M. Ikeda, *Chem. Commun.* p. 310 (1971).
127. Y. Tamura, Y. Miki, Y. Sumida, and M. Ikeda, *J. Chem. Soc. Perkin Trans. 1* p. 2580 (1973).
128. T. Sasaki, K. Kanematsu, and A. Kakehi, *Tetrahedron Lett.* p. 5245 (1972).
129. T. Sasaki, K. Kanematsu, and A. Kakehi, *J. Org. Chem.* **37,** 3106 (1972).
130. D. W. Adamson and J. W. Kenner, *J. Chem. Soc.* p. 286 (1935).
131. C. D. Hurd and S. C. Liu, *J. Am. Chem. Soc.* **57,** 2656 (1935).
132. G. L.'Abbé and G. Mathys, *J. Org. Chem.* **39,** 1778 (1974).
133. A. Ledwith and D. Parry, *J. Chem. Soc. B* p. 41 (1967).
134. J. L. Brewbaker and H. Hart, *J. Am. Chem. Soc.* **91,** 711 (1969).
135. A. Dornow and W. Bartsch, *Justus Liebigs Ann. Chem.* **602,** 23 (1957).
136. See footnote in refer. 135.
137. G. L. Closs, L. E. Closs, and W. A. Böll, *J. Am. Chem. Soc.* **85,** 3796 (1963).
138. G. L. Closs, W. A. Böll, H. Heyn, and V. Dev, *J. Am. Chem. Soc.* **90,** 173 (1968).
139. J. T. Sharp, R. H. Findlay, and P. B. Thorogood, *Chem. Commun.* p. 909 (1970).
139a. J. T. Sharp, R. H. Findlay, and P. B. Thorogood, *J. Chem. Soc. Perkin Trans. 1* p. 102 (1975).
140. J. Dingwal and J. T. Sharp, *Chem. Commun.* p. 128 (1975).
141. R. Grandi, W. Messerotti, U. M. Pagnoni, and R. Trave, *J. Org. Chem.* **42,** 1352 (1977).
142. J. N. Done, J. H. Knox, R. McEwan, and J. T. Sharp, *Chem. Commun.* p. 532 (1974).
143. R. K. Bartlett and T. S. Stevens, *J. Chem. Soc. C.* p. 1964 (1967).
144. W. Reichen, *Helv. Chim. Acta* **59,** 1636 (1976).
145. C. Wentrup as noted in footnote 2 refer. 144.
146. E. M. Burgess and H. R. Penton, *J. Org. Chem.* **39,** 2885 (1974).

147. T. Saito and S. Motoki, *J. Org. Chem.* **42**, 3922 (1977).
148. J. P. Freeman, D. L. Surbey, and J. E. Kassner, *Tetrahedron Lett.* p. 3797 (1970).
149. L. Wolff, *Justus Liebigs Ann. Chem.* **325**, 129 (1902).
150. O. Dimroth, *Justus Liebigs Ann. Chem.* **373**, 336 (1910).
151. O. Dimroth, *Justus Liebigs Ann. Chem.* **377**, 127 (1910).
152. O. Dimroth, *Justus Liebigs Ann. Chem.* **399**, 91 (1913).
153. B. R. Brown and D. L. Hammick, *J. Chem. Soc.* p. 1384 (1947).
154. T. Curtius and J. Thompson, *Chem. Ber.* **39**, 4140 (1906).
155. M. Regitz, *Tetrahedron Lett.* p. 3287 (1965); M. Regitz and H. Schwall, *Justus Liebigs Ann. Chem.* **728**, 99 (1969).
156. M. E. Hermes and F. D. March, *J. Am. Chem. Soc.* **89**, 4760 (1967).
157. R. E. Harmon, F. Stanley, Jr., S. K. Gupta, and J. Johnson, *J. Org. Chem.* **35**, 3444 (1970).
158. J. E. Leffler and S.-K. Liu, *J. Am. Chem. Soc.* **78**, 1949 (1956).
159. C. Wentrup, *Helv. Chim. Acta* **61**, 1755 (1978).
160. O. Dimroth, *Justus Liebigs Ann. Chem.* **364**, 183 (1909).
161. O. Dimroth and W. Michaelis, *Justus Liebigs Ann. Chem.* **459**, 39 (1927).
162. B. R. Brown, D. L. Hammick, and S. G. Heritage, *J. Chem. Soc.* p. 3820 (1953).
163. E. Lieber, C. N. R. Rao, and T. S. Chao, *J. Am. Chem. Soc.* **79**, 5962 (1957).
164. E. Lieber, T. S. Chao, and C. N. R. Rao, *J. Org. Chem.* **22**, 654 (1957).
165. G. Tennant and R. J. S. Ververs, *Chem. Commun.* p. 671 (1974).
166. T. Kindt-Larsen and C. Pedersen, *Acta Chem. Scand.* **16**, 1800 (1962).
167. J. Goerdeler and G. Gnad, *Chem. Ber.* **99**, 1618 (1966).
168. M. Regitz and H. Scherer, *Chem. Ber.* **102**, 417 (1964).
169. D. J. Brown and M. N. Paddon-Row, *J. Chem. Soc. C* p. 1856 (1967).
170. A. Albert, *J. Chem. Soc. C* p. 152 (1969).
171. M. Regitz and A. Liedhegener, *Justus Liebigs Ann. Chem.* **710**, 118 (1967).
172. R. Huisgen, J. Sauer, and H. J. Sturm, *Angew Chem.* **70**, 272 (1958).
173. R. Huisgen, J. Sauer, and M. Seidel, *Chem. and Ind. (London)* p. 1114 (1958).
174. W. Lwowski, A. Hartenstein, C. de Vita, and R. L. Smick, *Tetrahedron Lett.* p. 2497 (1964).
175. R. Puttner and K. Hafner, *Tetrahedron Lett.* p. 3119 (1964).
176. R. Huisgen and H. Blaschke, *Justus Liebigs Ann. Chem.* **686**, 145 (1965).
177. R. Huisgen, H. J. Sturm, and M. Seidel, *Chem. Ber.* **94**, 1555 (1961).
178. R. Huisgen, J. Sauer, H. J. Sturm, and J. H. Markgraf, *Chem. Ber.* **93**, 2106 (1960).
179. J. Sauer, R. Huisgen, and H. J. Sturm, *Tetrahedron* **11**, 241 (1960).
180. R. Huisgen and M. Seidel, *Chem. Ber.* **94**, 2509 (1961).
181. R. Huisgen, J. Sauer, and M. Seidel, *Chem. Ber.* **93**, 2885 (1960).
182. R. Huisgen, C. Axen, and H. Seidl, *Chem. Ber.* **98**, 2966 (1965).
183. See refer. 121, pp. 394–402.
184. R. N. Butler, *Adv. Heterocyclic Chem.* **21**, 365 (1977).
185. M. S. Gibson, *Tetrahedron* **18**, 1377 (1962).
186. M. S. Gibson, *Tetrahedron* **19**, 1587 (1963).
187. R. M. Herbst, *J. Org. Chem.* **26**, 2372 (1961).
188. C. Arnold, Jr. and D. N. Thatcher, *J. Org. Chem.* **34**, 1141 (1969).
189. A. Konnecke, R. Dörre, and E. Lippmann, *Tetrahedron Lett.* p. 2071 (1978).
190. H. Riemlinger, J. J. M. Vandewalle, G. S. D. King, W. R. F. Lingier, and R. Merenyi, *Chem. Ber.* **103**, 1918 (1970).
191. A. Könnecke and E. Lippmann, *Tetrahedron Lett.* p. 2187 (1977).
192. L. A. Burke, G. Leroy, M. T. Nguyen, and M. Sana, *J. Am. Chem. Soc.* **100**, 3668 (1978).
193. J. S. Meek and J. S. Fowler, *J. Am. Chem. Soc.* **89**, 1967 (1967).
193a. J. S. Meek and J. S. Fowler, *J. Org. Chem.* **33**, 985 (1968).

194. A. N. Nesmayanov and M. I. Rybinskaya, *Dokl. Akad. Nauk. SSSR* **166,** 1362 (1966).
195. F. P. Woerner and H. Riemlinger, *Chem. Ber.* **103,** 1908 (1970).
196. Y. Inagaki, R. Okazaki, and N. Inamoto, *Tetrahedron Lett.* p. 4575 (1975).
197. O. Dimroth, *Justus Liebigs Ann. Chem.* **335,** 1 (1904).
198. T. Curtius and J. Thompson, *Chem. Ber.* **39,** 4143 (1906).
199. C. Pedersen, *Acta Chem. Scand.* **12,** 1236 (1958).
200. W. Lwowski, *in* "Chemistry of the Azido Group" (S. Patai, ed.), pp. 507–520, Wiley (Interscience), New York, 1971.
201. L. Carlsen, A. Holm, J. P. Snyder, E. Koch, and B. Stilkerieg, *Tetrahedron* **33,** 2221 (1977).
202. L. A. Burke, J. Elguero, G. Leroy, and M. Sana, *J. Am. Chem. Soc.* **98,** 1685 (1976).
203. M. Tisler, *Synthesis* p. 123 (1973).
204. R. N. Butler, *Chem. Ind. (London)* p. 371 (1973).
205. C. Wentrup, *Tetrahedron* **26,** 4969 (1970).
206. A. Pollak, S. Polanc, B. Stanovnik, and M. Tisler, *Monatsh.* **103,** 1591 (1972).
207. A. Messmer, Gy. Hajos, P. Benko, and L. Pallos, *J. Heterocycl. Chem.* **10,** 575 (1973).
208. V. P. Krivopalov and V. P. Mamaev, *Izv. Akad. Nauk. SSSR Ser Khim.* p. 966 (1977).
209. A. F. Hegarty, K. J. Dignam, and D. M. Hickey, *Tetrahedron Lett.* p. 2121 (1978).
210. J. Plenkiewicz, *Tetrahedron Lett.* p. 341 (1975).
211. K. J. Dignam, A. F. Hegarty, and P. L. Quain, *J. Org. Chem.* **43,** 388 (1978).
212. F. Eloy, *J. Org. Chem.* **26,** 953 (1961).
213. E. Alcade, J. de Mendoza, and J. Elguero, *J. Heterocycl. Chem.* **11,** 921 (1974).
214. E. Alcade and R. M. Claramunt, *Tetrahedron Lett.* p. 1523 (1975).
215. E. Alcade and J. de Mendoza, *Chem. Commun.* p. 411 (1974).
216. R. Faure, J.-P. Galy, E.-J. Vincent, J.-P. Fayet, P. Mauret, M.-C. Vertut, and J. Elguero, *Can. J. Chem.* **55,** 1728 (1977).
217. K. Clusius and M. Vecchi, *Helv. Chim. Acta* **39,** 1469 (1956).
218. I. Ugi and R. Huisgen, *Chem. Ber.* **90,** 2914 (1957).
219. I. Ugi, H. Perlinger, and L. Behringer, *Chem. Ber.* **91,** 2324 (1958).
220. I. Ugi and R. Huisgen, *Chem. Ber.* **91,** 531 (1958).
221. C. D. Ritchie and D. J. Wright, *J. Am. Chem. Soc.* **93,** 2429 (1971).
222. I. Ugi, *Adv. Heterocycl. Chem.* **3,** 380 (1964).
223. I. Ugi, *Tetrahedron* **19,** 1801 (1963).
224. N. Zinin, *Chem. Ber.* **5,** 1104 (1872).
225. F. R. Japp and F. Klingemann, *J. Chem. Soc.* **57,** 669 (1890).
226. A. H. Blatt, *J. Org. Chem.* **15,** 869 (1950).
227. S. Lahiri, V. Dabral, M. P. Mahajan, and M. V. George, *Tetrahedron* **33,** 3247 (1977).
228. G. Maier and G. Wiessler, *Tetrahedron Lett.* p. 4987 (1969).
229. A. H. Blatt, *J. Am. Chem. Soc.* **56,** 2774 (1934).
230. J. P. Griffiths and C. K. Ingold, *J. Chem. Soc.* **127,** 1698 (1925).
231. R. L. Stern and J. G. Krause, *J. Org. Chem.* **33,** 212 (1968).
232. R. L. Stern and J. G. Krause, *J. Heterocycl. Chem.* **5,** 263 (1968).
233. P. Freche, A. Gorgues, and E. Levas, *Tetrahedron Lett.* p. 1495 (1976).
234. P. Freche, A. Gorgues, and E. Levas, *Tetrahedron* **33,** 2069 (1977).
235. K. Burger, W. Thenn, and A. Gieren, *Angew Chem. Int. Edit. Engl.* **13,** 474 (1974).
236. A. Gieren, P. Narayanan, K. Burger, and W. Thenn, *Angew Chem. Int. Edit. Engl.* **13,** 475 (1974).
237. K. Burger, W. Thenn, R. Rauh, H. Schickaneder, and A. Gieren, *Chem. Ber.* **108,** 1460 (1975).
238. K. Burger, H. Schickaneder, and A. Meffert, *Z. Naturforsch. B: Anorg. Chem., Org. Chem.* **30B,** 622 (1975).
239. K. Burger, H. Schickaneder and W. Thenn, *Tetrahedron Lett.* p. 1125 (1975).

240. S. Evans, R. C. Gearhart, L. J. Guggenberger, and E. E. Schweizer, *J. Org. Chem.* **42,** 452 (1977).

241. F. Yoneda, M. Higuchi, and T. Nagamatsu, *J. Am. Chem. Soc.* **96,** 5607 (1974).

242. F. Yoneda, T. Nagamatsu, T. Nagamura, and K. Senga, *J. Chem. Soc. Perkin Trans. 1* p. 765 (1977).

243. H. von Pechmann, *Chem. Ber.* **21,** 2751 (1888).

244. R. Stolle, *Chem. Ber.* **59,** 1742 (1926).

245. P. Grammaticakis, *C. R. Acad. Sci., Paris Ser. C.* **224,** 1509 (1947).

246. H. Bauer, G. R. Bedford, and A. R. Katritzky, *J. Chem. Soc.* p. 751 (1964).

247. K. B. Sukumaran, C. S. Angadiyavar, and M. V. George, *Tetrahedron Lett.* p. 633 (1971).

247a. K. B. Sukumaran, C. S. Angadiyavar, and M. V. George, *Tetrahedron* **28,** 3987 (1972).

248. K. B. Sukumaran, S. Satish, and M. V. George, *Tetrahedron* **30,** 445 (1974).

249. D. Y. Curtin and N. E. Alexandrou, *Tetrahedron* **19,** 1697 (1963).

250. S. Petersen and H. Heitzer, *Angew Chem. Int. Edit. Engl.* **9,** 67 (1970).

251. H. Bauer, A. J. Boulton, W. Fedeli, A. R. Katritzky, A. Majid-Hamid, F. Mazza, and A. Vaciago, *Angew Chem. Int. Edit. Engl.* **10,** 129 (1971).

252. C. S. Angadiyavar and M. V. George, *J. Org. Chem.* **36,** 1589 (1971).

253. A. R. Katritzky and J. M. Lagowski, "Chemistry of the Heterocyclic N-Oxides," pp. 106–107, 111–118, Academic Press, New York, 1971.

254. D. W. Russell, *Chem. Commun.* p. 498 (1965).

255. A. R. Katritzky, B. J. Ridgewell, and A. M. White, *Chem. Ind. London* p. 1576 (1964).

256. E. C. Taylor and E. E. Garcia, *J. Am. Chem. Soc.* **86,** 4721 (1964).

257. F. B. Mallory and C. S. Wood, *Proc. Nat. Acad. Sci. USA* **47,** 697 (1961).

258. G. Englert, *Z. Naturforsch. B* **16,** 413 (1961).

258a. G. Englert, *Z. Elektrochem.* **65,** 854 (1961).

259. R. K. Harris, A. R. Katritzky, S. Øksene, A. S. Bailey, and W. G. Patterson, *J. Chem. Soc.* p. 197 (1963).

260. P. Diehl, H. A. Christ, and F. B. Mallory, *Helv. Chim. Acta* **45,** 504 (1962).

261. F. B. Mallory, S. L. Manatt, and C. S. Wood, *J. Am. Chem. Soc.* **87,** 5433 (1965).

262. A. J. Boulton, P. B. Ghosh, and A. R. Katritzky, *J. Chem. Soc. B* p. 1011 (1966).

263. K.-I. Dahlquist and S. Forsen, *J. Magnet. Reson.* **2,** 61 (1970).

264. F. A. L. Anet and I. Yavari, *Org. Magnet. Reson.* **8,** 158 (1976).

265. A. J. Boulton, A. R. Katritzky, M. J. Sewell, and B. Wallis, *J. Chem. Soc. B* p. 914 (1967).

266. F. B. Mallory and A. Cammarata, *J. Am. Chem. Soc.* **88,** 61 (1966).

267. Y. Tominaga, Y. Miyake, H. Fujito, K. Kurata, H. Awaya, Y. Matsuda, and G. Kobayashi, *Chem. Pharm. Bull.* **25,** 1528 (1977).

268. A. Kakehi, S. Ito, T. Maeda, R. Takeda, M. Nishimura, and T. Yamaguchi, *Chem. Lett.* p. 59 (1978).

269. A. Kakehi, S. Ito, K. Uchiyama, and Y. Konno, *Chem. Lett.* p. 413 (1976).

270. A. Kakehi, S. Ito, K. Uchiyama, and K. Kondo, *Chem. Lett.* p. 545 (1977).

271. T. Toda, H. Morino, V. Suzuki, and T. Makai, *Chem. Lett.* p. 155 (1977).

272. A. Kakehi, S. Ito, K. Uchiyama, and K. Kondo, *J. Org. Chem.* **43,** 2896 (1978).

273. M. I. Komendantov, R. R. Bekmukhametov, and I. N. Domnin, *Tetrahedron* **34,** 2743 (1978).

274. T. L. Gilchrist, C. J. Moody, and C. W. Rees, *J. Chem. Soc. Perkin Trans. 1* p. 1871 (1979).

275. W. Sucrow, M. Slopianka, and A. Neophyton, *Chem. Ber.* **105,** 2143 (1972).

276. V. Bardakos, W. Sucrow, and A. Fehlauer, *Chem. Ber.* **108,** 2161 (1975).

277. J.-B. Cazaux, R. Jacquier, and G. Maury, *Bull. Soc. Chim. Fr.* p. 255 (1976).

278. A. Kakehi, S. Ito, T. Maeda, R. Takeda, M. Nishimura, M. Tamashima, and T. Yamaguchi, *J. Org. Chem.* **43,** 4837 (1978).

279. D. Borel, Y. Gelas-Mialhe, and R. Yessiere, *Can. J. Chem.* **54,** 1590 (1976).

280. J. J. Barr and R. C. Storr, *J. Chem. Soc. Perkin Trans. 1* p. 192 (1979).

281. S. Ito, Y. Tanaka, A. Kakehi, and K. Kondo, *Bull. Soc. Chem. Japan* **49,** 1920 (1976).

282. A. F. Hegarty, K. Brady, and M. Mullane, *Chem. Commun.* p. 871 (1978).

283. J. J. Barr and R. C. Storr, *J. Chem. Soc. Perkin Trans. 1* p. 185 (1979).

284. E. E. Schweitzer and S. Evans, *J. Org. Chem.* **43,** 4328 (1978).

CHAPTER 7
SIX ELECTRON–SIX ATOM SYSTEMS

The basic reaction of the six electron–six atom type, hexatriene to cyclo-hexadiene, is a modern reaction, first realized for this case in 1964.[1] Despite this many of the reactions which belong in this class were known early in this century, as for example, the ring opening reactions of pyridinium ions. Mechanistic detail, however, belongs to the Woodward–Hoffmann era.

I. cis-HEXATRIENES–CYCLOHEXADIENES

Predicted by Woodward and Hoffmann to be disrotatory, this electrocyclic process is revealed as a much more complex affair than the oft used analogy to the Diels–Alder reaction would lead one to expect. It is unfortunate, there-fore, that it has attracted so little attention from the theorists.

A. Theory

Lewis and Steiner[1] followed their experimental study of the rate of cycliza-tion of cis-hexatriene with a theoretical calculation of the activation param-eters. They used the method of Evans and Polanyi[2] but since the work preceded Woodward and Hoffmann's predictions, they chose a conrotatory description. Their calculation gave a reasonable activation energy (32–37 kcal/mol), even though the calculation relates to a symmetry forbidden route. Since the authors did not explicitly treat the question of delocalization in the transition state, the symmetry forbiddeness did not play a role.

Marvell[3] used a modified Hendrickson calculation in an attempt to define the transition state geometry for the allowed reaction and to ascertain the extent to which various factors influence the activation energy. A similar calculation for the forbidden reaction was also carried through. Neither process gave a transition state, the energy surface running smoothly downhill from hexatriene to cyclohexadiene. Possible reasons why this calculation did not lead to an energy barrier were considered and an artificial method of inducing a transition state was developed which led to interesting conclusions. The allowed transition state was ca. 14 kcal/mol lower in energy than the conrotatory state, but the difference resulted mainly from the difference in bond energies of the forming single bond. Delocalization energy was larger for the forbidden transition state. The disrotatory transition state had a boat-like ring with the C_1-C_2 distance of 2.0 Å, and the two cis-hydrogens of the terminal atoms (C_1, C_6) were separated by less than the sum of the van der Waal's radii.

Subsequently Komornicki and McIver[4] carried out a more sophisticated study of the allowed reaction using the MINDO/2 method. The $\Delta H^{\ddagger}$ obtained was ca. 5 kcal/mol lower than the experimental value of Lewis and Steiner,[1] but the transition state geometry was remarkably similar to that obtained by Marvell. In both cases the $C_1 C_2 C_3 C_4$ dihedral angle is close to $43°$ ($44°$ Marvell, $42°$ McIver), which suggests that some measure of radical character should be developed at C_2 and C_5 in the transition state. Both calculations indicate that severe non-bonded interactions should be developed between the substituents cis on the terminal double bonds of the triene if these are larger than hydrogen.

Tee and Yates[5] made a general study of the routes for cyclization of a series of electrocyclic reactions based on the principle of least motion. The results for the cyclopropyl cation–allyl cation reaction were in accord with the DePuy modified Woodward–Hoffmann disrotatory reaction mode. However, the hexatriene–cyclohexadiene reaction was found to constitute an exception to the direct correlation between symmetry conservation and least motion. The least motion principle clearly predicts a conrotatory process, and thus is not generally equivalent to the orbital symmetry conservation principle as a predictive tool.

Epiotis[6] has indicated that for substituent effects on the rate of hexatriene cyclization, the reaction may be considered an internal Diels–Alder reaction and an accumulation of donor substituents on one end and acceptor on the other end of the triene will enhance the rate. A rather simpler model was employed by Carpenter[7] who predicted that polar and conjugating substituents will *decrease* the rate of cyclization independent of the substitution position.

B. Stereochemistry

Contrary to all predictions based on steric effects, minimum disruption of the π system or least motion, the conservation of orbital symmetry indicates the electrocyclization should proceed in disrotatory fashion. Thus this case provides a specially demanding test of that theory. To be useful as a test case the reaction must be concerted and the stereochemistry of both reactant and product should be completely defined by methods independent of the orbital symmetry rule.

Of the examples which involve acyclic trienes the first useful case is the thermal conversion of vitamin D to a mixture of pyro-(143) and isopyro-calciferols (144) at 188°.[8] The stereochemistry of the thermal products was not completely determined until 1959.[9,10] Clearly the reaction of Eq. (7-1)

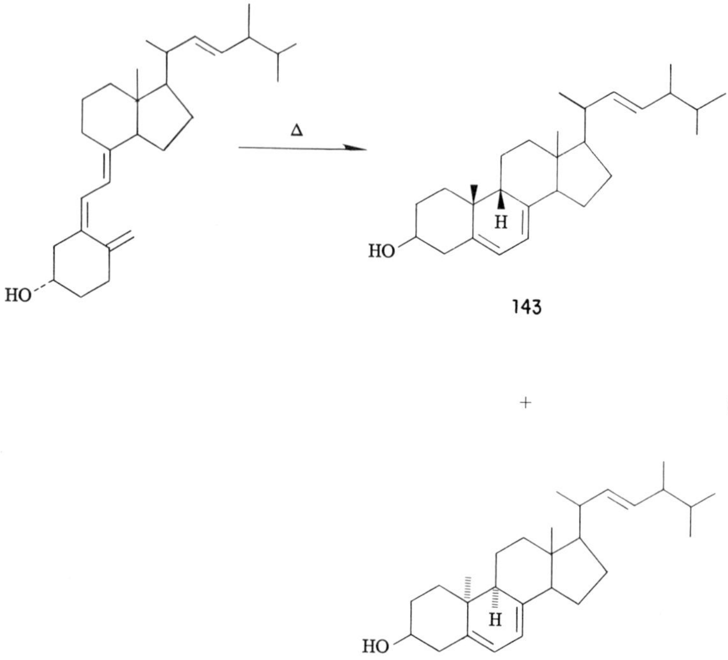

143

+   (7-1)

144

does not occur directly and the intermediacy of precalciferol[11] must be assumed. Given that assumption then this process constitutes reasonable evidence for the disrotatory reaction,[12] if the reaction is concerted.

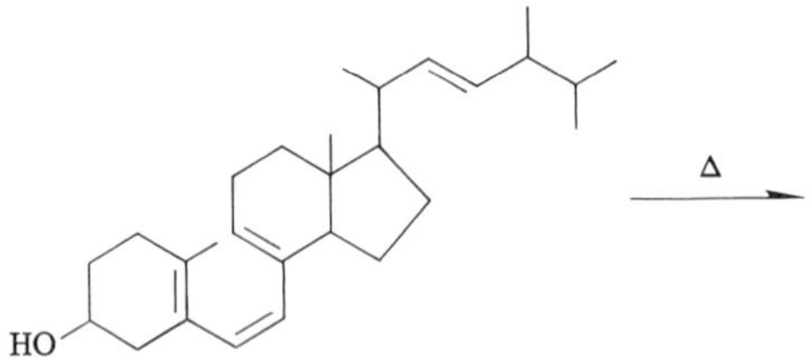

$\xrightarrow{\Delta}$ 143 + 144   (7-2)

The stereochemical question was investigated simultaneously and independently by two groups, both working with stereoisomers of 2,4,6-octatriene.[13,14,15] The trans, cis, trans isomer gave *cis*-5,6-dimethylcyclo-hexadiene in at least 99.5% purity at 140°. Concertedness was indicated since $\Delta H^{\ddagger}$ was 28.6 kcal/mol, which was at least 12 kcal/mol lower than a reasonable phase-independent path. Above 100° the all cis and cis, cis, trans stereoisomers are in mobile equilibrium and a kinetic analysis showed that

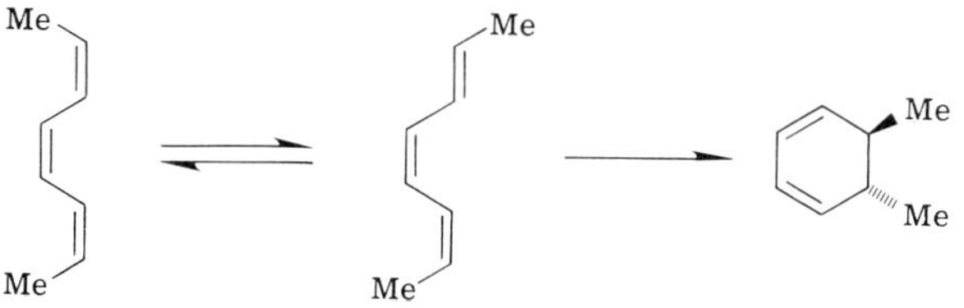

(7-3)

trans-5,6-dimethylcyclohexadiene found at 175° originated from the cis, cis, trans isomer.[14,15] Hilton[16] has prepared the trans, cis, trans form of 1-phenyl-1,3,5-heptatriene and has shown that less than 0.2% of any *trans*-5-phenyl-6-methylcyclohexadiene was formed in its thermal ring closure. Thus, stereoselectivity exceeds 99.8%.

A more heavily substituted set was examined tangentially by Padwa *et al.*[17] during an investigation of the photochemistry of some trienes [Eqs. (7-4) and (7-5)]. The stereochemistry of the reactants was determined purely from the mode of synthesis, and that of the products was ascertained by an inde-

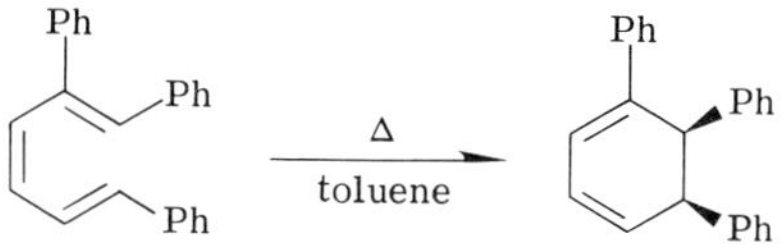

(7-4)

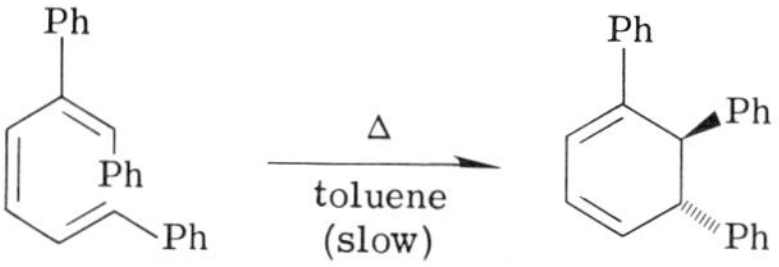

(7-5)

pendent synthesis from materials of known stereochemistry. The selectivity was not ascertained. One report of cyclization with apparent conrotatory stereochemistry attributed the result to a non-concerted process.[18] How-

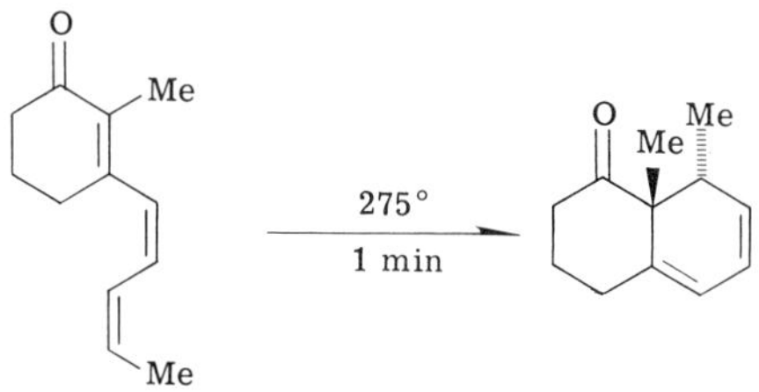

(7-6)

ever, the reaction is that of cis, cis, cis triene moiety and it appears most likely that the facile 1,7-hydrogen shift process converted the terminal double bond to a trans configuration, and ring closure occurred in accord with the orbital symmetry requirement.

The difficulty of preparing acylic trienes with established configurations at all three double bonds confined most of the early studies of this electrocyclization to medium ring trienes. All cis-1,3,5-cyclooctatriene, obtained by reduction of cyclooctatetraene, undergoes reversible formation of 2,4-bicyclo[4.2.0]octadiene at 80°–100°.[19] That the bicyclic ring fusion should be cis seems quite reasonable, but it has been confirmed since when the 145 is heated in the presence of acetic anhydride, it forms a stable tricyclic

145

anhydride [Eq. (7-7)].[20] However, the value of this case as a guide to stereochemical requirements of the reaction generally is lessened by questions about whether any other course could be followed.

Alder and Dortman obtained trans-2,4-bicyclo[4.3.0]nonadiene by Hofmann degradation of 10-methyl-10-aza-7-bicyclo[4.3.1]decene,[21] but the triene precursor was not isolated. Later Vogel et al.[13] prepared a cyclononatriene which cyclized at room temperature to give trans-bicyclo[4.3.0]-nonadiene. The all cis-triene was eliminated spectrally but it was not possible to exclude the trans, cis, trans isomer. However, Glass et al.[22] prepared all cis-cyclononatriene and showed that it gave only 146 [Eq. (7-8)].

146

Several groups developed a satisfactory route to all cis-cyclononatetraene by low temperature protonation of cyclononatetraenide ion.[23-26] Evidence for all the cis configuration of the tetraene was obtained from spectral studies[23-24] and direct formation of a Diels–Alder adduct.[24,27] Transannular ring closure of the tetraene occurs at or just above room temperature to give cis-8,9-dihydroindene [Eq. (7-9)].[23-26] No trans-8,9-dihydroindene

$$\text{(7-9)}$$

was found in the product[28] although 7% of an unidentified product was reported.[23] At least 97% of *cis*-dihydroindene was obtained in one case.[24] The same sequence of reactions was also observed by Schwartz,[28] but under less clearly defined conditions.

Photochemical formation of *N*-carbethoxyazonine permitted study of this aza analog of cyclononatetraene.[29,30,31] The all cis configuration was assigned mainly because of spectral resemblance to cyclononatetraene. This

$$\text{(7-10)}$$

azonine is not stable thermally but is converted readily to *cis*-8,9-dihydro-*N*-carbethoxyindole, whose stereochemistry was defined by complete reduction and comparison with an authentic *N*-carbethoxy-*cis*-perhydroindole.[29] Oxonin has been prepared by a similar photochemical route, and cyclizes thermally to *cis*-8,9-dihydrobenzofuran.[32,33,34] The oxonin was assigned an all cis configuration from spectral results and from the structure of its adduct with 4-phenyl-1,2,4-triazoline-3,5-dione.[35]

All of these examples, both acyclic and cyclic, provide overwhelming evidence of the predictive power of the Woodward–Hoffmann theory. However, there exist, albeit in the very special circumstance of steric confinement, two cases of conrotatory ring closure. The first of these is the remarkably facile conversion of **147** to *trans*-15,16-dimethyldihydropyrene [Eq. (7-11)].[36] The trans relation of the two methyl groups in the product

$$\text{(7-11)}$$

147

was established cleanly by X-ray crystallography of a precursor.[76] Thermal rearrangement occurs readily at 50° by a first order process.

The dihydropyrene skeleton is not necessary for this type of forbidden reaction since *trans*-9,10-dihydrophenanthrenes revert thermally very readily to *cis*-stilbenes.[266-269] Careful investigation of this reversion of the photochemically generated 9,10-dihydro-9,10-dimethylphenanthrene showed that the forbidden reaction occurred with $E_a = 22.5$ kcal/mol.[267,267a] An MO study of both the photo process and the thermal reversion showed that such forbidden thermal electrocyclic processes should occur with low activation energy if the reaction overall is highly exothermic and both reactant and product have low lying excited states of proper symmetry.[269]

A second conrotatory reaction was found with a bridged *cis, cis, trans*-cyclodecatriene [Eq. (7-12)].[36a] The twisted form of the triene is in accord

$$ \text{(7-12)} $$

X = O, S, SO$_2$

with ultraviolet spectral data but structure assignment was based otherwise on the mode of synthesis.

C. Energetics

The kinetic study of Lewis and Steiner[1] of the formation of 1,3-cyclo-hexadiene from *cis*-1,3,5-hexatriene provides a $\Delta H^{\ddagger}$ of 29.0 kcal/mol for this electrocyclic process. The heat of formation for 1,3-cyclohexadiene is 25.38 kcal/mol,[37] and that of *cis*-1,3,5-hexatriene is 40.58 kcal/mol.[28] Thus, this establishes $\Delta H_f^{\circ} = 69.6$ kcal/mol for the transition state of the electro-cyclic reaction, and also a $\Delta H^{\ddagger} = 44.2$ kcal/mol for the ring opening of 1,3-cyclohexadiene. Clearly unless strain factors intervene or the tempera-ture is sufficiently high cycloreversion will not occur. These expectations are clearly born out by the experimental results.

At temperatures above 60° 1,3,5-cyclooctatriene is in equilibrium with 2,4-bicyclo[4.2.0]octadiene,[19] and a study of this equilibrium at several temperatures gave $\Delta H^{\circ} = +1.1$ kcal/mol and $\Delta S^{\circ} = -1$ eu.[39,40] Direct analysis of the ΔH° values for the *cis*-hexatriene and cyclooctatriene exam-ples as follows leads to an interesting discrepancy. The strain energy normally given for cyclooctatriene is 3.5 kcal/mol,[38] and the value of ΔH for the hexatriene–cyclohexadiene reaction takes into account the delocalization energy difference and the strain energy of 1,3-cyclohexadiene. Bicyclo[4.2.0]-octadiene incorporates the added strain of the cyclobutane ring, 26.2 kcal/mol,[41] hence the calculated ΔH° for the cyclooctatriene cyclization is $+7.5$ kcal/mol. The discrepancy appears to be well beyond the error of these calculations. Arguments have been presented that the strain energy given for 1,3,5-cyclooctatriene is anomalously low, and that the true value could be as high as 9.3 kcal/mol.[38] With this value the $\Delta H°$ for the cyclo-octatriene electrocyclization becomes $+1.5$ kcal/mol. This agreement with the experimental value of $+1.1$ helps to confirm the suggested strain energy of ca. 9.3 kcal/mol. Huisgen *et al.*[40] have surveyed the influence of substituents on this equilibrium and have noted that it is difficult to provide any coherent rationale for the observed results.

That the retrocyclization can proceed at higher temperatures has been demonstrated by Baumann and Dreiding.[42] Automerization of **148** at 560° in the gas phase led to complete equilibration of the labels, and arguments

(7-13)

were adduced to show that this occurred by a repetition of the scheme shown in Eq. (7-13). In an earlier paper Spangler and Boles[43] had suggested that formation of 1,3-dimethylcyclohexadienes from 5,5-dimethyl-1,3-cyclo-hexadiene occurred via 1,5-methyl migration rather than through a retro-electrocyclization. Ring opening of a chemically excited 1,3-cyclohexadiene has also been observed.[44] Not all group migrations of 1,3-cyclohexadienes proceed via the ring opening and recyclization route since formyl groups have been shown to undergo 1,5-shifts,[77] and in some cases competition between the 1,5-shift and ring cleavage can occur.[78] Generally it appears likely that ring cleavage will be the dominant process if migration would necessarily involve an alkyl or an aryl group.[78] Spangler *et al.*[576] in a study intended to bolster the suggestions of Spangler and Boles,[43] have contended that 1,7-hydrogen shifts are slower than 6π electrocyclizations in some cases. Unfortunately, their work serves only to confirm their *assumption*, made to permit kinetic analysis, that a 1,7-hydrogen shift is slower than a 6π electro-cyclization.

No determination has been made of the $\Delta\Delta H^{\ddagger} = \Delta H^{\ddagger}(\text{forbid.}) - \Delta H^{\ddagger}$-(allowed) for the 6π electrocyclic reaction. However, the small amount of material which matched the glc peak for *trans*-5-phenyl-6-methylcyclohexane (0.18%) in the hydrogenated product from cyclization of *trans, cis, trans*-1-phenyl-1,3,5-heptatriene corresponds to a $\Delta\Delta H^{\ddagger}$ of 5.0 kcal/mol as a minimum value.[16]

D. Structure–Reactivity Relations

The picture of the transition state for the disrotatory cyclization which emerges from the theoretical studies permits predictions of structure–reactivity relations, and in general the experimental results collected in Tables 7-1 and 7-2 correlate very satisfactorily with those predictions. There are four possible monosubstituted isomers, and the results in these cases are probably the most revealing. Substituents trans at the terminal carbons would be expected to have little or no steric effects and should be partially disconnected from the conjugated system by the time the transition state is reached. The extent of this separation and the way in which substituents may influence the forming single bond is difficult to assess. The reality is rather unexpected in that substituents in this position show almost no effect on rate whether alkyl, aryl, or carbethoxyl. Perhaps the closest analogy to this type of process is the substituent effect at C_3 of the hexadiene moiety of the Cope rearrangement. In that case a methyl group increases the rate by a factor of about three,[80,81] while a phenyl group causes a seventeenfold rate enhancement.[82] Thus, while substituent effects on such bond conversions are small, those observed for the triene electrocyclization are abnormally low.

These results lead then to the prediction that terminal substituents in a cis orientation should have solely a steric effect, and since the spatial restrictions are more acute in the transition state rate retardation is to be expected. The anticipated result is observed but the magnitude is only moderate. Substituents at C_2 (or C_5) can act in two ways, ground state steric effects and stabilization of radical character generated at that carbon in the transition state because of the dihedral angle between the *p*-orbitals of C_2 and C_3.[3] Only the doubly *S-trans* conformation of *cis*-hexatriene is populated,[79] and the hydrogens at C_2 and C_5 apparently produce sufficient repulsive interaction to induce a torsion of $10°$ about the central double bond. Larger substituents should cause additional interaction, further deviation from planarity and a consequent rise in ground state energy. Since the transition state provides more space for these substituents, this should lead to a steric acceleration. No adequate test of this has been made, but to the extent that

TABLE 7-1

Kinetics and Activation Parameters for Acyclic Hexatriene Cyclizations

Compound	k (sec^{-1})	T (°)	$\Delta H^{\ddagger}$	$\Delta S^{\ddagger}$	Reference
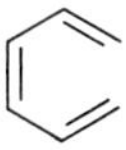 (hexatriene)	5.1×10^{-5}	132	29.3	-5	1
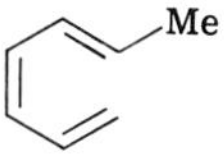 (Me-substituted)	2.39×10^{-5}	124	28.9	-4.9	45
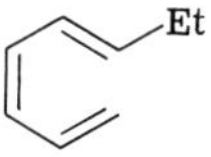 (Et-substituted)	5.56×10^{-5}	125	27.8	-6.2	45
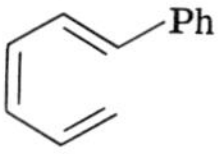 (Ph-substituted)	3.0×10^{-5}	125	28.2	-8	46
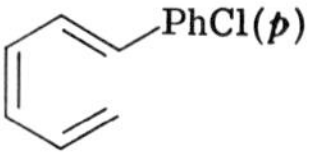 (PhCl(p)-substituted)	4.0×10^{-5}	125	24.5	-18	46
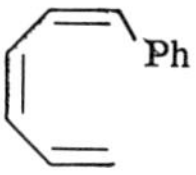 (Ph-substituted)	3×10^{-5}	195	—	—	46
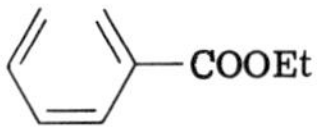 (—COOEt)	3.0×10^{-4}	75	—	—	47
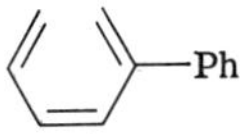 (—Ph)	4.42×10^{-5}	76.4		—	48
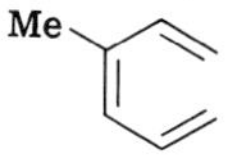 (Me-substituted)	6.43×10^{-5}	121	26.1	-12.4	45
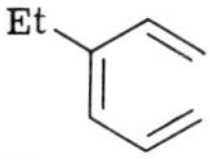 (Et-substituted)	1.32×10^{-4}	124	26.0	-11.3	45
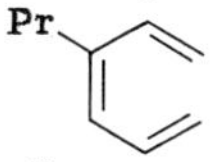 (Pr-substituted)	1.79×10^{-4}	125	23.5	-17.1	577
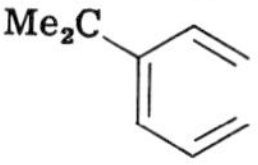 (Me$_2$C-substituted)	8.46×10^{-4}	125	26.7	-11.3	45
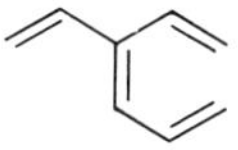 (vinyl-substituted)	2.26×10^{-4}	83.9	22.0	-13.8	49
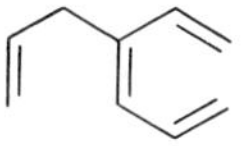 (substituted)	1.82×10^{-4}	125	24.5	-14.7	577

(*continued*)

TABLE 7-1 (Continued)

Compound	k (sec^{-1})	T (°)	$\Delta H^{\ddagger}$	$\Delta S^{\ddagger}$	Reference
(structure: Me, Me)	4.5×10^{-5}	132	28.6	-7	15
(structure: Me, Me)	2.2×10^{-5}	178	32	-5	15
(structure: Ph, Me)	2.1×10^{-4}	134	29.0	-6.8	16
(structure: Ph, Me)	5.0×10^{-5}	191			16
(structure: NMe$_2$, COPh)	2.69×10^{-5}	84	20.2	-23.3	578
(structure)	1.3×10^{-4}	132	28.2	-6	46
(structure)	3.68×10^{-5}	126.5	26.7	-12.1	46
EtOOC—⟨ ⟩—COOEt	5×10^{-4}	18	~ 20	—	50, 51
Me$_2$NOC—⟨ ⟩—CONMe$_2$	6×10^{-5}	18	—	—	51
(structure: O, CN)	3×10^{-3}	22	—	—	52
(structure: Me, Me)	6.9×10^{-6}	150	32	-13.8	53
(structure: Me)	1.3×10^{-4}	132	28.2	-1.0	46
(structure: Me)	2.6×10^{-5}	174	34	-1.0	46

TABLE 7-1 (*Continued*)

Compound	k (sec^{-1})	T (°)	$\Delta H^{\ddagger}$	$\Delta S^{\ddagger}$	Reference
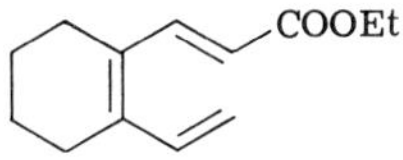 (structure, COOEt)	5.77×10^{-5}	130	26.2	−14	54
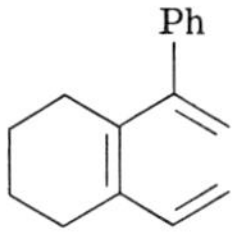 (structure, Ph)	4.84×10^{-5}	126.5	26.4	−12.1	55
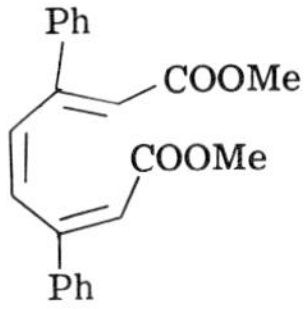 (structure, Ph, COOMe, COOMe, Ph)	3.4×10^{-6}	25	20.0	−17.5	56
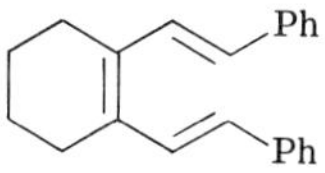 (structure, Ph, Ph)	2.3×10^{-4}	150	—	—	57
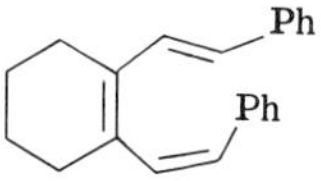 (structure, Ph, Ph)	1.1×10^{-5}	150	—	—	57
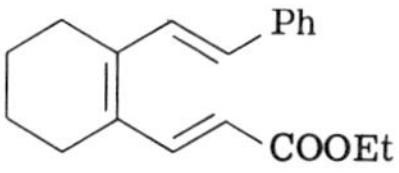 (structure, Ph, COOEt)	4.0×10^{-4}	135.7	27.8	−9.0	58
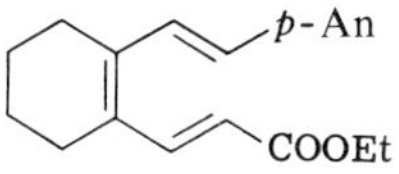 (structure, *p*-An, COOEt)	4.0×10^{-4}	135.7	27.8	−9.0	58
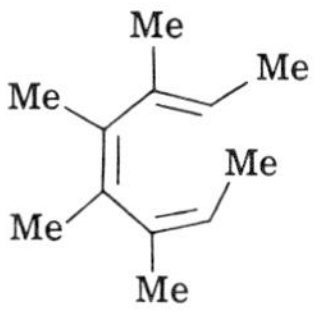 (structure, Me, Me, Me, Me, Me, Me)	3.7×10^{-4}	210	27.4	−19.0	59
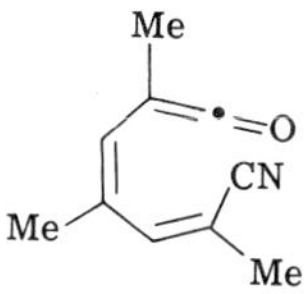 (structure, Me, =O, CN, Me, Me)	2.2×10^{-5}	−77	13	−3.5	52
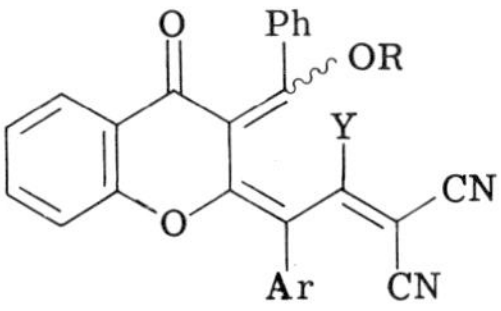 (structure, O, Ph, OR, Y, CN, CN, Ar, O)					

(*continued*)

TABLE 7-1   (Continued)

Compound	k (sec^{-1})	T (°)	$\Delta H^{\ddagger}$	$\Delta S^{\ddagger}$	Reference
R=H Ar=Ph Y=CN	8.8×10^{-2}	25			60, 61
R=Ac Ar=Ph Y=CN	6.2×10^{-6}	25			60, 61
R=H Ar=Ph Y=H	1.7×10^{-3}	25			61
R=H Ar=p—An Y=H	1.4×10^{-3}	25			61
R=H Ar=Me Y=H	4.2×10^{-4}	25			61
R=Ac Ar=Ph Y=H	$<1 \times 10^{-7}$	25			61

| | 2.7×10^{-1} | 25 | | | 61 |

a 1,2-tetramethylene group is indicative the effect is very modest. Conversely the groups known to stabilize radicals, phenyl and carbethoxyl, produce significant rate increases. However, the carbethoxyl is more effective than a phenyl group in contrast to their normal effects on radical stabilization.

Substituents at C_3 or C_4 may be expected to induce a steric acceleration as a result of ground state effects, and a transition state stabilization is expected of radical stabilizers, albeit of lesser magnitude than for the C_2 position. Reality again confirms the expectation. It is clear from the influence of a vinyl at C_3 and of phenyl and carbethoxyls at C_2 that there is little experimentally to support Carpenter's proposal that conjugating substituents should decrease the ring closure rate regardless of position. An attempt to test Epiotis' prediction that a combination of donor and acceptor substituents should produce a special rate enhancement found only insignificant acceleration, phenyl and carbethoxy gave a fourfold increase and β-anisyl and the carbethoxyl produced no further change.[58] The result may be attributed to the unresponsiveness of rate to substitution at the terminal centers. However, the results point to the conclusion that attempts to define substituent effects without full consideration of the transition state geometry are likely to meet with moderate success at best.

When substituents are sufficiently separated spatially so that no interference occurs in reactant or transition state, their rate effects are roughly additive in a number of cases, though this has not been given any serious test. Obviously synergism might be expected in some cases. That could be the case with the 1,6-dicarbomethoxy-2,5-diphenylhexatriene where a mixture of relief of ground state strain and the stabilizing effect of the phenyls in the transition state might account for the rather unexpectedly rapid rate. The extreme ground state strain induced by the heavily substituted s-cis

TABLE 7-2

Kinetics and Activation Parameters for Cyclic Hexatriene Electrocylizations

Compound	k (sec^{-1})	T (°)	$\Delta H^{\ddagger}$	$\Delta S^{\ddagger}$	Reference
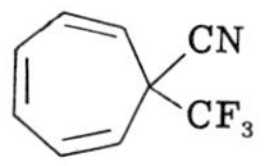 (cycloheptatriene) CN, CF$_3$	ca. 4×10^2	-85	—	—	62, 63
 (cycloheptatriene) O	ca. 6×10^3	-85	7	-2.3	63
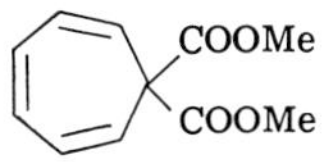 (cycloheptatriene) COOMe, COOMe	ca. 2×10^3	-85	7	-2.3	64
(cyclooctatriene)	5.3×10^{-7}	20	26.6	-1	65
	9.0×10^{-2}	125	24.7	-2	66
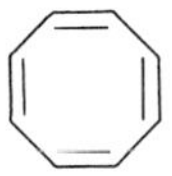 (cyclooctatriene)	0.17×10^{-7}	20	28.1	$+1$	65
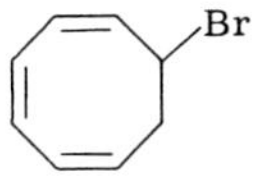 (cyclooctatriene) Br	6.0×10^{-6}	20	22.2	-7	65
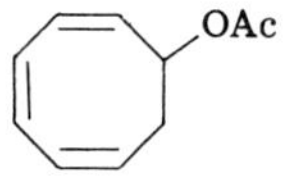 (cyclooctatriene) OAc	8.1×10^{-6}	20	24.6	$+2$	65
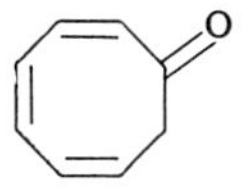 (cyclooctatriene) O	2.3×10^{-5}	20	22.7	-2	65
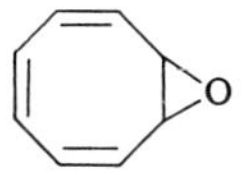 (cyclooctatriene epoxide) O	1.9×10^{-7}	20	24.6	-5	65
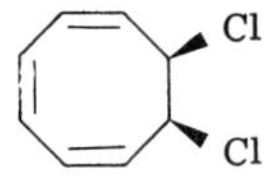 (cyclooctatriene) Cl, Cl	3.3×10^{-6}	20	24.7	$+1$	65
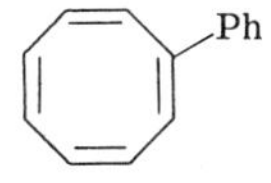 (cyclooctatriene) Ph	7.6×10^{-5}	60	25.0	-3	67
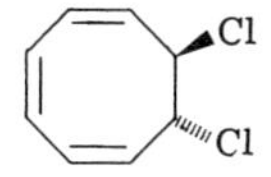 (cyclooctatriene) Cl, Cl	1.0×10^{-3}	20	—	—	65

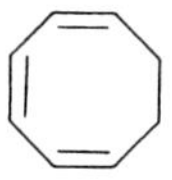

(*continued*)

TABLE 7-2 *(Continued)*

Compound	k (sec^{-1})	T (°)	$\Delta H^{\ddagger}$	$\Delta S^{\ddagger}$	Reference
(Br / Br structure)	1.5×10^{-4}	5	—	—	65
(bicyclic structure)	1.1×10^{-5}	20	—	—	231
(CH$_2$)$_3$	1.67×10^{-4}	50	23.0	−4.7	22
(CH$_2$)$_3$			20.1	−12.8	13
(bicyclic structure)	6.5×10^{-4}	35	21.4	−3.9	68, 69, 26
—CH$_2$OAc	4×10^{-3}	10	(CDCl$_3$)		70
(bicyclic structure)	1×10^{-2}	10	(CDCl$_3$)		70
X = O			19.4	−16	36
X = S			23.8	−12	36
X = SO$_2$			20.4	−15	36
(lactone structure)	$\sim 8 \times 10^{-4}$	25			71
(macrocyclic structure)	2.4×10^{-4}	−40	17.4 (E_a)		72, 73

TABLE 7-2   (*Continued*)

Compound	k (sec^{-1})	T (°)	$\Delta H^\ddagger$	$\Delta S^\ddagger$	Reference

R=X=Y=H	1.7×10^{-5}	30	22.4	-7	74
R=H X=Y=OAc	1.8×10^{-5}	30	22.4	-7	74
R=H X=NHAc Y=H	7×10^{-5}	30	22.1	-5	74
R=H X=CHO Y=H	9×10^{-3}	30	19.9	-7	74
R=H X=NHAc Y=CHO	2.7×10^{-2}	30	16.1	-13	74
R=Me X=Y=H	1.6×10^{-5}	30	23.9	-2	74
R=Me X=CHO Y=H	2×10^{-4}	30	20.6	-7	74

	3.3×10^{-6}	30	19.5	-19.2	75
Pyridinium salt of above	1.5×10^{-2}	4	19.8	4.8	75

s-trans conformation is probably largely responsible for the high rates of ring closure observed by Huffman *et al.*[60,61] The reduction in rate associated with acetylation of the hydroxyl group suggests the group is cis located on the terminal atom. The very high rates associated with dienyl ketenes is undoubtedly a ground state influence.

Confining the triene moiety within a ring quite clearly causes extremely large rate effects. The triene grouping must adopt a nearly perfect position for ring closure in the tub form of cycloheptatriene, and the terminal carbons are separated by less than the sum of their van der Waals' radii. The result is a ring closure rate which is competitive ratewise with the ring-flipping of the cycloheptatriene ring.[83] The situation is geometrically less favorable with cyclooctatrienes, cyclononatrienes, and cyclodecatrienes where individual valence isomers can be isolated but undergo very ready interconversion. The two cases of "forbidden" reaction are apparently studies in very diverse effects, the facility of ring closure with the bridged cyclodecatriene resulting exclusively from excessive strain in the ground state, but with the diencyclophane strain is supplemented by the decrease in energy separation between HOMO and LUMO resulting from the large conjugate system.

Rate changes produced by 7,8-substituents in the cyclooctatriene system indicate that electron attracting substituents all increase the rate of ring closure by about a factor of ten except for a keto group which is somewhat more effective. The rate depressing influence of an epoxy ring probably results from the developing ring strain of the 5-oxabicyclo[2.1.0]pentane moiety, and the quite significant increase which appears with trans-7,8-dihalo, but not cis-7,8-dihalo compounds, is certainly a steric effect of an endo halogen in the concavity of the cyclooctatriene tub.

E. Acyclic Examples

Excepting the early work on the formation of pyro- and isopyrocalciferols from vitamin D,[8] which was not completely clarified until 1959, all of the pre-1960 work on reactions which probably involve electrocyclizations initiated with trienes having a central trans double bond.[96–105] Reactions were carried out under various conditions including both liquid and vapor phase in the absence of catalysts and reactions in the presence of such catalysts as aluminum oxide. Goldblatt and co-workers[100–102] studied the simply pyrolyses of α-pinene, and the ring closure of one of the initial products of that pyrolysis, allo-ocimene [Eq. (7-14)]. Later Crowley[106] found

that 2,6-dimethyl-2,3,5-octatriene rearranged readily to two octatrienes, and at a higher temperature one of the pyronenes was formed [Eq. (7-15)]. How-

(7-15)

ever, the author did not convert the trienes directly to pyronene. Shortly thereafter Lewis and Steiner[1] made the first clean study of this electrocyclization. Finally the very complex series of thermal reactions which result from pyrolyses of α-pinene, and more significantly for this work, from allo-ocimene were unravelled by Crowley and Traynor.[107] They found 1,7-hydrogen shifts, electrocyclizations, 1,5-hydrogen shifts, central double bond stereomutation, and terminal double bond isomerizations with rates in decreasing order as listed.

Spangler *et al.*[140] and Spangler and Feldt[141] have studied the reactions which ensue when dienols are dehydrated over aluminum oxide, and the isolation of cyclohexadienes such as *cis*-5,6-dimethylcyclohexadiene has prompted him to suggest these cyclic products arise from an electrocyclic reaction. Orchard and Thrush[142] examined the high temperature cyclization of *trans*-1,3,5-hexatriene kinetically and found $E_a = 43.5$ kcal/mol, log $A = 12.65$. A survey of catalysts for the cyclization of eleostearic esters indicated that sulfur was very efficient and cyclization occurred readily at 160°.[143] Frater[144] obtained a cyclic product of appropriate stereochemistry by cyclization of a triene ester derived from β-ionone [Eq. (7-16)].

(7-16)

In certain cases cyclohexadienes will undergo photochemical ring opening to form trienes,[108] and studies of this process permitted access to a number of trienes whose thermal electrocyclizations were examined. Dauben and Coates photolyzed palustric acid and the triene product recycled readily [Eq. (7-17)].[109] Prinzbach and Hagermann[110] noted that 2,5-dicarbometh-

$$\qquad \qquad (7\text{-}17)$$

oxy-1,3,5-hexatriene reverted to 1,4-dicarbomethoxy-1,3-cyclohexadiene readily at 20°. Further work[47,50,51] included studies of the mono ester and other derivatives of the mono and dicarboxylic acids (Table 7-1). A number of complex examples were studied by Ullman and co-workers (Table 7-1).[60,61] Tetravinylethylene was prepared by photolysis of 2,3-divinyl-1,3-cyclohexadiene,[111] and ring closure was carried out in the vapor phase at 320° [Eq. (7-18)]. It is unfortunate that rate studies were not made since

$$\qquad \qquad (7\text{-}18)$$

3-vinyl-1,3,5-hexatriene reacts unexpectedly rapidly.[49] Courtot has prepared several trienes by photochemical ring cleavage and in some cases these have been thermally recyclized [Eq. (7-19)].[56,112] A less informative example

$$\qquad \qquad (7\text{-}19)$$

R = Me, COOMe

was provided by cyclization of 5-phenyl-cis,trans-1,3,5-heptatriene.[113]

Edmunds and Johnstone[114,115] reported cyclization of some trienes and polyenes in studies relating to formation of products found in cigarette smoke. When β-carotene was heated at 300° 1,1,6-trimethyl-1,2,3,4-tetrahydronaphthalene was obtained. They also reported formation of 1-phenyl-trans,cis-1,3,5-hexatriene and 1-phenyl-trans,cis,trans-1,3,5-heptatriene by a Wittig reaction and ring closure of these trienes at 170°–190°.[115] The identity of the phenylhexatriene is not directly clarified by the spectral data, but the preparation and cyclization have also been carried out by Marvell et al.[46] Hilton[16] has also prepared an authentic sample of the phenylheptatriene and has shown that the compound assigned that structure by Edmunds and Johnstone was actually 6-methyl-1-phenyl-1,3-cyclohexadiene. Padwa et al.[17] had better luck with the preparation of cis-trienes via the Wittig reaction and the cyclizations indicated that triphenyl trienes follow the Woodward-Hoffmann stereochemistry (see Section I,B).

Ramage and Sattar[18] found that the pyrolysis of the carbonate ester (149) for a short time gave a different product from that observed on prolonged heating [Eq. (7-20)]. Reversion of the kinetic product to a triene followed

$$(7\text{-}20)$$

by a 1,7-hydrogen shift and recyclization account for the second product. It has been shown that all three stereoisomers of 1,3,5-undecatriene give different thermal products.[116] The all trans isomer dimerizes via a Diels–Alder addition, while the cis-3,trans-5 isomer cyclizes normally and the dicis isomer undergoes solely a 1,7-hydrogen shift. A triene cyclization occurred unexpectedly following cleavage of a vinylcyclobutene [Eq.

$$(7\text{-}21)$$

(7-21)][117] where the more bulky group turned inward. A similar process has been observed with a benzocyclobutene [Eq. (7-22)].[118,119]

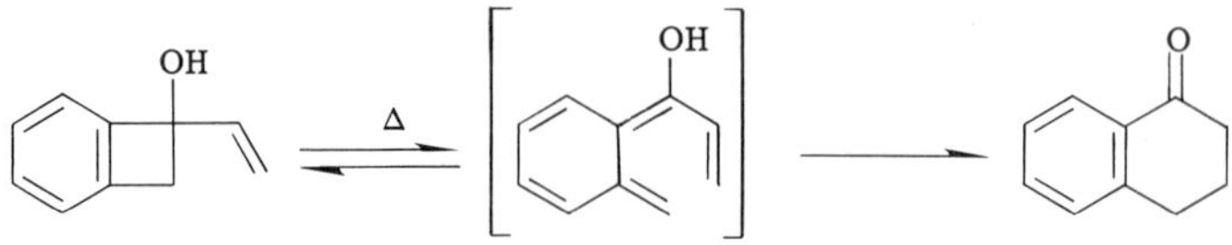

(7-22)

When the electrocyclization can lead to formation of an aromatic ring as in Eq. (7-22), the reaction should be facilitated. Indirect evidence that trans, trans **150** cyclizes thermally at 25° has been reported.[120] On the other

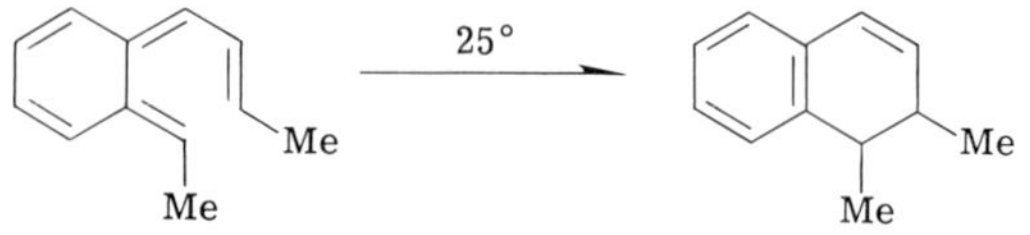

(7-23)

150

hand, when an aromatic ring bond participates in the electrocyclization, the

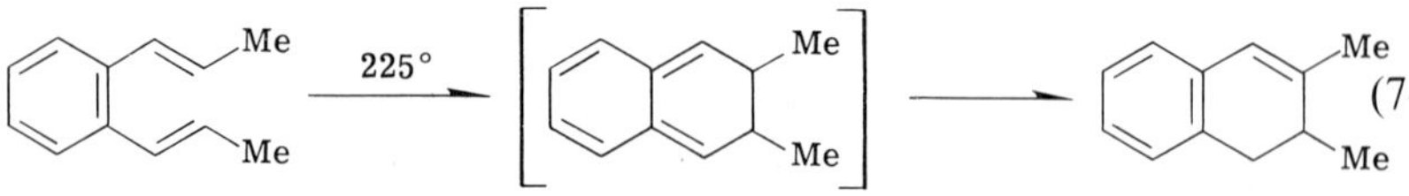

(7-24)

$k = 4.2 \times 10^{-6}$ sec.$^{-1}$ $\Delta H^{\ddagger} = 32.3$ $\Delta S^{\ddagger} = -14$

reaction is retarded,[121] and the retroelectrocyclization is facilitated [Eq.

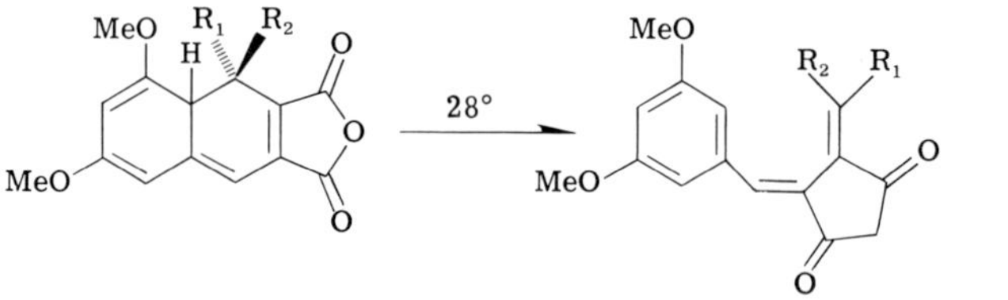

(7-25)

$R_1 = R_2 = Me$ $k = 2.37 \times 10^{-5}$ sec^{-1} $E_a \simeq 14$ kcal/mol
$R_1 = R_2 = Et$ $k = 0.91 \times 10^{-5}$ sec^{-1} $E_a \simeq 19$ kcal/mol

(7-25)].[122] However, a number of similar examples where ring closure occurs have also appeared.[579,580] In at least one case a cyclization which aromatizes one ring while dearomatizing another also proceeds at a surprisingly low temperature.[123] Absence of the rearomatizing ring returns the reaction to a

$$(7\text{-}26)$$

high temperature process.[124] Pyrolysis of a methylated hexahelicene leads to a complex rearrangement which apparently involves an unusual electrocyclization [Eq. (7-27)].[125]

$$(7\text{-}27)$$

When one double bond is part of a ketene moiety the electrocyclization rate is greatly enhanced,[52,126,127] (see Table 7-1). Generally the dieneketenes are formed photochemically, and unless trapped by alcohols or amines will revert to cyclohexadienones [Eq. (7-28)].[127,128] In one case the diene-

$$(7\text{-}28)$$

ketene was obtained via thermal ring opening of a vinylcyclobutenone.[129] When the diene–ketene grouping is heavily substituted an unusual cycliza-

$$Y = Me^{130,131,132} \qquad Y = OAc^{133}$$

tion competes with the electrocyclic reaction. Normally a photochemical reaction this double cyclization is favored by polar solvents, reaction being rapid at $-100°$ in ethanol. In view of the importance of the heavy substitution in aiding the double cyclization the observation of the same competition between normal electrocyclization and double cyclization with an alkyl-amino triene [Eq. (7-30)] must be considered most unusual.[134]

Substitution of one or both of the terminal double bonds by triple bonds does not appear to prevent ring closure, though no complete understanding of the details (concertedness, etc.) of the reaction has been reached. Hopf and Musso[135] found that both *cis*- and *trans*-1,3-hexadiene-5-yne were stable in the gas phase to 224°, but at 274° cis-trans interconversion and benzene formation were active. Toluene can be produced by heating 1,6-heptadiyne in the presence of potassium *tert*-butoxide,[136] and the reaction was shown to involve *cis*-1,3-heptadien-5-yne.[137] Purely thermal cyclization occurred

Ar = H	R = H	274°	90 min	50%
Ar = H	R = Me	274°		
Ar = Ph	R = H	300°	15 sec	~ quant.

at 274°, but in the presence of 5-butoxide reaction proceeded readily at 80°. This might suggest the slow step is the hydrogen migration which, if concerted, would be a symmetry forbidden 1,3-shift. Moy[138] found that 2-phenyl-1,3-hexadien-5-yne gave approximately a quantitative yield of biphenyl in 15 sec at 300° in a flow system. Jones and Bergman[139] indicated by deuterium labelling that *cis*-3-hexen-1,5-diyne will undergo an automerization [Eq. (7-32)]. Evidence favoring the intermediacy of *p*-benzenediyl

$$ \text{(7-32)} $$

was provided by the formation of *p*-dichlorobenzene when carbon tetra-tetrachloride was present. The enthalpy of activation was estimated at 32 kcal/mol.

Cyclization of a dienallene does occur but because the relevant double bond was initially trans the temperature for rearrangement (490°) is not indicative of the ease of cyclization.[582]

F. Cyclic Examples

One of the fascinating problems associated with the electrocyclic reaction is the cycloheptatriene–norcaradiene question. Evoked initially by the famous Buchner acids obtained by treating benzene with diazoacetic ester,[145] the structural uncertainties remained until in 1956 Doering and co-workers showed that all the acids have the cycloheptatriene structure.[146] The early work has been elegantly reviewed by Maier[147] and this section will thus concentrate on more recent studies. The results prior to 1965 can be readily summarized as follows. All simple substitution products of cycloheptatriene had been shown to have the monocyclic structure (CHT). The norcaradiene (NC) structure could be stabilized in two ways, bridging C_1 and C_6 by a short

$$ \text{(7-33)} $$

$$ \text{(7-34)} $$

chain of atoms (normally three or less), and incorporation of one or more of the double bonds of the NC form into an aromatic ring.

This simple pattern was completely altered by the discovery that 7,7-dicyanonorcaradiene was a stable compound showing no tendency to form a cycloheptatriene isomer.[62] The discovery triggered a series of studies of this valence isomerization with a wide variety of substituted molecules. Molecules for which an equilibrium constant has been ascertained are listed

TABLE 7-3

Equilibrium Constants for Substituted Cycloheptatriene–Norcaradienes

Compound			K	$T\,(°)$	Reference
(CF₃, CN)-cycloheptatriene			0.25	−85	62, 163
(COOMe, COOMe)			0.326	−30	64, 149
(Ph, Y)	Y=COOEt		0.42	25	84
	Y=PO(OMe)₂		3.78	25	84
(COOR, aryl-Y)	R	Y			
	Me	H	0.510	1.0	85, 162
	Me	NO₂	0.296	−1.5	85, 162
	Me	OMe	0.435	−1.5	85, 162
	Et	H	0.686	−1.5	85
(COY, PO(OMe)₂); Y = HN–aryl(NO₂), MeOOC			0.23	37	86
			0.30	37	86
(C⁺R₁R₂)	R₁	R₂			
	NH₂	OMe	0.431	−70	87
	NH(CH₂)₂–	O	0.214	−70	87
	NMe₂	OMe	0.737	−70	87
	NMe₂	OEt	0.628	−70	87
spiro lactone			1.23	36	88
(Me, CN), R	R=Ph		5.7	−120	89
	R=Me		0.25	−120	89
(Me, COOMe)			0.85	−120	89
(Me, COOMe), R₁, R₂	R₁	R₂			
	H	H	0.32	−120	89
	Br	H	>19	−120	89
	Me	H	>19	−120	89
	H	Me	0.67	−120	89

TABLE 7-3 (*Continued*)

Compound			K	$T (°)$	Reference
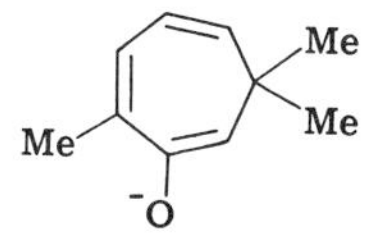			~2	28	90
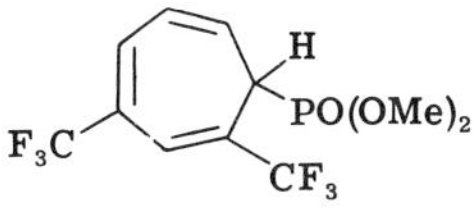			0.097	37	86
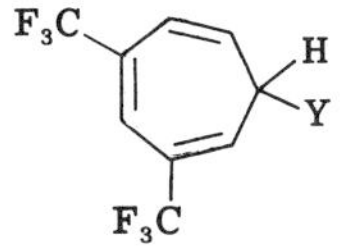	Y=PO(OMe)$_2$		1.25	25	86
	Y=COOEt		0.25	25	86
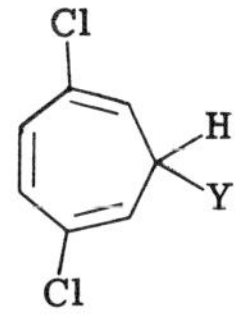	Y=PO(OMe)$_2$		0.65	25	86
	Y=COOEt		0.76	25	86
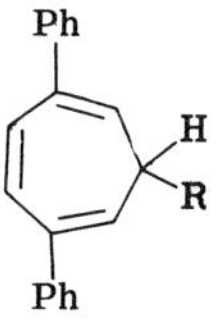	R = 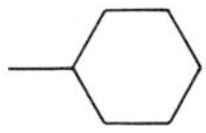		~0.3	—	91
	R = 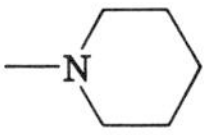		~0.7	—	91
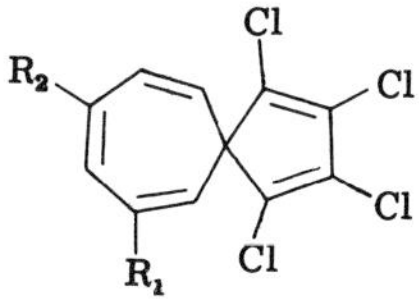	R_1	R_2			
	H	H	0.24	37	92, 93
	H	CF$_3$	0.41	37	92, 93
	CF$_3$	CF$_3$	2.1	37	92, 93
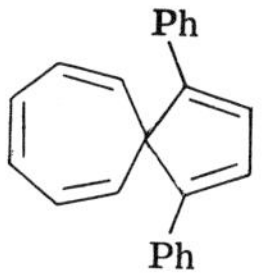			0.25–0.43	37	93
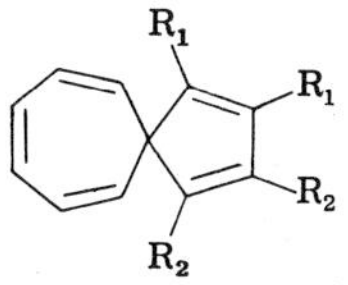	R_1	R_2			
	Ph	Ph	0.053	37	93
	H	benzo	~1.0	37	93
	Ph	benzo	0.24	37	93
	benzo	benzo	0.33	37	93

(*continued*)

TABLE 7-3   (Continued)

Compound	K	T (°)	Reference
	0.28	37	93
	0.25	−90	94
	0.031	20	95

in Table 7-3. In Tables 7-4 and 7-5 are compounds which have been shown
to exist primarily in the cycloheptatriene or the norcaradiene form, respectively. Norcaradiene itself can be calculated to be about 10.5 kcal/mol less
stable (ΔH) than is cycloheptatriene, and since 7,7-dicyanonorcaradiene is
about 6 kcal/mol more stable than the CHT form,[83] each cyanide group
stabilizes the NC form by about 8 kcal/mol. Hoffmann[160] and Günther[155]
have indicated that this stabilization results from electron transfer from the
cyclopropane to a π acceptor group.

Any attempt to set up some scale of stabilizing effect on the NC form must
take into account the topomerization of norcaradienes via ring flipping of
the cycloheptatrienes [Eq. (7-35)].[83,164] Since this interconverts exo and

$$\text{} \tag{7-35}$$

endo positions at C_7 in the norcaradienes and the spatial character of these
positions is very different, steric effects of substituents will differ depending
on whether they exist to a greater or lesser extent in an exo or an endo
position. Generally, however, it appears that the series stable C^+, CN,
$PO(OMe)_2$, COOR show decreasing ability to stabilize the NC form substituted at C_7. Two vinyl groups present in a spiro-fused five- or six-numbered
ring stabilize the NC form, and this result has been attributed to possible
cross-conjugation with the cyclopropane.[64] The bond angle between the
two C_7 external bonds can exert an influence and is undoubtedly important

TABLE 7-4

Compounds Existing as Cycloheptatrienes[a]

Compound	Reference
1,2,5-Triphenyl	152
1,3,5-Triphenyl	151
1,4,5-Triphenyl	152
7-Methyl	148
7-Isopropyl	148
7-*tert*-Butyl	148
7-Phenyl	148
1,4,7-Triphenyl	152
7,7-Dimethoxy	64
7-Methyl-7-cyano	89
1,7-Dimethyl-7-carbomethoxy	89
7,7-Dimethyl-1,4-diphenyl	161
1,7,7-Trimethyl-6-phenyl	161

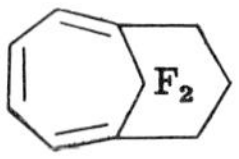

155

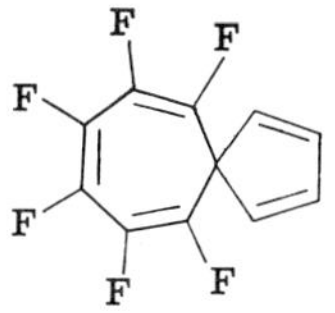

150

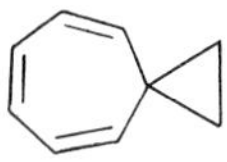

154

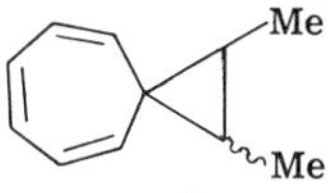

(both cis and trans)   153

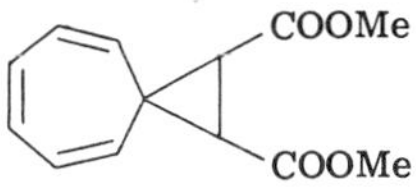

(trans)   154

[a] Reported since Maier's review in 1967.

in keeping the spiro[6.2]nonatrienes in CHT form, and by steric interference in the endo position which again favors the CHT form.

Substitution at positions other than C_7 can also play a significant role in determining the equilibrium position. A 4,5-diaza compound exists purely as an NC form because the nitrogen–nitrogen double bond required in the CHT form [Eq. (7-36)] is exceedingly weak.[164] Phenyls at C_2 and C_5 exert a

TABLE 7-5

Compounds Existing as Norcaradienes[a]

Compound	Reference
1-Methyl-7,7-dicyano	156
2-Methyl-7,7-dicyano	156
3-Methyl-7,7-dicyano	156
2,5-Bisdifluoromethyl-7,7-dicyano	83
7-Phenyl-7-cyano	163
7-Carbomethoxy-7-cyano	163
7,7-Dimethyl-2,5-diphenyl	161
2,5,7-Triphenyl	151
2,5-Diphenyl-7-cyano	160
2,5-Diphenyl-7-(2-phenylethynyl)	160
2,5-Dichloro-7,7-dicarbomethoxy	86
2,5-Dichloro-7-phenyl-7-(dimethylphosphonate)	84
2,5-Dichloro-7-carbomethoxy-7-(dimethylphosphonate)	86
2,5-Dichloro-7-(N-p-nitrophenylcarboxamido)-7-(dimethylphosphonate)	86
2,5-Dimethoxy-7-phenyl-7-(dimethylphosphonate)	84

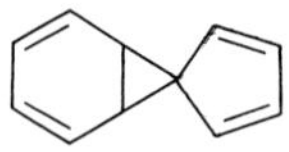

157

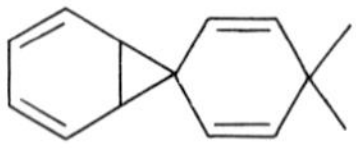

150

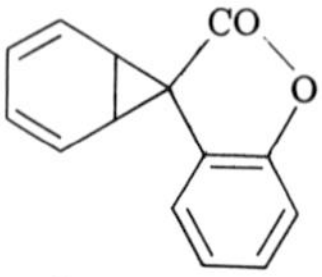

158

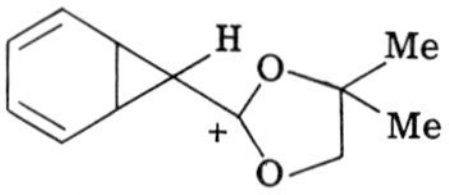

88

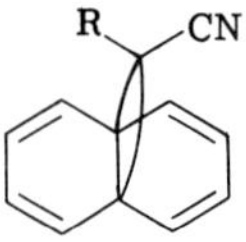

159

R = Me, PhCH$_2$

$$\text{(7-36)}$$

much larger influence than might be expected, and halogens in these positions also stabilize the NC form.

The 1,6-methano bridged [10] annulenes present a particularly interesting case since the CHT form is now stabilized by the ten electron aromatic system. For $R_1 = R_2 = H$, Me Cl, Br the [10] annulene is the stable form.[165, 166,159] The annulene is also more stable when $R_1 = Me$ $R_2 = COOMe$, or

$$\text{(7-37)}$$

$R_1 = Cl$ $R_2 = CN$, but the norcaradiene is the stable form when $R_1 = Me$ or $PhCH_2$ $R_2 = CN$.[159]

An especially fascinating type of norcaradiene, the bisnorcaradiene [Eq. (7-28)], was apparently first postulated as an intermediate in the thermolysis of benzo[6]cycloheptatriene.[167] Further èvidence indicating the presence of the bisnorcaradiene in photolysis of spiro[6.4] undecaptentaenes was ad-

$$\text{(7-38)}$$

duced from the reaction of a series of substituted compounds [Eq. (7-38)].[168] Another thermal route into the same manifold was noted by Mitsuhashi and Jones,[169] while Vedejs and Wilbur invoked the bisnorcaradiene in an additional case.[170] Finally Dürr et al.[171] obtained spectral evidence for the bisnorcaradiene ($R_1 = R_4 = Ph$, $R_2 = R_3 = H$) by low temperature photolysis. Kinetic studies on the thermal rearrangement of a group of bisnorcaradienes gave $\Delta H^{\ddagger} = 15$–20 kcal/mol and $\Delta S^{\ddagger} = -17$ to -30 eu.[172] Further examples have been uncovered.[581] However, rearrangement via the bisnorcaradiene is not always observed, since the reaction of Eq. (7-39)

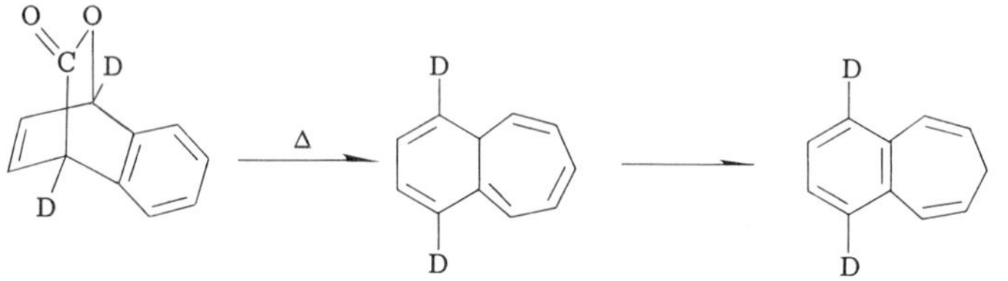

$$(7\text{-}39)$$

leads to a product completely devoid of deuterium in the cycloheptatriene ring.[173]

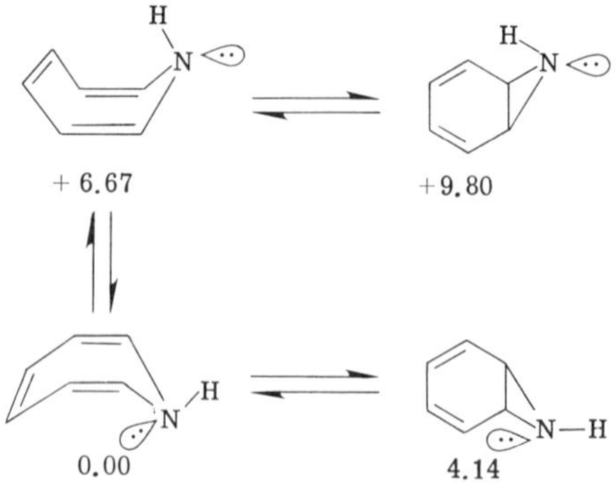

Presence of a heteroatom at C_7 introduces in those cases where the heteroatom has an unshared pair of electrons the question of antiaromaticity. Oxepins and azepins are eight π electron heterocycles, hence either non- or antiaromatic. Added interest in the oxepins has been generated by the formation of arene oxides in the metabolism of aromatic hydrocarbons and the implication that these are involved in the carcinogenetic activity of polycyclic aromatic compounds. An interesting theoretical treatment of the valence tautomeric behavior of oxepins and azepines has appeared.[174] This extended Hückel calculation indicated that benzene oxide should be more stable than oxepin by 3.9 kcal/mol. The results shown below for the azanorcaradiene-azepine system are more interesting and a similar pattern was found to hold for the tropilidene anion. However, protonated oxepin was not expected to follow suit, while the protonated azepin was calculated as 6.2 kcal/mol less stable than the protonated azanorcaradiene form.

Two excellent reviews of oxepin-benzeneoxide chemistry have been published,[175,176] so only a brief summary of the salient data on the valence isomerism will be given here. Benzene oxide is 1.7 kcal/mol more stable than oxepin, but the entropy gain in forming the oxepin (10.5 eu) leads to its presence in excess at room temperature.[63] The rate of interconversion of isomers is exceedingly rapid even at $-85°$ and the activation parameters for the oxepin ring closure are given in Table 7-2. For the reverse step $E_a = 9.1$ kcal/mol and $\log A = 12.1$.[63] A methyl group at C_1 tips the equilibrium more toward the oxepin form with $\Delta H° = 0.4$ kcal/mol (oxide $\rightleftarrows$ oxepin) and $\Delta S° = 5$ eu. The activation energies for forward and reverse reactions are almost equivalent at 9.0 kcal/mol.[177] Stabilization of the appropriate valence tautomers by incorporation of one or more double bonds into aromatic rings occurs as expected. Valence isomerization is slowed sufficiently in 1,2-naphthalene oxide to permit isolation of the optically active compound.[178] Bridging the 1,6-positions by carbocycles of four carbons or less locks the system in the benzene oxide form, but longer chains permit the oxepin form to exist in equilibrium with the oxide. A very interesting structural change occurs when **151** is acidified, though **151** itself shows no tendency toward ring closure.[193]

$$(7\text{-}40)$$

151

Azepins present a strongly contrasting behavior to the oxepins and indeed to cycloheptatrienes. Thus, while norcaradiene cannot be identified in cycloheptatriene by spectral methods, reaction does occur via the norcaradiene form, but azepins are so strongly favored that reaction generally occurs via that form only.[179–182] Some of the chemistry of the 1 H-azepines has been reviewed,[183] but the valence tautomerism question has been considerably clarified since then. Azanorcaradienes can be stabilized by the usual 1,6-bridging but a chain of three or fewer atoms is required.[183] However, the usual annelation technique does not appear to be successful, at

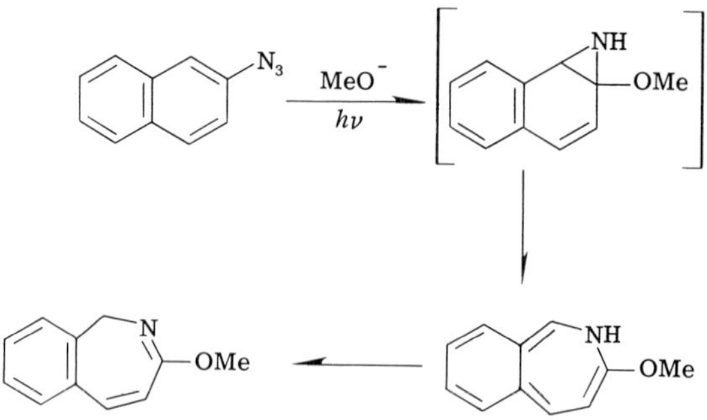

(7-41)

least when hydrogen migration from the nitrogen can occur.[184] Generally 1-H-azepines are not stable when a hydrogen migration is possible [Eq. (7-42)].[185,186] This migration is an extremely rapid reaction with a rate

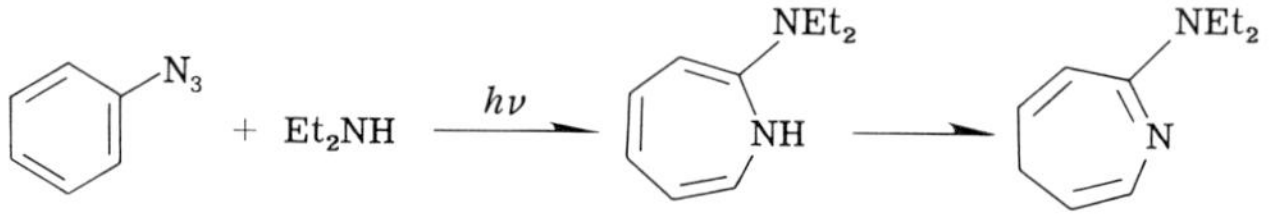

(7-42)

constant of 11 sec^{-1} at 21.7°.[187] Annelation to a thiophene ring does not

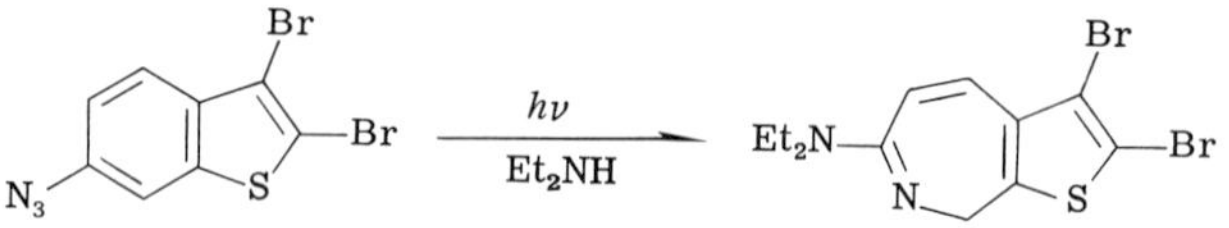

(7-43)

stabilize the azanorcaradiene form sufficiently to prevent the ring expansion.[188]

N-carbethoxy-1H-azepines are stable thermally[183] and 1H-azepines annelated to a benzene ring are also stable.[189] These latter may be prepared by ring expansion of a cyclobutene derivative, and a number of substituted

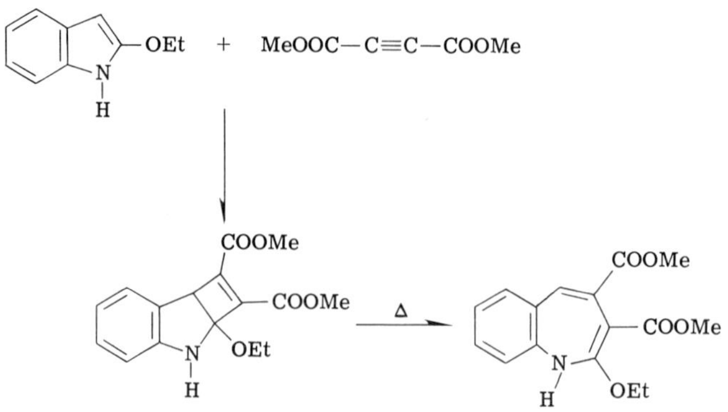

(7-44)

azepines have been prepared in this way.[190] Chemical evidence for valence isomerization of N,4,5-tricarbethoxyazepine has been reported,[191] and an nmr study of **152** showed a rapid valence tautomerism ($\Delta G^{\ddagger} \sim 12$–16

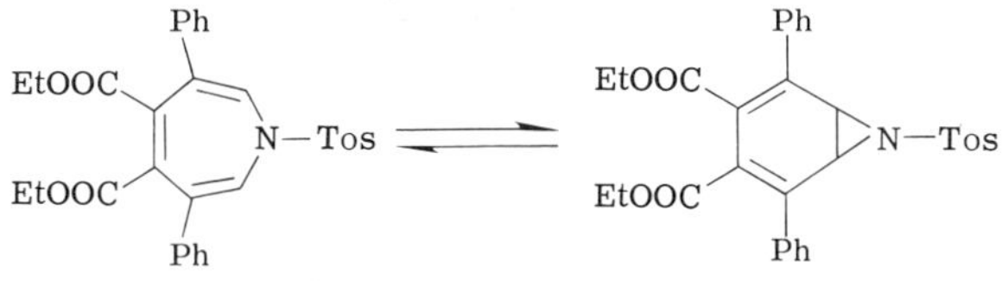

(7-45)

152

kcal/mol) with about 3% of the azanorcaradiene form present at equilibrium at 20°.[192]

A novel problem appears with the thiepin thianorcaradiene system where no stable thianorcaradiene has been isolated. In fact no hard evidence has been presented which indicates the presence of such structures even as transient intermediates, but the circumstantial evidence strongly favors the view that thianorcaradienes generally extrude sulfur thermally with great facility. A common example is shown in Eq. (7-46).[194,195] The chemistry relating

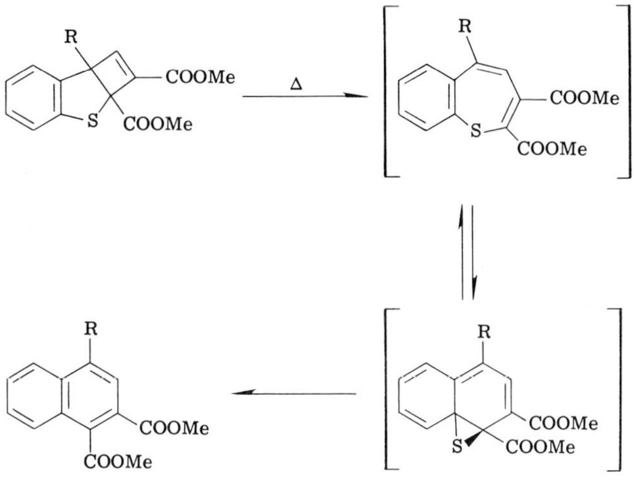

(7-46)

to annelated thiepins has been reviewed[196,197] and only specially pertinent material will be noted here. The postulated formation of the benzothiepins has been confirmed by their isolation and subsequent thermal desulfurization.[198,198a] Thiepins fused to heterocyclic rings also lose sulfur under mild conditions giving benzene rings.[199] The reaction can be useful in synthesis. Monocyclic thiepins have provided quite a synthetic challenge. An attempt to prepare thiepin itself via dehydrohalogenation of 3,4-dibromocyclohexene episulfide gave only benzene.[200] A thermally stable thiepin with spatially demanding substitution at C_2, C_4, C_5, and C_7 has been prepared.[201] However, the most valuable evidence was uncovered by low temperature studies of the dimethyl acetylenedicarboxylate adduct of an aminothiophene.[202,202a]

$$R_1 = H, Me$$

(7-47)

At 25° only the dimethyl phthalate derivative was obtained, but at −30° the initial adduct slowly rearranged into the thiepin which was identified spectrally. Loss of sulfur occurred slowly at −30° but no evidence for the presumed intermediacy of the thianorcaradiene was obtained.

Unlike the thiepins, their sulfone derivatives are relatively stable thermally.

Mock[203] has isolated the parent compound and has shown that it evolves sulfur dioxide at temperatures above 100°. Substituted sulfones are more stable yet.[204] An attempt to isolate a benzoselenepin led directly to a naphthalene derivative presumably via the selenanorcaradiene tautomer.[205]

Borepin, a heterocyclic electrolog of tropylium ion, might well be expected to exist as a stable substance. The heptaphenyl derivative has been prepared and appears to be stable as the borepin valence isomer.[206] However, spectral data indicated that the ammonia complex probably existed in a norcaradiene form. A 1-phenyl-2,3,4,5,6,7-hexa-p-tolyl analog showed six methyl signals in its nmr spectrum, and at elevated temperatures these coalesce first to five (57°) and then four signals (113°). The authors suggested that the borepin–boranorcaradiene equilibrium could account for these results. It is not easy to see what thermal process is represented by the 57° spectral change, and further information on this system would be useful. A number of benzo annelated borepins have been prepared,[207] but these would be expected to exist primarily, if not exclusively, as borepins.

A silicon analog of cycloheptatriene has been obtained and thermolysis led to extrusion of the silicon-bearing moiety.[208] Extrusion may very well

$$(7\text{-}49)$$

occur from the norcaradiene valence isomer, but the silepin is clearly the dominant form. The two phenyl substituents undoubtedly stabilize the silepin form. Benzo annelated silepins have been prepared[208–211] but in all cases annelation favored the silepin valence isomer and no silanorcaradienes have appeared.

Diazepines with proper substitution on nitrogen can be obtained by photolysis of pyridinium ylides.[212,213,214] The reaction is generally useful

$$(7\text{-}50)$$

Y = COOEt, COPh

for the preparation of substituted diazepines.[215] Some evidence which indicates the diazanorcaradiene tautomer can be an intermediate in the thermal decomposition of the diazepine has been presented, but no evidence (via variable temperature nmr) could be obtained for its presence in equilibrium with the diazepine form.[215] Evidence relating to the diazanorcaradiene form as an intermediate in the photochemical reaction, and simultaneously showing that the ring closure and opening to the diazepine are thermal processes has been presented.[216,217] Photolysis of the 2,6-dideuterio ylide[216] and an ylide enriched at C_2 with ^{13}C[217] showed an inverse isotope effect in each case. Thus, the rate controlling step must be thermal and must convert the original ylide C_2 from sp^2 to sp^3 hybridization. Only the initial step of Eq. (7-50) fits the hybridization requirement and it must thus occur via excitation and conversion to a vibrationally excited ground state prior to the ring closure. These results support the suggestion that the base catalyzed formation of amino-pyridines from dihydrodiazepinones proceeds via the

$$(7\text{-}51)$$

enolate and a diazanorcaradiene [Eq. (7-51)].[218]

Heterocyclic amine N-oxides undergo a photochemically induced reaction sequence analogous to, but much more complex than, that of the pyridinium ylides and the rather extensive studies of this process have been reviewed.[219] A novel feature of the amine oxide photolysis is that one product is a 3-oxazepine [Eq. (7-52)].[220] The structure of this product has been confirmed by

$$(7\text{-}52)$$

X-ray analyses, and a series of analogs has been obtained. This reaction is strongly solvent dependent, best yields are generally obtained in acetone and the process is of some synthetic value.[221] While the first step may be photochemical, it has been suggested[219] that the oxaziridine ring rearrangement and ring cleavage are thermal. However, in view of the more recent results with the pyridinium ylides ring closure to the oxaziridine could be a ground state process. It could be presumed that the oxygen walk was necessitated by the reversibility of Eq. (7-53) and the instability of the 2-oxazepine

$$(7\text{-}53)$$

form lacking the aromatic ring. However, the reaction of pyridine N-oxides having proper substitution with aryl or cyano groups leads to the same 3-oxazepine type product.[222] No evidence for an equilibrium content of the azoxanorcaradiene form was found using variable temperature nmr, though the substitution pattern was not favorable.[222]

The first example of a 2-oxazepine was found with a doubly annelated system [Eq. (7-54)].[223] When treated with silica gel this oxazepine rearranged

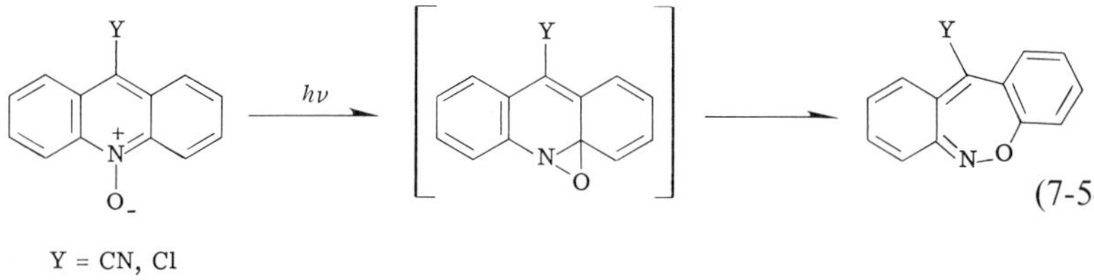

$$Y = CN, Cl$$

(7-54)

to a series of products apparently derived from the oxazanorcaradiene form. A 3,6-diazaoxepine derivative has been obtained [Eq. (7-55)],[224] but photo-

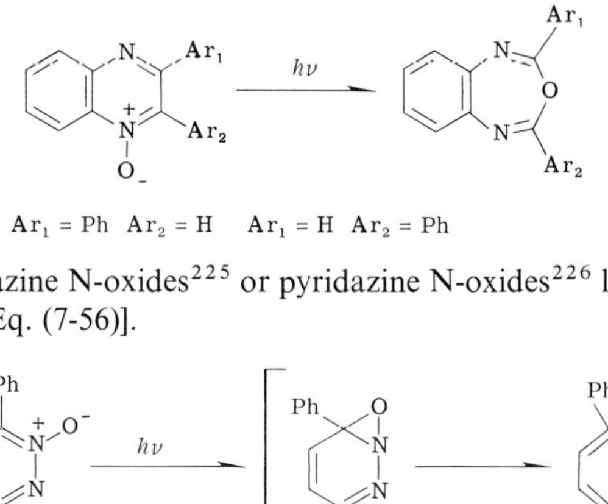

(7-55)

$$Ar_1 = Ph \quad Ar_2 = H \qquad Ar_1 = H \quad Ar_2 = Ph$$

lyses of phthalazine N-oxides[225] or pyridazine N-oxides[226] lead to very different results [Eq. (7-56)].

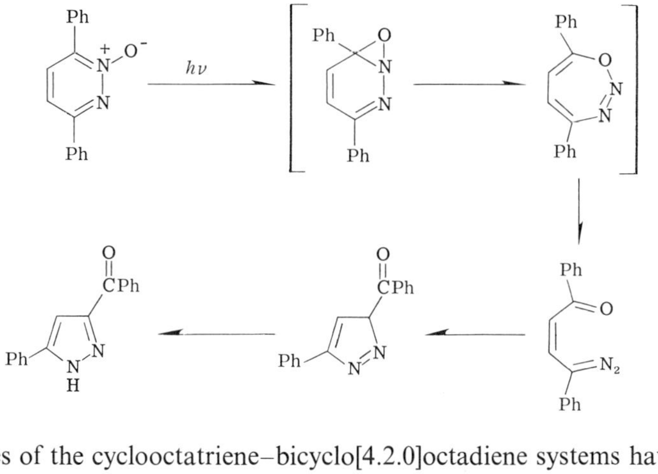

(7-56)

Studies of the cyclooctatriene–bicyclo[4.2.0]octadiene systems have been far less pervasive than for cycloheptatrienes. The rates of ring closure and their variation with substitution are listed in Table 7-2. Brookhart and co-workers[229,230] have shown that tricarbonyl iron complexes of the valence isomers can be isolated and the complex of cyclooctatriene is converted thermally ($\Delta G^{\ddagger} = 29.3$ kcal/mol) completely to the bicyclic complex.[234] The pure bicyclic valence isomers are readily obtained from the complexes by low temperature oxidation, and these have been used to determine rates of some retro-electrocyclic reactions.[230] Thus, reversion rates are tricyclo-[4.3.0.0^{7,9}]nona-2,4-diene, 2.2×10^{-4} sec^{-1} at 30°, its 8-methyl derivative, 1.7×10^{-4} sec^{-1} at 30°, and tricyclo[4.4.0.0^{2,5}]deca-7,9-diene, 4.2×10^{-5} sec^{-1} at 45°.

Equilibrium constants for the cyclooctatriene bicyclo[4.2.0]octa-2,4-diene sytstem have been measured for a number of derivatives and these results have been gathered in Table 7-6. Probably the most striking feature of the table is the enormous variation in the equilibrium constants which range over at least four orders of magnitude. Certain trends are clearly evident, as for example fusion of a small ring to the 7,8-positions of cyclooctatriene introduces an added increment of strain in the bicyclic isomer. The added ring has been shown to be fused anti in the tricyclic system.[231] The addition of a cyclopentane by contrast stabilizes the tricyclic form, probably because the cyclopentane accomodates the eclipsed form of the cyclobutane ring very conveniently. Two groups larger than hydrogen at C_7 or trans-7,8 introduce more steric problems with the tublike cyclooctatriene than with its valence isomer. The bond angle problem with accommodating an sp^2 carbon in a cyclobutane ring is certainly responsible for the preference of cyclooctatrienone for the monocyclic form. However, as Huisgen has noted,[65] the electronic features which control the effects of bromo, hydroxy, acetoxy, azido, etc., groups on the equilibrium are not consistently interpretable.

TABLE 7-6

Equilibrium Data for Cyclooctatrienes

Compound	% Bicyclic isomer	Temperature	Reference
Cyclooctatriene	10.8	60	65
7-Bromo	35	60	65, 227
7-Hydroxy	90	35	228
7-Acetoxy	53	60	65
	75	35	228
7-(7'-Cycloheptatrienyl)oxy	90	35	228
7-Azido	75	35	228

TABLE 7-6 (*Continued*)

Compound	% Bicyclic isomer	Temperature	Reference
7-Keto	6.6	60	65
	0.35	0	229
7,8-Epoxy	0.40	60	65
7-Methyl-7-hydroxy	95	35	228
7,7-Dimethoxy	>95	60	65
cis-7,8-Dichloro	80	60	65
trans-7,8-Dichloro	99	−30	65
cis-7,8-Dimethyl	81	60	65
trans-7,8-Dimethyl	94	60	65
trans-7,8-Diacetoxy	>95	60	65
(cyclooctatriene fused cyclopropane)	~0.02	50	230
(cyclooctatriene fused cyclopropane, H_3C, H)	~0.02	50	230
(cyclooctatriene fused cyclobutane)	2.5	45	230
(cyclooctatriene fused cyclopentane)	~97	58	231
(cyclooctatriene fused cyclohexane)	50	58	231
NC, Me_2C ... CN, CMe_2	~100	25	232

(azepinone structure with R, R′, N, O, R)

R	R^1			
H	H	97.6	60	233
Me	H	80.5	35	233
Me	Me	~100	35	233
H	SO_2Ar (or Me)	~100	35	233

Both Maier *et al.*[236] and Wilson and Warrener[237] have found that 3,4-diaza analogs have a very strong preference for the bicyclic structure, based no doubt on the weakness of the —N=N— bond. Maier has shown that

$$(7\text{-}57)$$

two stereoisomeric 3,4-diazabicyclo[4.2.0]octadienes can interconvert at elevated temperatures ($E_a = 35$ kcal/mol) and the valence isomerization step is probably rate determining. When only a single nitrogen is present as with a 3-aza compound, the azacyclooctatriene is the stable form.[238]

Cyclooctatetraene undergoes electrocyclization to bicyclo[4.2.0]octa-2,4,7-triene [Eq. (7-58)]. Huisgen and Mietzsch[67] have made a kinetic

$$(7\text{-}58)$$

analysis of the reaction scheme of Eq. (7-58) which permitted measurement of the rate of the cyclization. At 100° only 0.01% of the bicyclic isomer is present at equilibrium.[65] The bicyclic isomer was isolated in high purity by Vogel *et al.*[239] by low temperature dehalogenation of dibromide [Eq. (7-59)]. Reversion to cyclooctatetraene occurs at 0° and the rate was meas-

$$(7\text{-}59)$$

$$t_{1/2} = 14 \text{ min}$$

ured (E_a = 18.7 kcal/mol).[239] Later Askani[240] prepared the bicyclic triene by loss of nitrogen from 7,8-diazatricyclo[4.2.2.0^{2,5}]deca-3,7,9-triene. Monosubstituted cyclooctatetraenes should be capable of four different bicyclic valence isomers. In a number of cases the form isolated in the Diels–Alder adduct has the substituent on an sp^2 carbon of the cyclobutene ring.[67,241] That this may not give reliable information about the possible valence isomeric structures has been detailed in a few cases. Bromocyclooctatetraene reacts with TCNE to form one single adduct, but 4-phenyltriazolidine-3,5-dione gives two different adducts [Eq. (7-60)].[242] Despite this difference in

$$(7\text{-}60)$$

adduct formation, the rate of formation of the 1-bromobicyclo[4.2.0]-octatriene is almost twice that of the 7-bromo. The former rate (k_1 in 7-60) is 1.11×10^{-2} sec^{-1} at 120° in acetic acid with $\Delta H^{\ddagger}$ = 23.1 kcal/mol and $\Delta S^{\ddagger}$ = −9.5 eu.[243] Rate of the analogous ring closure with chlorocyclooctatetraene is 1.5×10^{-3} sec^{-1} at 100°.[244]

Normally the bicyclic valence isomer is not isolable at ambient temperatures and must be identified indirectly, but a number of stable bicyclic isomers have been uncovered. Conceptually this can arise because the bicyclic molecule is the more stable thermodynamically, or because the barrier to interconversion becomes high enough to prevent isomerization at room temperature. Examples of both types are known. The earliest case, octamethylbicyclo[4.2.0]octatriene, apparently cannot be converted thermally to the monocyclic isomer,[235] though the reverse reaction goes readily. Paquette has found that fusion of a chain of four methylene groups from C_1 to C_6 stabilizes the bicyclic valence isomer which shows no variation in its nmr

$$(7\text{-}61)$$

spectrum from 40° to 165°[245,246] However, extension of the chain to five methylenes tips the equilibrium to the opposite extreme.[246] Properly situated annelation to an aromatic ring can raise the activation energy for conversion of the bicyclic to the monocyclic form by enforcing participation of the

$$(7\text{-}62)$$

annelated ring [Eq. (7-62)].[247] An interesting variant on this principle stabilizes the bicyclic isomer of Eq. (7-63), where ring opening requires

$$(7\text{-}63)$$

$E_a = 30$ kcal/mol.[248]

The influence of vicinal substitution on this valence isomerization is of considerable interest. A relatively stable bicyclo[4.2.0]octatriene having four adjacent bulky substituents was isolated [Eq. (7-64)].[249] At 65° in deuteriochloroform this rearranges to an equilibrium mixture containing the dimethyldiphenylcyclooctatetraene and a new symmetrical bicyclic isomer. Reversion to the initial bicyclic compound was not observed. Both

$$(7\text{-}64)$$

the retarding influence of the four substituents and the double bond orientation in the cyclooctatetraene show that this isomerization is a true six electron retro-electrocyclic reaction and not a cyclobutene ring opening. This is, of course, precisely what the Woodward–Hoffmann theory would predict. The second bicyclic isomer can be isolated in pure form as a solid, but at temperatures above zero exists in an equilibrium mixture with the cyclooctatetraene form. At $-60°$ the equilibrium constant is about unity. Above $120°$ another pair of isomers appear and can be identified spectrally and by adduct formation. This study produced the first directly observable equilibrium between the valence isomers of a cyclooctatetraene–bicyclo-[4.2.0]octatriene pair. A further example of the same type of behavior was observed with the tetrachloro derivative.[250] Phenyl groups on the cyclo-butene double bond and at C_2 and C_5 stabilize the bicyclic isomer. Thus, 1,2,4,7-tetraphenylcyclooctatetraene exists in equilibrium with its bicyclic form with $K = 1.0$ at $25°$.[253]

Monoazacyclooctatetraenes exist primarily as the monocyclic isomer,[251] but the bicyclic form can be stabilized by fusing a short chain of atoms across the 1,6-positions [Eq. (7-65)].[252] When the chain has four methylene units or less the bicyclic isomer is the sole stable form, but with five methyl-

$$(7\text{-}65)$$

enes in the chain an equilibrium mixture appears. When unsubstituted 1,2-diazacyclooctatetraene is stable in the monocyclic form with a single bond between the nitrogens.[583] At $140°$ it decomposes with formation of benzene ($\Delta H^{\ddagger} = 34.4$ kcal/mol, $\Delta S^{\ddagger} = 4.8$ eu) and pyridine ($\Delta H^{\ddagger} = 30.6$ kcal/mol, $\Delta S^{\ddagger} = -4.1$ eu), presumably via the bicyclic valence tautomers. On the other hand a heavily substituted example exists solely as one bicyclic form.[236] Evidence that the valence isomerization may intrude at higher temperatures

$$(7\text{-}66)$$

R	H	Me	Me
R'	Ph	Ph	Me
E_a	32.8	29.8	28.2

is again suggested by the loss of nitrogen which occurs with formation of a benzene derivative [Eq. (7-66)].[236]

The facile cyclizations of cyclononatrienes and cyclononatetraenes were discussed in detail in Section I,B of this chapter. The trans, cis, trans stereo-isomer of a substituted cyclodecatriene was prepared thermally at 140°, but cyclized to give *trans*-bicyclo[4.4.0]deca-2,4-diene at 220°.[254] Methyl groups at both C_1 and C_6 of *trans,cis,trans*-1,3,5-cyclodecatriene markedly reduce the rate of ring closure, presumably because one methyl must pass through the ring before the disrotatory transition state can be achieved.[255] The cis, cis, trans isomer of 1,3,5-cycloundecatriene undergoes a disrotatory ring closure at 150°.[256]

The annulenes from cyclodecapentaene through [16]annulene provide a fertile area for electrocyclizations which are generally restricted to the sterically favorable 6π disrotatory process. Research prior to 1971 on $(CH)_n$ series has been reviewed.[257] All cis cyclodecaptentaene can be formed photochemically at low temperature and at $-10°$ undergoes electrocyclic ring closure to *cis*-9,10-dihydronaphthalene.[258-261] The *cis,cis,cis,cis,trans*-cyclodecapentaene has also been observed at low temperature as a photolysis product from *cis*-9,10-dihydronaphthalene, and it cyclizes to *trans*-9,10-dihydronaphthalene at $-25°$.[261] This isomer was also obtained by reaction of cyclononatetraenide ion with methylene chloride at low temperature, and ring closure was observed at $-20°$.[262,263] Though not isolated this latter pentaene isomer has been given a role in the thermal conversion of bicyclo[6.2.0]decatetraene to *trans*-9,10-dihydronaphthalene [Eq. (7-67)].[264]

$$(7\text{-}67)$$

A halogenated derivative has been implicated in the thermal reaction of 9-fluoro-10-chlorobicyclo[6.2.0]decatetraene.[265]

Normally [12]annulene exists as the configuration with alternating cis and trans double bonds, but conversion at low temperature to other isomers has been postulated to account for the facile thermal ring closure [Eq. (7-68)].[72] The initial product of that ring closure can react further just above

$$(7\text{-}68)$$

room temperature to give a tricyclic molecule which eventually decomposes into benzene.

No electrocyclic ring closures of [14]annulene itself have been reported, but numerous examples of electrocyclic ring closure leading to the bridged [14]annulene, *trans*-dihydropyrene are known.[74,75] These were discussed in Section I,B. Conversely, [16]annulene is readily converted thermally by a double 6π electrocyclization to a tricyclic isomer, and both closures follow

$$(7\text{-}69)$$

the expected disrotatory pattern.[271,272] The configuration of [16]annulene which leads to the cyclization is not the most stable form but isomerizations occur readily at low temperature.[272]

II. HETEROATOMIC SIX ATOM SYSTEMS

The presence of heteroatoms in a conjugated chain perturbs the π system markedly and raises the question of their influence on the electrocyclic process. Is the stereochemical result altered, or the rate subject to major change, or is their influence of little consequence? Some of these and other questions are answered, but some remain open to future experimentation.

A. One Oxygen Atom: α-Pyrans vs. *cis*-Dienones

Examination of a comprehensive text on organic chemistry such as Rodd's "Chemistry of Carbon Compounds" shows that α-pyrans are not well known. α-Pyran itself has never been observed or isolated, surely the simplest unstrained monocycle yet resistant to synthesis. The first report[273] of a simple α-pyran, the 2-methyl derivative, had a short literary half-life before it was shown to be 2-vinyl-2,5-dihydrofuran.[274] Pyryllium salts have been well known since before 1900 and their reactions with nuclephiles have been of considerable synthetic utility. Pseudobases obtained by treating these salts with aqueous alkali were sometimes formulated as α-pyrans.[275-278] Treatment of pyrylium salts with alcoholic ammonia gave pyridines suggesting a ring opening and reclosure process.[280,281] A large body of work (see below) on pyrylium salts and α-pyrones added strong support to the basic supposition that α-pyrans were transient intermediates in many of these reactions. However, no direct evidence for the α-pyran or direct observation of a ring opening reaction appeared in the literature prior to 1950.

An important contributor to this lack of direct information about α-pyrans and their ring opening proclivities was the difficulty in proving the structure of an α-pyran prior to the general availability of adequate spectral evidence. Any chemical test not proceeding notably faster than a ring opening, or closure if the dienone structure is considered, could give misleading information. Table 7-7 covers the spectral properties of some α-pyrans and dienals or dienones, and these show that even spectral results need to be viewed with care and caution in assigning structures. Generally the ultraviolet maxima appear at similar wavelengths for equivalently substituted pyrans and dienones and the lower extinction coefficient for the pyran can be more indicative than the λ_{max}. In the infrared region the carbonyl band of a dienone often appears at a slightly higher frequency than the enol ether band of the pyran. Dienones generally have three peaks in the 1550–1700 cm^{-1} region whereas α-pyrans have but two, but aromatic substituents can interfere with this test. Pyrans may show a C—O—C stretch in the 1070–1215 cm^{-1} region which may be useful in diagnosis. When applicable, however, the proton nmr permits the most direct and clear evidence of structure. In the absence of strong perturbing features no proton on a ring carbon of an α-pyran appears further downfield than ca 5.75 ppm, and such protons rarely are shifted below about 6.5 ppm. On the other hand, olefinic protons on the dienone system rarely appear upfield from 6.2 ppm unless terminal. Numerous instances of improper structural assignments have resulted from failure to appreciate the small differences in spectral properties of these valence isomers (see below).

The first stable α-pyran isolated and adequately characterized was obtained by Büchi and Yang[282] by irradiation of trans-β-ionone. One assumes that cis-β-ionone should be an intermediate in this process. To study experimentally the possibility that the presumed ring cleavage reactions of α-pyrans and closures of cis-dienones might be an electrocyclic reaction, Marvell and his students had attempted to prepare α-pyrans or cis-dienones of established structure.[300,301] As a part of this study the Büchi pyran was used as a model

$$\qquad\qquad\qquad\qquad\qquad\qquad\qquad\qquad\qquad (7\text{-}70)$$

for the spectral properties of an α-pyran. The nmr spectrum contained a set of weak bands not associated with the pyran which were appropriate for the cis-β-ionone. Variable temperature studies showed that an equilibrium mixture of dienone (17.8% at 54°) and α-pyran (82.2%) was present. This fortuitous result permitted a direct study of the equilibrium and the rate of its attainment.[302] The ring closure of cis-β-isonone is a rapid unimolecular

TABLE 7-7

Spectral Data for α-Pyrans and Dienones

Part 1. Ultraviolet Spectra

Compound	uv Max.	Reference
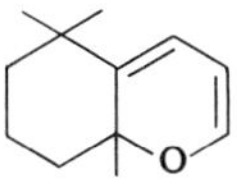	286 (3.68)	282
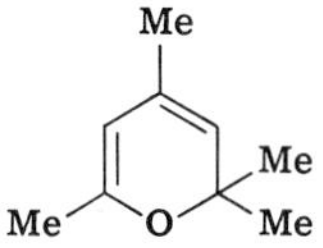	278 (3.22)	283, 284
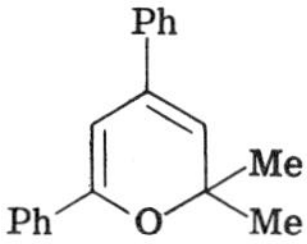	337 (3.94)	283
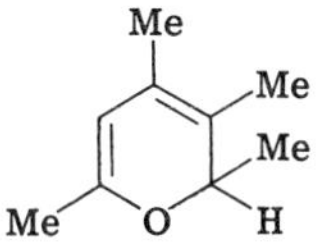	275	285
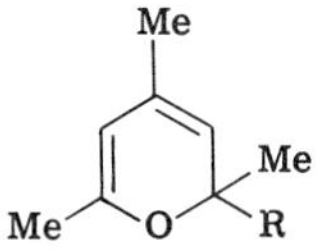 R = Me, Et, *i*-Pr	282–284	294
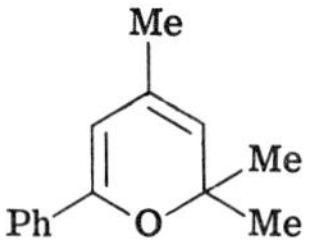	325 (7500)	287
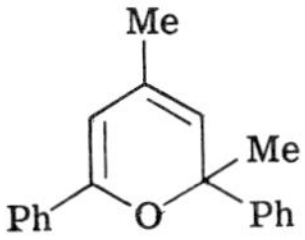	323	288
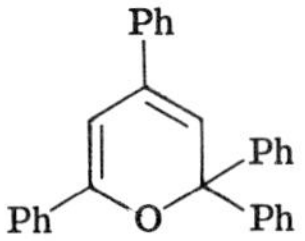	353 (13,500)	287

(*continued*)

Part 1. Ultraviolet Spectra (*continued*)

Compound	uv Max.	Reference
CH$_2$=CH—CH=CCl—CHO	271 (4.25)	289
$\begin{array}{c} \text{Ph} \\ \quad \text{C=CH—CH=CH—CHO} \\ \text{Me} \end{array}$	319	290
	276 (13,500)	291
Me$_2$C=CH—CH=CCl—CHO	299 (4.58)	292
PhCO—CH=CH—CH=CMe$_2$	315 (25,000)	287
PhCO—CH=CH—CH=CHPh	342 (4.5)	293

Part 2. Infrared Spectra

Compound	1550–1700 cm^{-1} bands		Reference
	1668	1610	282
	1680	1632	285
R Me Pr t-Bu	1670 1665 1678	1610 1605 1615	294
	1660	1590–1575	286
	1640	1590–1560	287

Part 2. Infrared Spectra (*continued*)

Compound	$1550-1700 \text{ cm}^{-1}$ bands	Reference
	1650 1600–1580	287
	1640 1590–1540	287
$Me-CO-CH=C-CH=CH-Me$ $\quad\quad\quad\quad\quad\mid$ $\quad\quad\quad\quad\quad Me$ (cis) (trans)	1680	295
	1695	289
	1660	291
$MeCO-CH=CH-CH=Me_2$ (cis) (trans)	1675 1620 1570 1670 1635 1590	286 286
$\quad\quad\; Ph\; Ph\; Ph$ $\quad\quad\;\; \mid\;\; \mid\;\; \mid$ $Ph-CO-C=C-C=CH-Ph$ (cis) (trans)	1660	296

Part 3. nmr Spectra

Compound	Chemical shifts (δ) and J's	Reference
	H_3 4.88 H_4 5.56 $J_{34} = 6$ Hz	283
	H_2 4.58 H_5 4.86 Me_2 1.22 Me_6 1.75	285
	H_2 4.60 H_4 5.5 H_5 4.90 $J_{45} = 6$ Hz Me_2 1.28 Me_3 1.76 Me_6 1.67	297

(continued)

Part 3. nmr Spectra (*continued*)

Compound	Chemical shifts (δ) and J's	Reference
	H_3　4.6–4.7　H_5　4.63–4.81 J_{35} = 1.0–1.5 Hz	294
	H_3　4.90　　H_5　5.59	286
	H_2　5.0　　H_3　5.3 H_4　6.5　　J_{23} = 3.5 H_2 J_{34} = 10 Hz	298
	H_a　7.37　H_b　5.35 H_c　5.45	291
$Me_2C{=}CH{-}CH{=}CCl{-}CHO$ (cis)	H_3　6.53　H_4　7.51 J_{34} = 11.2 Hz	292
$\underset{Ph}{\overset{MeO}{>}}C{=}CH{-}CH{=}CH{-}CO{-}Ph$	H_2　6.6　　H_3　6.88 H_4　7.4　　J_{23} = 10.9 Hz J_{34} = 10.9 Hz	299
$Me_2C{=}CH{-}CH{=}CH{-}CO{-}Ph$	H_2　6.79　H_3　7.61 H_4　6.06	287
$Me_2{-}C{=}CH{-}CH{=}CMe{-}CO{-}Ph$	H_3　6.98　H_4　6.23	287

reaction which shows relatively little sensitivity to solvent[304] and it is certainly an electrocyclic reaction.

$$(7\text{-}71)$$

$k_1 = 1.3 \times 10^{-3}$ sec^{-1} (18°)

TABLE 7-8

Electrocyclic Reaction Rates for Dienones and α-Pyrans

Reaction	Solv.	T (°)	Rate (sec^{-1})	Reference
	TCE	0	k_1 1.58×10^{-4} k_{-1} 8.6×10^{-6}	303
	Py	0	k_1 0.68×10^{-4} k_{-1} 6.2×10^{-6}	
	Et$_2$O	15	1.6×10^{-4}	303
	Et$_2$O	15	5.35×10^{-4}	303
	pentane	13	3×10^{-3}	305
		58	4.35×10^{-5} $\Delta H^\ddagger = 22.4$ $\Delta S^\ddagger = -11.5$	334
		59	7.37×10^{-5} $\Delta H^\ddagger - 21.8$ $\Delta S^\ddagger = -12.1$	334

The rates for electrocyclic ring closures of dienones and of ring opening of α-pyrans are listed in Table 7-8. These are all notable for the contrast in rates between *cis*-trienes and *cis*-dienones. For the closure of *cis*-β-ionone $\Delta H^\ddagger = 20$ kcal/mol and $\Delta S^\ddagger = -5$ eu, while for *cis,cis*-1-cyclohexenyl-1,3-pentadiene the $\Delta H^\ddagger$ is 34 kcal/mol and $\Delta S^\ddagger = -5$ eu.[46] Thus, replacement of a terminal CH_2 by an oxygen reduces the free energy of activation by 14 kcal/mol. No theoretical study of the possible causes of this remarkable rate enhancement has been made, but three causes might be considered. The carbonyl group with its strong polarizing influence increases contributions of the dipolar resonance contributor with a negative charge on oxygen and a positive charge on the terminal carbon. Second, the unshared electron pair on oxygen might participate in the process as theory suggests for the azidoimine–tetrazole case.[306] However, there appears to be rather severe stereoelectronic problems for such participation in the present case, which are absent with azidoimines. Evidence suggesting that the unshared pair does not play a role has been obtained with dienimines (see Section II, E, 1). Finally, the oxygen atom might perturb the π system sufficiently to permit the disallowed conrotatory process to intervene. The similarity between the $\pi \rightarrow \pi^*$ bands in trienes and dienones suggests this is most unlikely. Most likely then, largely by default, is the polar effect of the electronegative oxygen. Some theoretical treatment of this would certainly be helpful.

The hydrolysis of a series of pyrylium salts has been studied by Williams,[307] and he has concluded that the reaction cannot proceed via the process of Eq. (7-72) because the reverse reaction would require initial formation of the

$$(7\text{-}72)$$

enol. At pH 7 the enol formation rate can be calculated from the data of Bell[308] as ca 1×10^{-6} sec^{-1}, and this is too slow for the reverse reaction which has $k = 4 \times 10^{-4}$ sec^{-1}. Though Williams views this as a hemiacetal ring opening, his results would invalidate an electrocyclic process as well. However, at pH 7 and 25° the reaction below has a rate constant of 1.1×10^{-2} sec^{-1},[309] which is quite fast enough to account easily for the reverse reaction viewed as

$$Me \cdot CO \cdot CH_2 - CO - Ph + H_2O \rightarrow Me - \bar{C}O - CH - CO - Ph + H_3O^+$$

an electrocyclic (or hemiacetal formation) process.

The equilibrium between dienones and α-pyrans is exceedingly closely balanced as far as enthalpy is concerned. From the group contributions of Benson[310] the $\Delta H°$ for the equilibrium between cis-pentadienal and α-pyran can be calculated as -0.1 kcal/mol. Thus in the absence of other strongly interactive substituents, the equilibrium can be expected to be dominated by entropy. Since cyclization normally leads to a loss of entropy, unrestricted cis-dienones are generally favored over α-pyrans. Duperrier and Dreux[311] have considered some of the structural features which control the equilibrium position, but their treatment is complexed by consideration of the added equilibrium between *cis-* and *trans*-dienones which does not normally occur under purely thermal conditions. They note that the normal conformation for *cis*-dienones ($R_1 \neq H$) is that of Eq. (7-73) and steric interactions be-

$$\text{(7-73)}$$

tween R_3 and R_5 or R_4 and the C=O are important steric features which can tip the scale in favor of the pyran. Groups which interact conjugatively in the dienone but not the pyran, namely, R_5 and/or $R_6 = $ Ar favor the dienone. As is often the case with equilibria, the modest energy changes needed to cause major alteration in the equilibrium constant (ca. 3 kcal/mol from $K = 0.1$ to $K = 10$), it is not yet possible to pinpoint the exact causes in all cases. Table 7-9 lists the majority of α-pyrans for which reasonable evidence of structure is available.

TABLE 7-9

Compounds Existing Mainly as Monocyclic α-Pyrans at Equilibrium

R_1	R_2	R_3	R_4	R_5	R_6	Reference
Me	H	Me	H	H	Me	297
Me	Me	H	Me	H	H	313
Me	Me	H	Me	H	Me	283, 294
Et	Et	H	Me	H	Me	283, 312
Me	Et	H	Me	H	Et	312
Pr	Pr	H	Me	H	Me	283
Bu	Bu	H	Me	H	Me	283
Me	H	Me	Me	H	Me	285
Me	Me	H	Me	H	Et	286
Me	Me	H	Et	H	Me	286
Et	Me	H	Me	H	Me	286, 294
Pr	Me	H	Me	H	Me	286, 294
Bu	Me	H	Me	H	Me	286, 294
i-Bu	Me	H	Me	H	Me	286, 294
t-Bu	Me	H	Me	H	Me	286, 294
Me	Me	H	Me	H	Ph	286

(continued)

TABLE 7-9 (*Continued*)

R_1	R_2	R_3	R_4	R_5	R_6	Reference
Me	Me	H	Ph	H	Me	286
Me	Me	H	Ph	H	Ph	283, 287, 314
Ph	Me	H	Me	H	Ph	283, 315
Ph	Ph	H	Me	H	Ph	287
Ph	Ph	H	Ph	H	Ph	287
PhCH$_2$	Ph	H	Ph	H	Ph	316
i-Pr	Me	H	Me	H	i-Pr	317
Ph	OH	H	Ph	H	Ph	279
Ph	H	CHO	Ph	H	Ph	318
Me	Me	Me	H	Me	Me	286, 319
Me	Me	Me	Me	H	Me	286
Me	H	Me	Et	Me	Me	320
Me	Me	Me	H	Ph	Me	311, 319
F	F	H	Ph	CF$_3$	F	321
F	F	CF$_3$	Ph	H	F	321
Me	Me	Cl	Cl	Cl	Cl	322
Et	Et	Cl	Cl	Cl	Cl	322
Pr	Pr	Cl	Cl	Cl	Cl	322
Me	Me	Cl	Cl	Cl	H	322
Me	H	Cl	Cl	Cl	Cl	323
Et	H	Cl	Cl	Cl	Cl	323
Pr	H	Cl	Cl	Cl	Cl	323
Bu	H	Cl	Cl	Cl	Cl	323
Pent	H	Cl	Cl	Cl	Cl	323
Ph	Cl	Cl	Cl	Cl	H	324
Me	Me	Cl	Cl	Cl	Ph	322
Et	Et	Cl	Cl	Cl	Ph	322
Pr	Pr	Cl	Cl	Cl	Ph	322
Me	Me	Cl	Cl	Cl	p-Tol	322
Et	Et	Cl	Cl	Cl	p-Tol	322
Me	Me	Cl	Cl	Cl	Mes	322
Me	Me	Cl	Cl	Cl	p-ClPh	322
Et	Et	Cl	Cl	Cl	p-ClPh	322
Me	Me	Cl	Cl	Cl	α-Nap	322

The rapid interconversion rate, the delicately balanced equilibrium and the general difficulty in making clean structural assignments has led to a number of erroneous reports of the formation of α-pyrans. Schinz and his students assigned α-pyran structures to the semireduction products of several eny-nones,[325] but the mixture of products contained only *cis*-dienones. Köbrick undoubtedly obtained 2-phenyl-2,4,6-trimethyl-α-pyran but formation of a dinitrophenylhydrazone suggested the product was the valence isomeric dienone. Though Gompper and Christman[283] obtained a series of α-pyrans their spectral data are confusing since neither the 2,2-diethyl-4,6-dimethyl or 2,2-dipropyl-4,6-dimethylpyrans were reported to have a λ_{max} at longer wavelength than 257 nm. The mode of synthesis also indicates that neither of these is a pure compound and Rouiller and Dreux[312] ascertained the composition of the diethyl derivative. Gompper and Christman assigned the structure of 2,2-diphenyl-4,6-dimethylpyran to a compound having $\lambda_{max} = 323$, which was shown to be 2,6-diphenyl-2,4-dimethylpyran by Köbrick and Wunder.[315] The product from the reaction of Eq. (7-74) was assigned a pyran structure but the data are in good agreement with 1-acetyl-2-hydroxy-4-phenyl-1,3-cyclohexadiene.[327] Sarel and Rivlin suggested that

$$\text{PhC}{\equiv}\text{C}-\text{CH}_2-\text{CH}-(\text{COMe})_2 \quad \xrightarrow[\text{HOAc}]{\text{HBr}} \quad \text{(structure with COMe, Ph, O, Me)} \qquad (7\text{-}74)$$

they had isolated 5-chloro-α-pyran, but the ultraviolet spectrum ($\lambda_{max} = 240$ nm) does not support the assignment.[328]

A substance $C_{17}H_{14}S_2O$ tentatively assigned the structure 3,6-di(phenylthio)-α-pyran is probably α,5-di(phenylthio)-2-methylfuran.[329] Another interesting case is the assignment of the structure 3-hydroxy-2-(p-nitrobenzoxymethyl)-α-pyran to an optically active substance.[330] It is not obvious what the substance is, but the nmr data also show that it cannot be an α-pyran. Similarly, the pyran structures postulated by Jankowski and Luce[331] and by Anschütz and Datta[332] are not in agreement with the spectral data provided.

Despite the growing number of cases where evidence shows that either a pyran or a dienone is the sole identifiable form, there exist a few cases where evidence for the presence of both forms in equilibrium is clear. Aside from the case of Büchi's pyran discussed earlier in this section, Kluge and Lillya[297] showed that 2,3,6-trimethyl-α-pyran exists at room temperature in equilibrium with 13% of the dienone. Maier and Wieseler[335] found $K \sim 1$ for 2,3,4-trimethyl-6-phenyl-2-carbomethoxy-α-pyran. Gosink[320] showed that 2,3,5,6-tetramethyl-4-ethyl-α-pyran contained ca. 25% of the dienone form, while both 5-methyl-4,6-nonadien-3-one and 2,2,5,8,8-pentamethyl-4-6-nonadien-3-one contain ca. 20% of the relevant pyran forms. Finally, de Groot and Jansen[298] found that 3-butenyliden-2,4-pentanedione contains 20% of the pyran form at equilibrium under normal conditions. However, it should be noted that the number of cases subjected to careful examination has been quite limited.

The rapid equilibration of dienone and α-pyran is expected to induce rapid cis-trans isomerization of appropriately substituted cis,cis-dienones [Eq. (7-75)]. Kluge and Lillya[297] have attributed the lack of cis-γ,δ isomers in the

$$(7\text{-}75)$$

photochemical isomerization of dienones to this process. Schiess and his students have investigated some examples of the reaction and have made use of the process to obtain both isomers of a dienone.[291,333,334] Maier and Wiessler observed double bond isomerization via this mechanism during their study of cyclobutene ring opening when carbonyl substituents were present.[335]

The most interesting consequence of the rapid equilibration was observed by Roedig and his collaborators with polychlorinated dienals and dienones [Eq. (7-76)].[336] Labeling studies (^{14}C)[337] and use of an unsymmetrically

$$(7\text{-}76)$$

substituted dienal ($R_1 = Cl, R_2 = Me, R_3 = R_4 = Br, R_5 = H$)[338] were used to establish the mechanism. When $R_1 = Br$ both bromine and chlorine could migrate,[339] and kinetic studies have shown that $\Delta H^{\ddagger} \sim 22$ kcal/mol and $\Delta S^{\ddagger} \sim -10$ eu.[340] Roedig suggests that ring closure to the pyran is the rate determining step, but the rate is rather sensitive to solvent,[337] which might be interpreted as indicating that neither ring closure nor chlorine migration is fully rate determining. When $R_1 = R_2 = R_3 = R_4 = Cl$, $R_5 = Me$, Et, Pr,i-Pr, C_6H_{11}—, Ph, p-Tol the half-life in boiling carbon tetrachloride varies from 35 min(Me) to 365 min (p-Tol) with a marked change occurring from aliphatic to aromatic substituents.[341] When $R_1 = Ph, R_2 = R_3 = R_4 = Cl, R_5 = H$ ring closure and chlorine migration lead to a stable α-pyran which does not open.[342]

Pyrans have been prepared most commonly by treating pyrylium salts with organometallic reagents,[286,294,315,326,343–345] and in most cases both α- and γ-pyrans can be formed. An attempt to devise a theoretical explanation for the regioselectivity has been published.[345] It is obvious that the final isolation of α-pyrans or *cis*-dienones depends on the substitution, and when 2,4,6-trisubstituted pyrylium salts are treated with borohydride only dienones are obtained.[295,296,346] Although α-pyrans may be obtained when 2-pyrones are treated with organometallic reagents,[283,287,288,312,317,319,347] the reaction is complex proceeding via pseudobases and ring opened intermediates and a number of different products are obtained depending on substitution patterns and conditions. Dihydropyrones have been used as precursors for α-pyran preparation,[313,314] and in some cases photochemical procedures have been used to obtain the *cis*-dienones.[282,297] An interesting and certainly unexpected route to α-pyrans involves photoinduced rearrangement of γ-pyrans having aromatic substituents.[348,349]

Attempts to prepare α-pyrans from dihydrofurans by ring expansion with dichlorocarbene gave only dienones,[289,292,328] but properly substituted dihydrofurans should give α-pyrans. Both Surber *et al.*[325] and Marvell[300,301] attempted to prepare proper *cis*-dienones for ring closure by semihydrogenation of enynones. The procedure should succeed with a properly tetrasubstituted double bond, though overhydrogenation could provide problems. Pyrans have been reported as products of addition to diynols [Eq. 7-77)],[350]

$$\text{Me}_2\text{C}-\text{C}\equiv\text{C}-\text{C}\equiv\text{C}-\text{Br} \ + \ \text{R}_2\text{NH} \longrightarrow \qquad\qquad\qquad (7\text{-}77)$$
$$\underset{\text{OH}}{|}$$
$$R = \text{Et, Pr}$$

but it seems rather surprising that an α-pyran should be more stable than a dienamide. A pyran has also been proposed as the stable form rather than a diencarboxylate [Eq. (7-78)].[351] The first case of a pyran stable under

$$(\text{CF}_3)_2\text{C}=\text{CF}_2 \ + \ \text{H}_2\text{C}\underset{\text{COOEt}}{\overset{\text{CN}}{<}} \ \xrightarrow[\text{BF}_3]{\text{Et}_3\text{N}} \qquad\qquad (7\text{-}78)$$

anhydrous conditions but existing purely in open chain form in water has been reported.[352]

Valence isomerization of pyrans plays an important role in the phenomenon of thermochromism, i.e., the color alteration resulting from heating and reversed by cooling. Spiropyrans often form colored melts but are

colorless at room temperature, the dienone form being one resonance contributor to the colored compound [Eq. (7-79)]. The subject has been reviewed

$$\text{(7-79)}$$

colorless   purple

by Day.[353] Photochromism is a related process where the reaction is forced
in one direction by irradiation and reversion occurs thermally. In many of
these cases chromenes or related compounds are involved and the regenera-

$$\text{(7-80)}$$

tion of an aromatic ring renders reversion rapid even at $-40°$.[354,355,356,357]

$$\text{(7-81)}$$

Any process which leads to the o-quinomethide form [Eq. (7-80)] provides
a route to chromene formation and several processes based on this idea have
been developed. Schmid and his students utilized a set of thermal reactions

$$\text{(7-82)}$$

with a sequence composed of a Claisen rearrangement, a 1,5-hydrogen shift
and the electrocyclization.[358,359,360] Cycloaddition to a benzyne followed
by thermal ring cleavage [Eq. (7-83)] has also proved successful.[361–364]

$$(7\text{-}83)$$

$$(7\text{-}84)$$

A simple route to 2-methylchromenes involves heating *cis*-(*o*-hydroxyphenyl) butadiene.[365] In DMSO the reaction kinetics gave $\Delta H^{\ddagger} = 23$ kcal/mol and $\Delta S^{\ddagger} = -17$ cu, a surprisingly low $\Delta H^{\ddagger}$ for a reaction which

$$+ \quad Me_2C=CH-COMe \quad \longrightarrow \qquad (7\text{-}85)$$

interrupts the aromatic ring. Carbonyl condensation reactions have also proved useful [Eqs. (7-85) and (7-86)],[366–369] and the closure is reversible

$$+ \quad Me_2C=CH-CHO \quad \longrightarrow \qquad (7\text{-}86)$$

since with unsymmetrical resorcinols the two possible chromenes are interconvertible.[370]

B. *cis*-Enoneketenes and 2-Pyrones

An electrocyclic process closely related to the α-pyran type forms 2-pyrones from ketenes having a *cis*-enone substitutent [Eq. (7-87)]. Though the process

$$\rightleftharpoons \qquad (7\text{-}87)$$

is reversible the equilibrium lies far on the side of the 2-pyrone. The reaction appears to have been invoked initially to rationalize an unexpected rearrangement [Eq. (7-88)].[371] The mechanism [Eq. (7-89)] was verified by ^{18}O

$$(7\text{-}88)$$

$$(7\text{-}89)$$

labeling studies.[372] Ring closure may be invoked to rationalize the pyrone formation of Eq. (7-90).[373]

$$(7\text{-}90)$$

Low temperature photolysis of 2-pyrone permits identification of the ketene valence isomer which reverts readily to the pyrone.[374,375] This has permitted a kinetic study of the electrocyclization reaction which has $E_a = 9.4$ kcal/mol, log $A = 12.4$.[376] At 26.5° the ketene lifetime is 2.4 μsec, thus making this reaction one of the most rapid electrocyclic reactions studied. The same type of reaction functions with isocyanates[377] and a ketene having an imino function.[378] Syntheses of coumarins by a photochemical process has been shown to utilize this same ketone cyclization.[379]

$$(7\text{-}91)$$

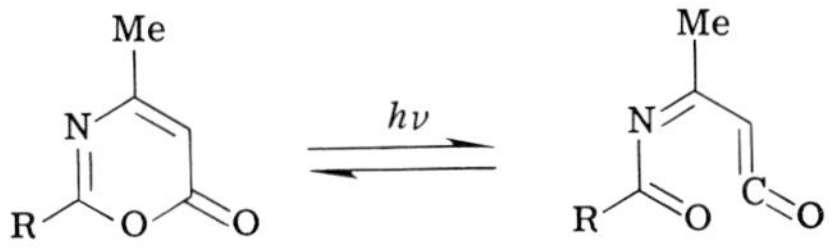

$$\text{(7-92)}$$

R = CF$_3$, *tert*-Bu

C. Two Oxygen Atoms

The interesting but unknown 1,2-dioxin would be expected to undergo a particularly facile and exothermic retroelectrocyclic reaction [Eq. (7-93)].

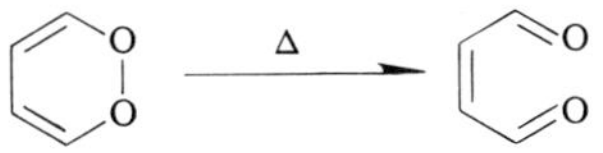

$$\text{(7-93)}$$

The reaction is calculated to be exothermic by about 72 kcal/mol. Goldschmidt and Wessbecher[380] prepared several quinones of structure **153**,

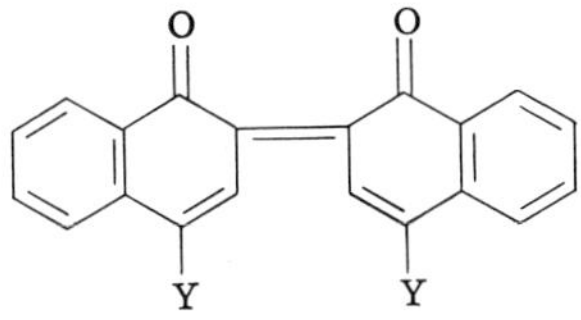

Y = OMe, Br

153

deep blue compounds which were relatively unstable and showed no indication of cyclization to the 1,2-dioxin form. However, a 1,2-dioxin was obtained when 2-*tert*-butyl-4-methoxyphenol was oxidized with potassium ferricyanide or lead dioxide [Eq. (7-94)].[381] This colorless compound formed

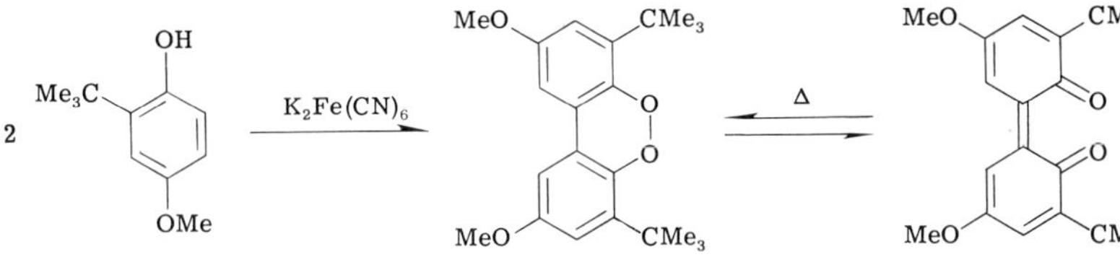

$$\text{(7-94)}$$

154

a deep blue solution in non-polar solvents such as benzene, the blue material being assigned the structure **154**. Newman and Childers[382] prepared 4,5-dihydroxyphenanthrene but were unable to obtain either a 1,2-dioxin or a dione on oxidation.

Evidence indicating the presence of an *o*-xylylene peroxide as an intermediate in the decomposition of **155** was confirmed by trapping the peroxide with maleic anhydride.[383] In the absence of maleic anhydride and with exclusion of oxygen the peroxide rearranges to *o*-dibenzoylbenzene [Eq.

$$(7\text{-}95)$$

(7-95)]. The diketone is produced in an excited state since chemiluminescence is produced when biacetyl is present. Michl[384] has suggested that a similar process is responsible for the chemiluminescence of luminol, and he theorizes that decomposition occurs via bond elongation to produce a biradicaloid doubly excited state of the dicarbonyl product. It is not clear what prevents a rotational process in the ring scission. A very surprising feature is the relatively slow thermal decomposition of the xylylene peroxide which apparently requires a $\Delta G^{\ddagger}$ approaching 20 kcal/mol.

D. One or More Sulfur Atoms

Thiapyrans, unlike their oxygen analogs, are relatively stable and show little tendency to form dienthiones. Thus formation of 2H-thiapyrans by electrocyclization is a reasonable synthetic approach. This method was developed by Brandsma and his students,[385–388] although the necessary dienthione moiety was formed indirectly [Eq. (7-96)]. The R groups surveyed

$$(7\text{-}96)$$

included methyl and ethyl groups in various positions. When R_1 is methyl or ethyl partial rearrangement to an exocyclic double bond occurred.[386] The reaction is successful when the double bond is part of a thiophene ring.[387,388] Reaction in the presence of a tertiary amine diverted part of the reactant to a thiophene.[386,388] Some 1,2-dithins have been prepared and shown to exist in the cyclic rather than the open chain dithioketone form.[584-586]

E. One Nitrogen Atom

Three isomeric aza-trienes are to be expected and while all appear to participate in electrocyclic reactions, the 1-aza isomers are the most common participants and have been most carefully studied.

1. 1-Azatrienes

In the absence of any strongly perturbing substituents 2-dihydropyridines are thermodynamically more stable than 1-azahexatrienes, hence electrocyclization should be an observable process. The well known ring-opening reactions of certain pyridinium ions[389-392] and the ring closure reactions of König's salts,[393] are the initial examples of reactions which could conceivably utilize this mechanism. All of these reactions are accomodated by the general scheme of Eq. (7-97), but the ring-opening–closure

$$\text{(7-97)}$$

process has been visualized as either a nucleophilic addition to an imine [Eq. (7-98a)], or an electrocyclic reaction [Eq. (7-98b)].[394] However, these two processes differ only in the movement of electrons and thus become a matter of personal preference depending only on the particular resonance contributor one chooses to write. Further details of the mechanism are, however, of considerable importance since the influence of the nitrogen on the electrocyclic reaction is of interest.

$$\text{(7-98)}$$

Ring closure of König's salts is mechanistically simpler and more directly revealing of the electrocyclic process. Zincke[390] first noted that **155** forms

$$\left[\text{Ph}\diagdown\text{N}\diagup\diagdown\diagup\diagdown\text{N}\diagup\text{Ph} \atop \text{H} \qquad \text{H} \right]^{+} \longrightarrow \left(\text{pyridinium} \right) + \text{PhNH}_2 \qquad (7\text{-}99)$$

N-phenylpyridinum ion and aniline when heated with excess aniline or ethanolic hydrogen chloride. Marvell and his students have studied the reaction in some detail.[395–398] The red salt **155** certainly exists as the extended all-trans form since both nmr spectra of **155** and X-ray studies of analogous salts[399] generally confirm this structure. Ring closure requires at least one double bond in the cis form, and it has been shown that such trans $\rightleftarrows$ cis interconversions in salts such as **155** correspond to conformational changes rather than configurational, i.e., the interconversion barrier in **155** is 6 ± 2 kcal/mol.[400,401] Thus, neither reactant nor product permits examination of the stereochemistry of the ring closure.

Compound **155** acts as a weak acid in methanol and ring closure proceeds only from the neutral base.[394,395] Though loss of aniline from the direct ring closure product requires addition of a proton this step does not become rate determining under conditions which maintain complete ionization of **155**. Ring closure rate is not altered by added salts, and lowering the polarity of the solvent increases the rate to a modest degree. In methanol containing sufficient triethylamine to convert **155** to its free base $\Delta H^{\ddagger} = 22.7$ kcal/mol and $\Delta S^{\ddagger} = 0.0$ eu with $k_{obs} = 3.5 \times 10^{-4}$ sec^{-1} at 40°. It is obvious from the rate observed that the substitution of a terminal nitrogen for a CH group in the triene has enhanced the rate of the electrocyclic reaction. However, since measurement was made in terms of disappearance of **156t**, the measured rate constant $k_{obs} = K_{tc}k_{el}$ where K_{tc} must be less than 1×10^{-2}. Thus, the effect of the nitrogen on the electrocyclization rate appears to be comparable with that of an oxygen (cf. Section II,A). However, the influence of p-substituents on the rate is minimal, i.e., fivefold from NMe$_2$ to NO$_2$.[396]

Kavalek and Sterba[402] and Kavalek *et al.*[407] have examined the kinetics of ring opening of 2,4-dinitrophenylpyridinium ion (**157**) with aniline in aqueous ethanol buffered to maintain constant pH and at constant ionic strength. The rate was followed by monitoring the appearance of **155** via uv spectroscopy. The observed rate was first order in substrate, first order in aniline and an inverse function of the hydrogen ion concentration. Meas-

(7-101)

urements were made from pH $\sim$ 7 to pH $\sim$ 13 and a plot of the second order rate constant k_2 vs. $[H^+]^{-1}$ showed a change of slope at ca. pH = 10. In region from pH 10–13 the reaction showed general base catalysis. The authors interpreted the change in slope at pH 10 as indicating a change in mechanism and suggested that at low pH the ring opening step (k_3) is rate determining, while at high pH the proton removal step k_2 becomes rate determining. This seems to contradict chemical good sense because the ring opening step is independent of pH while the proton removal step can only become faster as the base concentration increases (at higher pH). Possibly a more reasonable interpretation is that the first two steps occur together being general base catalyzed and rate determining, while the change in slope is a medium effect. In any event, the authors calculate that the ring closure rate $k_3 = 1 \times 10^2$ sec^{-1}, under the conditions they assume make the ring closure rate determining.

Two earlier studies of the mechanism of this ring opening of **157** with aromatic amines were less detailed but indicated that some step subsequent to addition of the amine was rate determining.[403,404] The complex reaction of **158** in basic solution has been studied by Johnson and Rumon.[405]

The reaction to form pyridine is not subject to general base catalysis and the pH-rate profile has a maximum at 11.8. A mechanism involving a nucleophilic addition to a carbonyl group was proposed.

The ring scission of **157** by bases is a general phenomenon, and the kinetics of the reaction with hydroxide ion has also been studied.[406] The reaction was carried out in aqueous ethanol which apparently introduced some complications. At low pH the reaction is reversible and as the pH is increased the rate increases until at high pH (ca. 12 in 50% aqueous ethanol) the rate becomes effectively independent of pH. As the percent of ethanol in the solvent increases the independent region occurs at lower pH's and the rate for that region decreases. The rate determining step is preceded by two rapid

$$(7\text{-}103)$$

reactions, one very fast reaction with $t_{1/2} \ll 1$ μsec, and a slower reaction which appears to lead to product formation. The authors suggest the faster reaction is the reversible addition of ethoxide to the substrate, while the slower is the addition of hydroxide. Again the removal of a proton from the added hydroxide is considered to be rate determining, which necessitates an equilibrium constant for the addition of hydroxide of less than 1×10^{-8}, while the constant for addition of ethoxide is measured at 2.5×10^3.

The influence of substituents on the N-aryl ring was also measured.[406] As would be expected the reaction was very sensitive to these substituents, with electron attracting groups providing strong acceleration ($\rho = +5.05$). Second order rate constants (1 mol^{-1} sec^{-1}) and the substituents on the phenyl ring are listed under Eq. (7-104). It was indicated that for the ring

$$(7\text{-}104)$$

Ar	2,4-diNO$_2$	4-NO$_2$	3-NO$_2$	3-Cl
k	4.5×10^7	1.53×10^2	0.97	1.85×10^{-2}

Ar	4-Cl	H	4-Me
k	4.15×10^{-3}	2.65×10^{-4}	5.0×10^{-5}

closure step a value of rho equal to about two could be estimated, and based on this the authors dismiss an electrocyclic reaction in favor of a nucleophilic addition process.

The acid catalyzed ring closure of **155** to N-phenylpyridinium ion has been studied by Marvell *et al.*[397] As Zincke[390] had indicated, added hydrochloric acid will enhance the rate of ring closure in methanol.[395] In anhydrous acetic acid containing sulfuric acid the first order ring closure rate is approximately linear in H_0 with a slope of -0.62. In DOAC/D$_2$SO$_4$ $k_D/k_H = 2.4$, which indicates a mechanism involving a prior acid-base equilibrium followed by a rate determining reaction of the substrate where no exchangeable proton is removed. The mechanism shown in Eq. (7-105)

$$(7\text{-}105)$$

was considered the most probable alternative. No evidence for direct ring closure of **155** was obtained, and it was assumed that the high resonance energy of **155** prevented the direct reaction. If the mechanism of Eq. (7-105) is correct, this result suggests that the unshared pair of electrons on the nitrogen is not responsible for the accelerating influence of the nitrogen on the electrocyclization. Thus, the electronegativity must be the important feature.

Zincke[390] also noted that when methylamine was allowed to react with **157** the product was **159** [Eq. (7-106)] where the dinitrophenyl moiety has

$$\text{157} + \text{Me}-\text{NH}_2 \longrightarrow \left[\begin{array}{c} \underset{Ar}{\overset{}{}} \underset{H}{\overset{}{N}} = \diagdown \diagup \diagdown \underset{H}{\overset{}{N}} \diagup \text{Me} \end{array} \right]^{+} \qquad (7\text{-}106)$$

157   159

not been extruded from the product. The further reactions of **159** were studied by Marvell and Li.[408] In DMSO at 40° **159** forms N-methylpyridinium ion and 2,4-dinitroaniline by a first order process, $k_1 = 2.25 \times 10^{-5}$ sec^{-1}. The reaction is inhibited by added acid. However, the most surprising feature is that the nmr spectrum of **159** in DMSO-d_6 is that of an equilibrium mixture of **159** and **157** plus methylamine ($K \sim 0.5$). This spectrum appears immediately upon solution of a pure sample of solid **159**, or of an equimolar mixture of **157** and methylamine. Conversely, the spectrum of **159** in trifluoroacetic acid is quite normal. The ultraviolet spectrum showed

$$\text{159} \;\rightleftharpoons\; \underset{N}{\overset{Ar}{}} = \diagdown \diagup \diagdown \underset{\underset{H}{\overset{}{N}}}{\overset{}{}} \diagup \text{Me} \; + \; \text{H}^+$$

$$(7\text{-}107)$$

that **159** loses a proton in DMSO and with the assumption that that proton comes from the dinitroaniline terminus it is clear that the sequence of reactions of Eq. (7-107) must be very rapid. The rate determining step was not ascertained, but it is certain that ring closure is far faster when a 2,4-dinitrophenyl group is on the ring nitrogen than when a phenyl is present.

A more detailed study of the reactions between **157** and piperidine has been made.[409,410] In this example no ambiguity about the proton removed or added can exist, but ring formation from the open chain product can only take place by reversion to reactants. The reaction was run in buffered 50% aqueous ethanol which added complications because of the presence of hydroxide ion [Eq. (7-108)]. Formation of **160** + **160H**$^+$ showed general

(7-108)

base catalysis and the removal of a proton from the piperidine addition product was assumed to be rate determining. Conversion of **160** to **161** was accompanied by a more rapid reversion of **160** to **157** and piperidine. This reversion showed general acid catalysis except in a phosphate buffer, in which case the authors list the ring closure as rate determining, $k_{obs} = 1 \times 10^{-2}$ sec^{-1}.

The kinetics of the reaction between **157** and the methyl cyanoacetate anion [Eq. (7-109)] have also been examined.[411] In methanol containing

$$(7\text{-}109)$$

added methoxide the rate depends on methoxide up to a certain concentration and is independent of base after. At that point the rate $k/K + 1 = 0.56$ sec^{-1} where $K \times$ **162/163**. Once again the ring opening step is a very fast reaction.

In view of the rapidity of the above ring closures, the thermal reversal rates of some photochromic dihydroquinolines is certainly unexpected. Thus, for example, the monocyclic form [Eq. (7-110)] can be generated

$$(7\text{-}110)$$

Y = COOEt, CN
Z = CN, OH

photochemically at $-196°$, but the color of this highly conjugated molecule persists for "some time" at room temperature.[457]

Pyridine *N*-oxide reacts with Grignard reagents giving dienoximes.[412,413] van Bergen and Kellogg[413] showed by nmr that the oxime obtained with

$$(7\text{-}111)$$

phenylmagnesium bromide was the trans, cis, anti form **164**. They proposed that the phenyl entered from the axial side giving a cis dihydropyridine which underwent a retro-electrocyclic reaction faster than any conformational change. However, Schiess and Ringele[414] found the initial adduct was stable at $-50°$ but at $0°$ it reacts to form the oxime. The initial adduct was considered to be the *trans*-dihydropyridine which rearranged to the cis isomer before ring cleavage.[415] The dienoxime recloses to an isomeric dihydropyridine when heated,[414] and the acetate is converted thermally to 2-phenylpyridine [Eq. (7-111)].[413] Dienoximes having a central cis double bond generally undergo a reversible electrocyclic closure to dihydropyridines which irreversibly lose water to form substituted pyridines.[416] However, when a good leaving group is at C_2 the *N*-oxide can be isolated [Eq. (7-112)].[342]

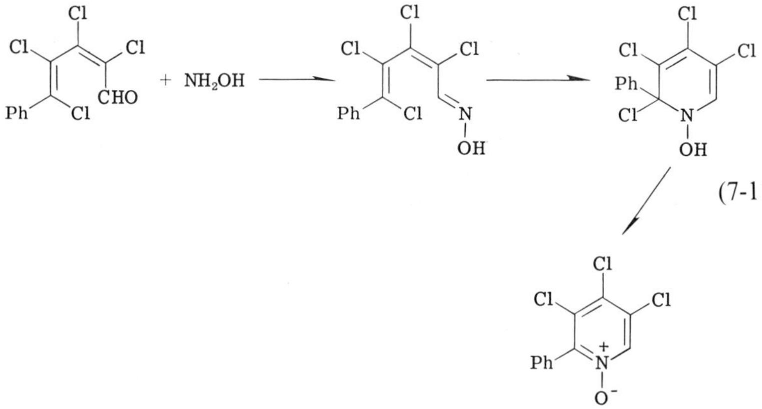

(7-112)

O-Substituted N-oxides will also undergo the ring opening process when treated with bases.[417-423] The mechanism of this reaction does not appear to have attracted much attention but Sliwa and Tartar[423] showed that ring opening occurred under very mild conditions and produced an oxime derivative with a cis central double bond [Eq. (7-113)]. Ring closure of an un-

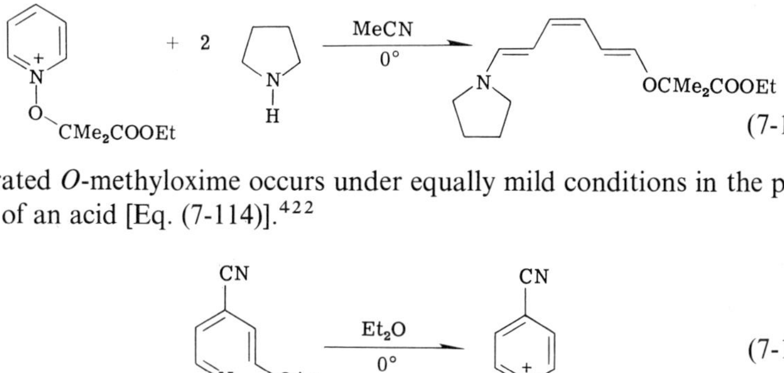

(7-113)

saturated O-methyloxime occurs under equally mild conditions in the presence of an acid [Eq. (7-114)].[422]

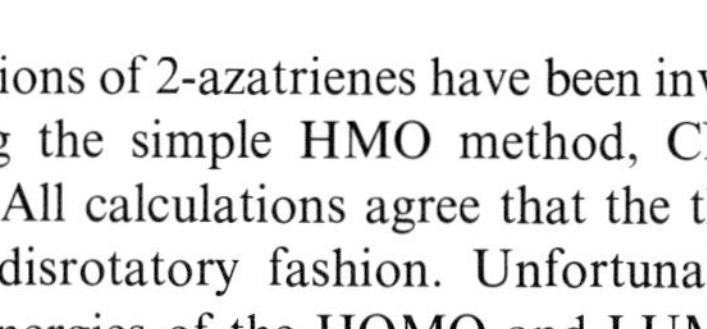

(7-114)

2. 2-Azatrienes

Electrocyclic reactions of 2-azatrienes have been investigated theoretically by Neiman[424] using the simple HMO method, CNDO/2 and extended Hückel procedures. All calculations agree that the thermal process should occur normally in disrotatory fashion. Unfortunately, Neiman did not record the relative energies of the HOMO and LUMO for hexatriene and 2-azahexatriene, so no conclusions vis-a-vis the trend for $\Delta\Delta G^{\ddagger} = \Delta G^{\ddagger}$

(forbidden) $- \Delta G^{\ddagger}$ (allowed) can be reached. It is equally unfortunate that no consideration was given to the alternate 5-atom ring closure (see Chapter 6, Section VI).

Problems associated with the synthesis of 2-azatrienes with the proper double bond configurations have limited the study of this electrocyclic reaction severely. However, Hassan and Fowler[425] discovered serendipitously a means to investigate reasonably directly an example of this reaction

$$(7\text{-}115)$$

$$R = (CH_2)_3CH{=}CH_2$$

[Eq. (7-115)]. Deuterium labeling was used to trace the route shown in Eq. (7-115). Further study with an *N*-ethyl group enabled them to isolate a 5,6-dimethyl-5,6-dihydropyridine which was shown by nmr to have a *trans* configuration. Thus, a disrotatory ring closure must have occurred.

$$(7\text{-}116)$$

All of the other sequences which may involve a 2-azatriene electrocyclization are more speculative. Perhaps the most convincing uses a cationic nitrogen atom [Eq. (7-117)].[426] Another case involving a 2-azatriene having

$$(7\text{-}117)$$

a positive charge leads to an aromatic product by loss of a proton [Eq.

$$(7\text{-}118)$$

(7-118)].[427] The rather extensive changes of Eq. (7-119) apparently encompass a 2-azatriene electrocyclization,[428] and closure of a dienisocyanate

$$(7\text{-}119)$$

was invoked to account for the formation of a pyridone from **165** [Eq. (7-120)].[429]

$$(7\text{-}120)$$

3. 3-Azatrienes

Neiman[424] has calculated that insertion of a nitrogen in place of a CH group at C_3 does not alter the predictions of orbital symmetry theory. There are no simple experimental examples which would test this conclusion. Inevitably the electrocyclic reactions of 3-azatrienes appear as one step in a lengthy sequence proposed to explain an isolated product. A typical example

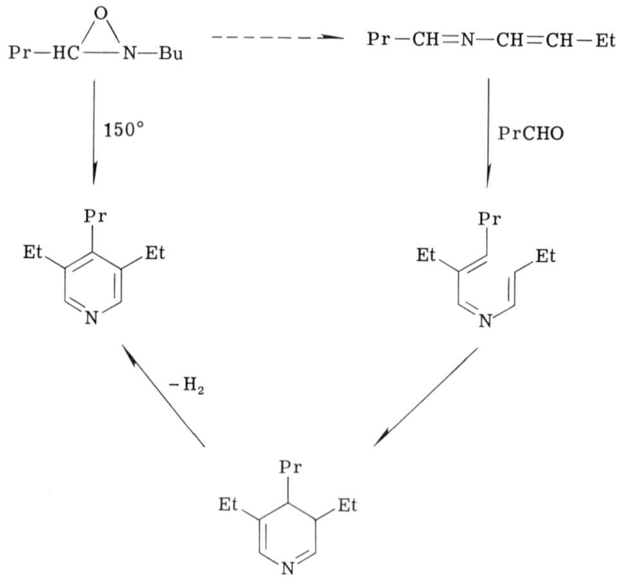

(7-121)

is shown in Eq. (7-121).[430] Further examples are illustrated by Eqs. (7-122)

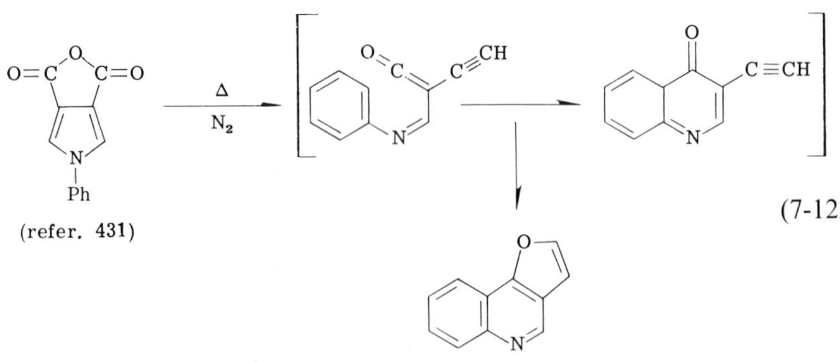

(refer. 431)

(7-122)

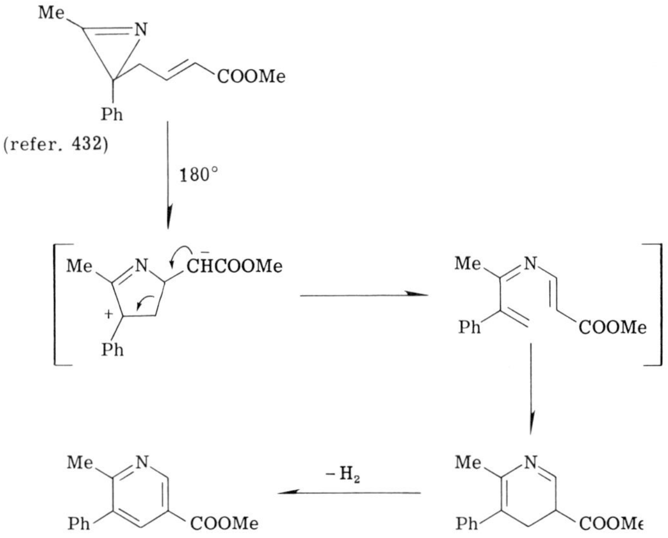

(refer. 432)

180°   (7-123)

and (7-123). An interesting reaction for which no mechanisms was supplied[453] is shown with a suggested route in Eq. (7-124). The Conrad–Limpach syn-

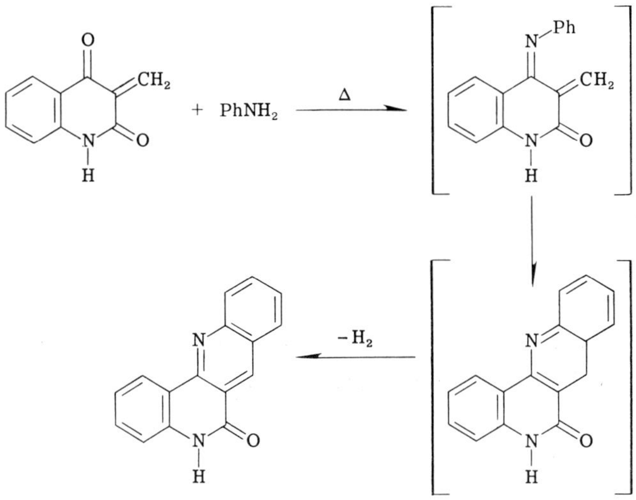

(7-124)

thesis of quinolines may also involve a similar electrocyclic reaction (see Section III,B).

F. Two Nitrogen Atoms

Calculations by Neiman[424] indicated that replacement of two CH groups at C_2,C_4 and C_2,C_5 by nitrogen leaves unaltered the preferred disrotatory thermal electrocyclization. This prediction has been adequately confirmed in the case of the 2,5-diazatrienes. However, these 2,5-diazatrienes can also undergo a 5-atom ring closure (see Chapter 6, Section VI) to form imidazolidines. The six atom ring closure is readily observed when diimines derived from o-diamines are heated.[434] More direct observation of this ring closure has been made by Padwa and co-workers.[435-441] Thus, **166**, which is ob-

$$(7\text{-}125)$$

tained photochemically in several ways, undergoes a facile electrocyclic reaction to give a cis-5,6-diphenyl-5,6-dihydropyrazine [Eq. (7-126)].[435]

$$(7\text{-}126)$$

The cis, cis, trans isomer of **166** is converted to **166** faster than it will cyclize.[435] The six atom electrocyclization proceeds nicely in benzene solution, but in

$$(7\text{-}127)$$

methanol the cyclization takes an entirely different course.[436] Similar cyclizations had been observed earlier.[443,444]

When 1,4,8-triphenyl-2,5-diazaoctatetraene is formed photochemically cyclization to a dihydropyrazine occurs readily [Eq. (7-128)].[437,438] On the

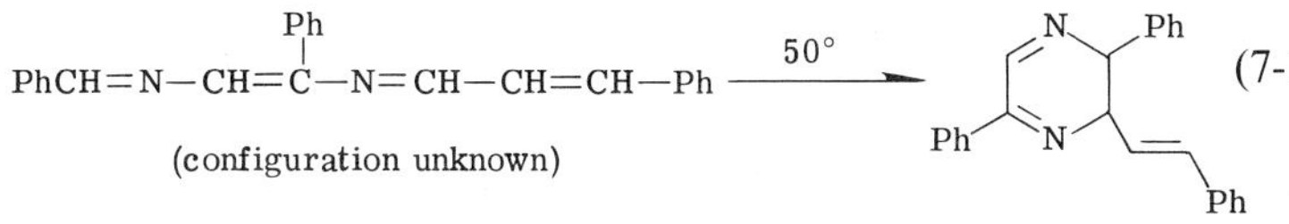

$$\text{PhCH}{=}\text{N}{-}\text{CH}{=}\overset{\overset{\displaystyle \text{Ph}}{|}}{\text{C}}{-}\text{N}{=}\text{CH}{-}\text{CH}{=}\text{CH}{-}\text{Ph} \xrightarrow{\;50°\;} \qquad\qquad (7\text{-}128)$$

(configuration unknown)

other hand, 4,5-diphenyl-2,7-dimethyl-3,6-diaza-2,4,6-octatriene does not cyclize to a dihydropyrazine but undergoes instead a 1,2-hydrogen shift.[441]

Yoneda and co-workers[445,446] have provided an extention of the reaction of Eq. (7-125) which leads very readily to condensed pyrazines [Eq. (7-129)].

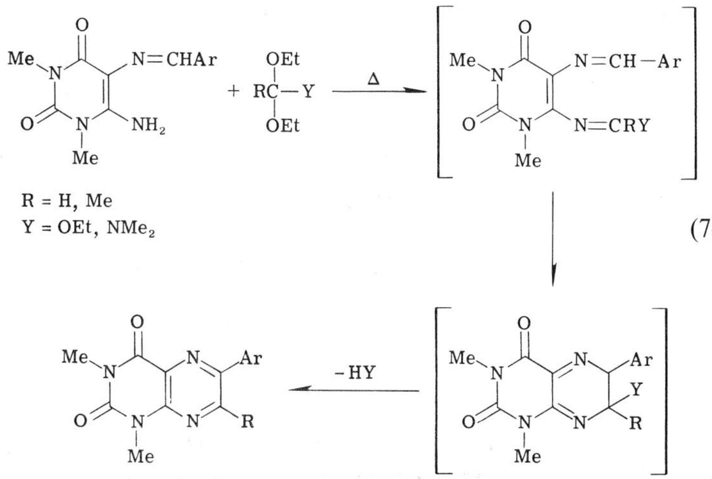

R = H, Me
Y = OEt, NMe₂

$$(7\text{-}129)$$

There is little evidence concerning the cyclization of 2,4-diazatrienes, although several examples which involve either isocyanates or isothio-

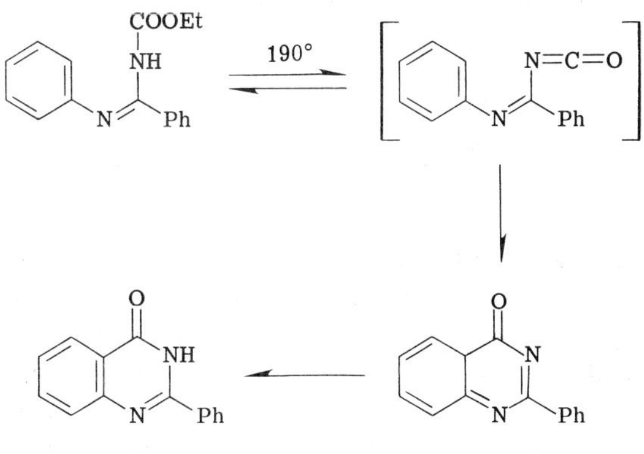

$$(7\text{-}130)$$

$$(7\text{-}131)$$

cyanates have been reported.[447–449] Gilchrist *et al.*[450] utilized an electrocyclization of a 2,4-diazatriene to rationalize a minor product in a thermolysis. One example of a 1,5-diazatriene cyclization has been published [Eq. (7-132)].[451] The direct cyclization product loses water to give an aro-

$$(7\text{-}132)$$

R = Ph, *p*-An, , COOEt

matized product. Formation of pyrimidine *N*-oxide from the oxime of 3-formamido-*cis*-acrolein may also involve an electrocyclic reaction of this type.[456]

Baldwin and Basson[452] suggested a retro-electrocyclic reaction of a 1,2-dihydro-1,2-diazine as one step in the rationalization of an unexpected

$$(7\text{-}133)$$

product from the reaction of *trans*-1,2-diacetyl-4-cyclohexene with dimethylhydrazine. Generally, however, 1,2-dihydro-1,2-diazines are thermally stable,[453] though that ring opening seems quite reasonable.

Recently van der Plas and his students have uncovered an unusual and unexpected nucleophilic aromatic substitution process with pyrimidines called the $S_N(ANRORC)$ process.[454,455] Work on the mechanism including

$$(7\text{-}134)$$

^{15}N labeling studies has been reviewed.[454] Initial addition followed by electrocyclic ring opening appears to be a general opening gambit, but several routes may pertain subsequently [Eq. (7-134)]. In many cases a pyrimidyne route or a direct addition-elimination process may compete. However, the scheme serves to emphasize once again the enormous facility of the electrocyclic ring opening and closure reactions with these nitrogen containing rings.

G. Trienes with Three Nitrogens

An interesting reaction which can formally be written as an electrocyclic reaction, but which ultimately must involve an unshared pair of electrons on one nitrogen is shown in Eq. (7-135). The literature of this reaction has

$$(7\text{-}135)$$

been reviewed.[458] Conceptually the simplest route involves addition of an unshared pair from the oxime to the diazonium ion.

III. USES IN SYNTHESIS

The potential for the triene electrocyclic reaction in synthesis should be great, but the general requirement for a central cis double bond has been the major stumbling block. Stereospecific synthesis of trienes with predetermined configurations of all double bonds has proved difficult, particularly via the very appealing semi-hydrogenation of dienynes.[459] Thus, the most important search has been for solutions to the problem of generating the required *cis*-trienes.

A. Carbocyclic Synthesis

Synthesis of cyclohexadienes having a specified configuration of the two sp^3 hybridized centers can be achieved by this electrocyclization. However, this method has never been seriously exploited, though Frater used the reaction in a synthesis of chamigrene.[470] One problem which often causes problems is a 1,5-hydrogen shift, and generally when a terminal cis double bond is present the 1,5-hydrogen shift becomes competitive with cyclization.[15] A similar problem arises with the cyclization of *o*-divinylbenzenes, which constitutes a potentially useful synthesis for 1,2-dihydronaphthalenes.[121]

Probably the simplest and most obvious method for bypassing the need to prepare a central cis double bond is to operate at a temperature high enough to permit double bond stereomutation. As a route to cyclohexadienes of known stereochemistry this procedure is of no value. Despite the rather general observation that the central double bond isomerizes faster than terminal ones, the 1,5-hydrogen shift also intervenes. The early studies of Goldblatt[96,97,98] on the formation of pyronenes is an example of the use of this technique. Both Woods *et al.*[100], Fleischaker and Woods,[101] Woods and Viola,[102] Pines and Kozlowski,[104] and Pines and Chen[105] modified the method via the introduction of a solid catalyst such as aluminum oxide. Under appropriate conditions this led to the formation of aromatic products.[104,105] Workers at the Northern Regional Research Laboratory in Peoria, Illinois, devised a procedure for cyclizing non-conjugated polyene acids such as linolenic by heating with potassium hydroxide in ethylene glycol.[460–465] The first step appears to be an isomerization to a conjugated triene, which is followed by cyclization or by a Diels–Alder dimerization depending on the conditions. Eleostearic acids behave similarly under the same conditions. Nayak *et al.*[143] used a sulfur catalyst to cyclize the methyl esters of eleostearic acids at 160°.

Edmunds and Johnstone[114] obtained methylated tetralins from pyrolysis of β-carotene. A triene ester derived from β-ionone cyclized stereospecifically at 240° with no apparent 1,5-hydrogen shift [Eq. (7-16)].[144] Pyrolysis of 1-arylbutadienes or their precursors has been used to prepare both dihydro- and fully aromatic products [Eq. (7-136)].[466,467] The reaction also has been tested in examples where the aryl moiety is a pyridine,[466,468] furan, thiophene, or *N*-methylpyrrole group.[469]

$$(7\text{-}136)$$

Conrotatory ring-opening of properly substituted cyclobutenes can give rise to *cis*-trienes, and in the event that a *trans*-triene is formed the reversibility of that ring opening may permit the proper *cis*-triene to form more slowly but cyclize to a cyclohexadiene irreversibly. Because of the ease with which reversion occurs this latter possibility has been most utilized with benzo-

$$(7\text{-}137)$$

167

cyclobutenes. Thus, 9-phenylanthracene can be prepared conveniently from **167**.[471] A dihydro product was isolated from tetraphenylbenzocyclobutene because ring closure occurs at a low temperature [Eq. (7-26)].[123] Photochemical synthesis of vinylbenzocyclobutenes led to the development of a

$$(7\text{-}138)$$

useful approach to the preparation of dihydronaphthalenes.[472] α-Tetralone can be obtained from 1-vinylbenzocyclobutenol [Eq. (7-22)][118,119] and polynuclear aromatics have been obtained from chlorinated naphthocyclo-

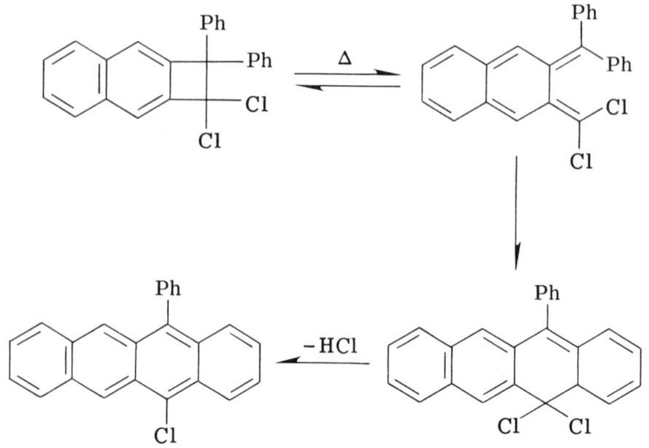

(7-139)

butenes [Eq. (7-139)].[473,474] An example which uses a vinylcyclobutene is shown in Eq. (7-21).[117]

In some cases the problem of the cis double bond does not arise at all, i.e., Eqs. (7-26) and (7-139). While a limited case at best it does serve in some

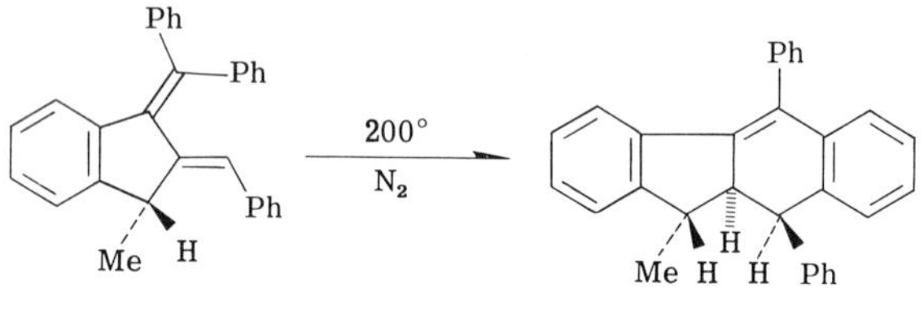

(7-140)

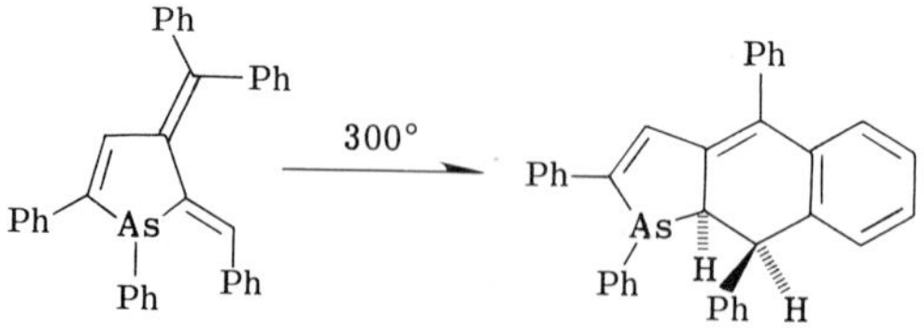

(7-141)

instances.[475,476] Where the central double bond lies in a normal ring the problem of the cis configuration is automatically solved. Schiess has developed a very useful scheme which is nicely adapted to the preparation of hexahydronaphthalenes [Eq. (7-142)].[51,54] The process serves best when a

$$(7\text{-}142)$$

stabilized Wittig reagent is used so that the last vinyl group has a trans configuration. However, double bond positions in the product cannot be assumed to be as shown.[54,55,16]

Photochemical opening followed by thermal closure can be used to alter the stereochemistry of cyclohexadienes. While this is not a directly coupled pair of reactions, the method was employed to convert ergosterol to pyro- and isopyrocalciferols.[9,10,11,12] Vogel et al.[13] added an example with a cyclononatriene intermediate, while Dauben has investigated the process more thoroughly.[255,256]

The readily reversible ring opening of cyclobutenones (see Chapter 5, Section III,C) and the rapidity of 6π cyclization of dienketenes[52] combine to make a very interesting route to naphthols and phenols. A nice example

$$(7\text{-}143)$$

is illustrated in Eq. (7-143),[477] and other cases of a similar type have been carried out.[478,479] Preparation of a phenol is shown in Eq. (7-144).[129]

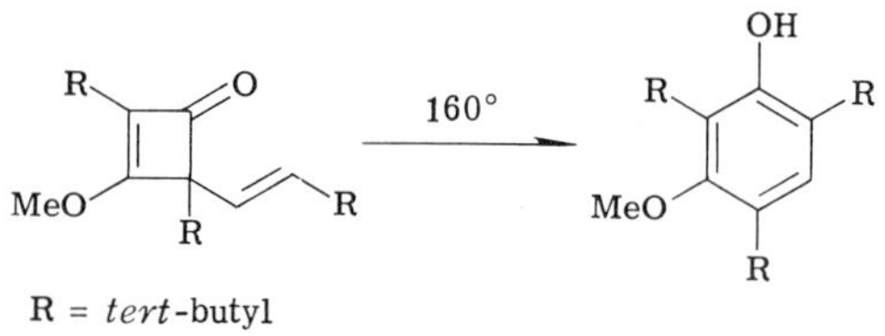

$$
\text{(7-144)}
$$

R = *tert*-butyl

Breslow *et al.*[480] obtained α-naphthol from cyclopropenone via a reaction which might be useful for some special cases. A more generally valuable route to β-naphthols has been devised [Eq. (7-145)].[481,587]

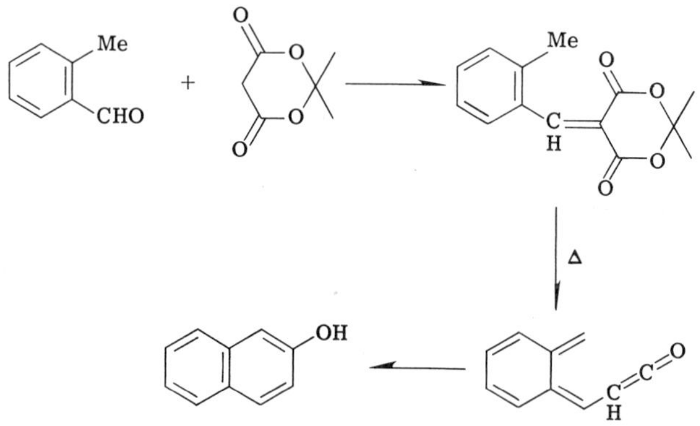

$$
\text{(7-145)}
$$

The widest ranging and most important solution to the problem of the cis double bond is the use of 1-dialkylaminotrienes which interconvert geometric isomers readily under conditions which lead to cyclization. This is the electrocyclic–elimination sequence leading to fully aromatized rings, a procedure based on the Ziegler–Hafner azulene synthesis (see Chapter 8, Section II), but developed as a general process by Jutz and Wagner.[482] The very broad exploitation of this scheme [Eq. (7-146)] by Jutz and his

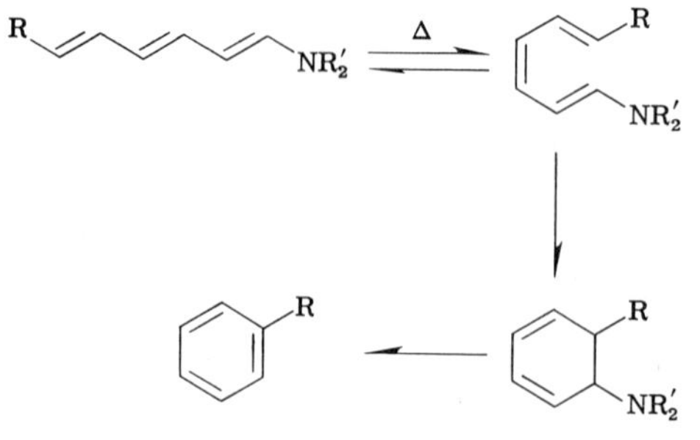

$$
\text{(7-146)}
$$

students has been given a complete review.[483] Typical examples are illus-

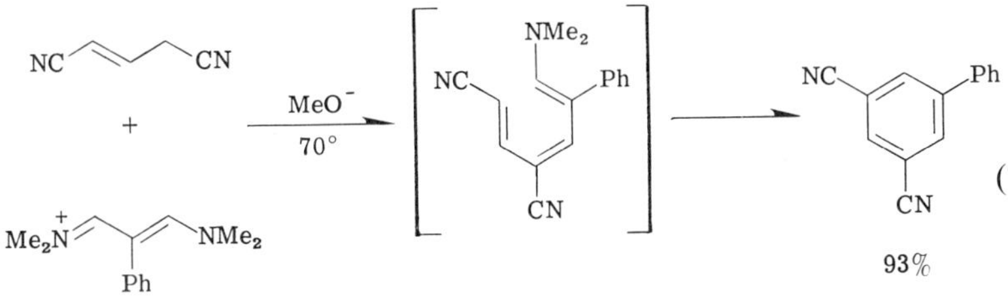

$$(7\text{-}147)$$

93%

trated in Eqs. (7-147) and (7-148). An anthranilate derivative was prepared

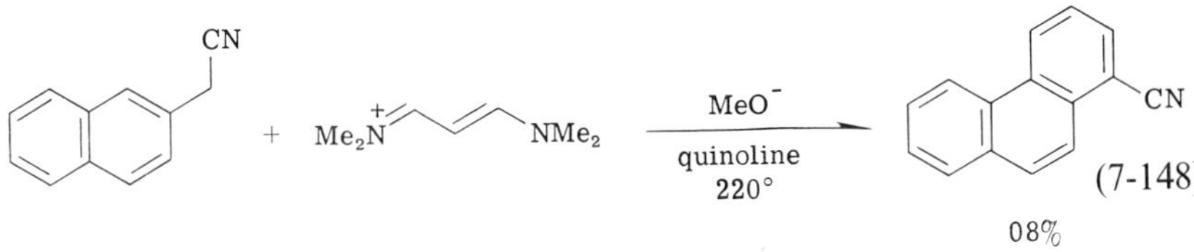

$$(7\text{-}148)$$

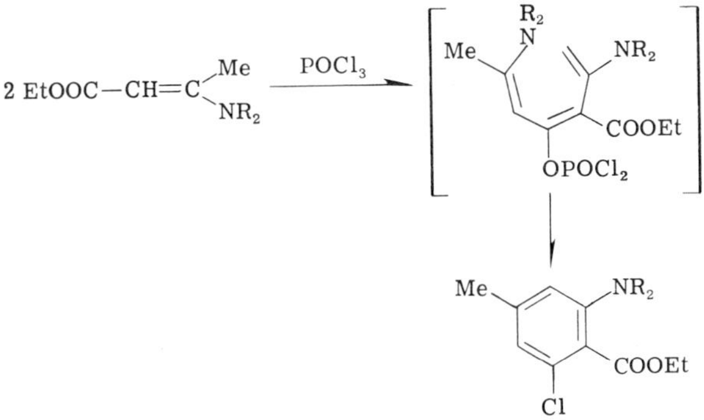

08%

$$(7\text{-}149)$$

by a closely related reaction.[484] Jutz[483] has proposed that the well known
Elbs reaction[485] is an example of this same electrocyclic–elimination sequence. The initial step is a 1,5-hydrogen shift, and the sequence is illustrated
in Eq. (7-150).

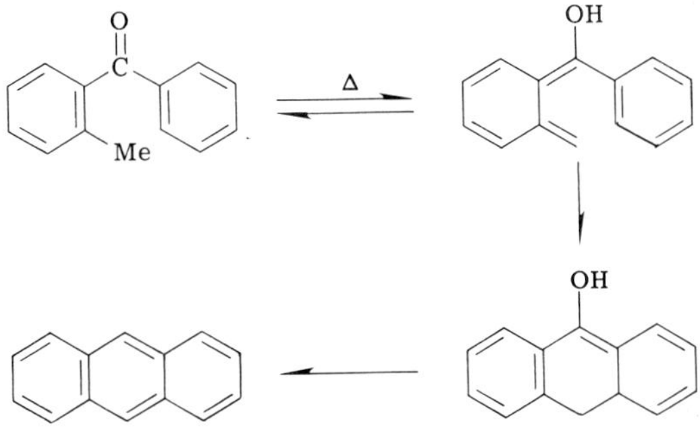

$$(7\text{-}150)$$

Prior to the development of this electrocyclic–elimination sequence of Jutz, a preparatively useful procedure for the conversion of pyrylium salts to aromatic compounds had been developed.[486-495] The most common sequence is shown in Eq. (7-151), where Nu: can be OH^-, R_2NH, $RMgX$,

$$(7\text{-}151)$$

or $CH(CN)_2^-$. A second common sequence involves the shift of a hydrogen from the added nucleophile rather than from an alkyl group already present on the pyrylium ring. Nucleophiles which participate in that process include

$$(7\text{-}152)$$

$^-CH_2NO_2$ and $ArCH_2MgX$.[348] A more recent development utilizes a vinylpyran which opens and recloses.[494,495] The alternate eliminations apparently depend on the nature of the R groups on the nitrogen.[494] Triene electrocyclization proceeds more rapidly than normal which could be the result of push–pull substitution as suggested by Epiotis. The early work on formation of aromatic compounds from pyrylium salts has been reviewed.[496]

Extension of this method to pyridinium ions having a 2-methyl substituent has been reported.[497,498] A more unusual reaction was noted when 2-methyl-5-nitropyridine was treated with methylamine.[498]

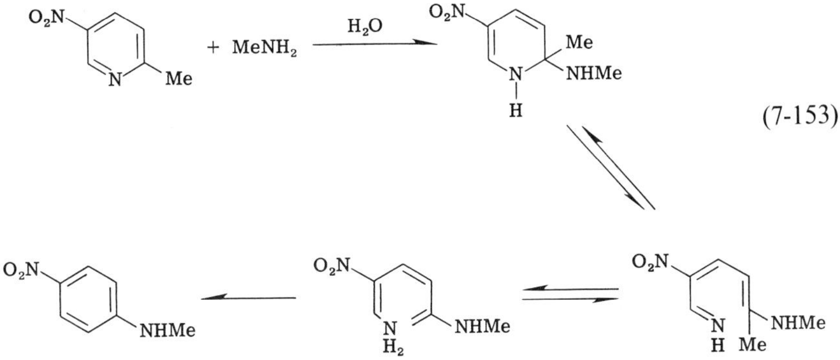

(7-153)

A potentially very valuable synthesis which generates two rings at one time has been developed by Hilton.[16] The two rings could be formed sequentially or simultaneously [Eq. (7-155)]. The stereochemistry has not been

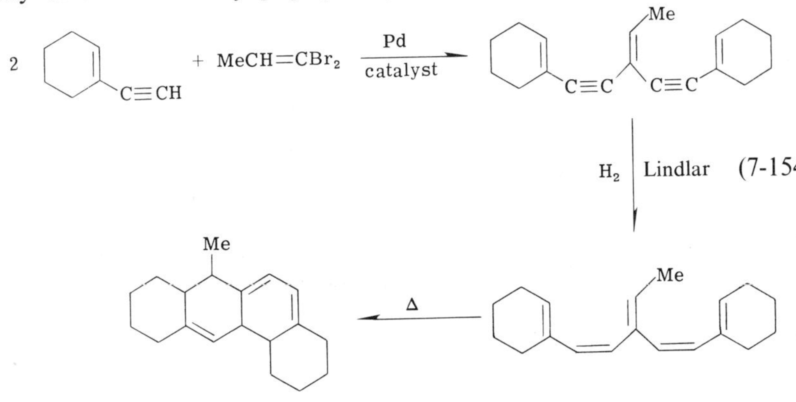

H_2 | Lindlar   (7-154)

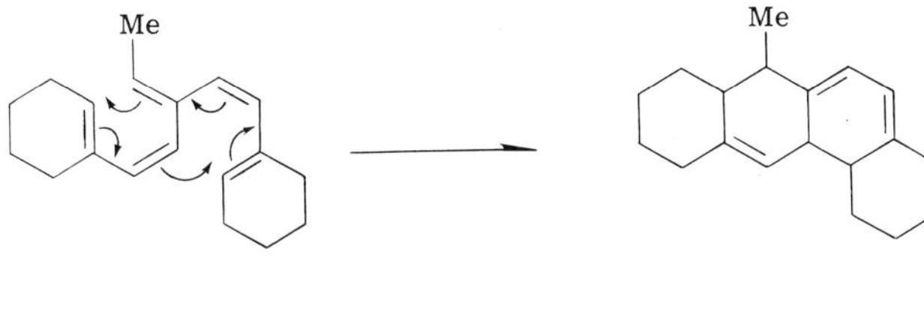

(7-155)

examined, but if the closures are sequential both should be disrotatory, but if simultaneous both could be conrotatory. This latter would be an example of a ten π electron electrobicyclization. As long as the bicyclization constitutes a single cycle process, the rules about nodes will hold, so for this system an even number of nodes is required.

The strain in 2,4-bicyclo[4.1.0]heptadienes and 2,4-bicyclo[4.2.0]octadienes generally makes the monocyclic trienes more stable than the bicyclic tautomers, a result which makes this retroelectrocyclization useful in synthesis. Vogel et al.[20] used this approach to prepare dimethyl-1,6-cyclooctatrienedicarboxylate. Knox et al.[499] prepared **168** by this procedure, and

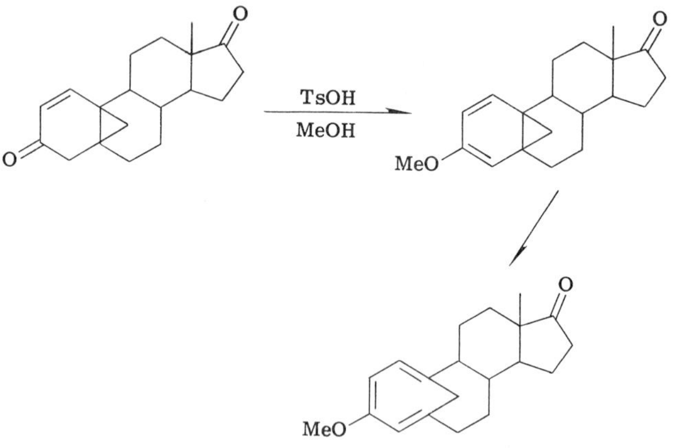

(7-156)

168

Darms et al.[500] used the same route to dimethyl 1,6-cycloheptatrienedicarboxylate. Vogel has made extensive use of the technique in a very elegant fashion during his classic studies of the aromaticity of bridged annulenes.[501] Paquette and Phillips[245] found this route the most appropriate for their study of bridged cyclooctatetraenes. Cyclooctadienones have been prepared successfully by this procedure.[502]

B. Synthesis of Heterocycles

Very broad use has been made of electrocyclic reactions in the synthesis of heterocyclic molecules, both in a direct sense and via retro-electrocyclic reactions employed to prepare reactants leading to heterocycles. The literature relating to this usage is so large that we can only suggest the general routes without any attempt at comprehensive coverage. It is fortunate,

therefore, that van der Plas has provided a very thorough survey of that part
of the area which relates to conversion of one heterocycle to another.[503]

1. One Oxygen Atom

Electrocyclization with the formation of oxygen heterocycles has not been
an important synthetic process. Use in the investigation of α-pyans was
reviewed in Section II,A and will not be considered further here. Schroth
and Fisher[504] used the electrocyclic–elimination route to prepare pyrylium
salts from enamines (see also refer. 578). Chromenes are readily prepared
via an electrocyclic process with either propargyl phenyl ethers[358-360] or
benzoxetenes[361-364] serving as primary reactants. Crombie and co-workers

$$(7\text{-}157)$$

have studied the formation of chromenes from phenolates and α,β-un-
saturated aldehydes as part of their work on citrans.[367,369,370,505-507]
Flindersine has been synthesized by an electrocyclic ring closure of the pyran
unit.[298,508]

2. One Nitrogen Atom

Electrocyclic reactions have been very important in the synthesis of
pyridines, pyridinium ions, pyridones and pyridine N-oxides and polynuclear relatives of these. Formation of pyridines from pyrylium salts by
treatment with alcoholic ammonia has been known since the early years
of this century.[280,281] A general scheme for the reaction is shown by Eq.
(7-158) where ring closure to a dihydropyridine is written as an electrocyclic

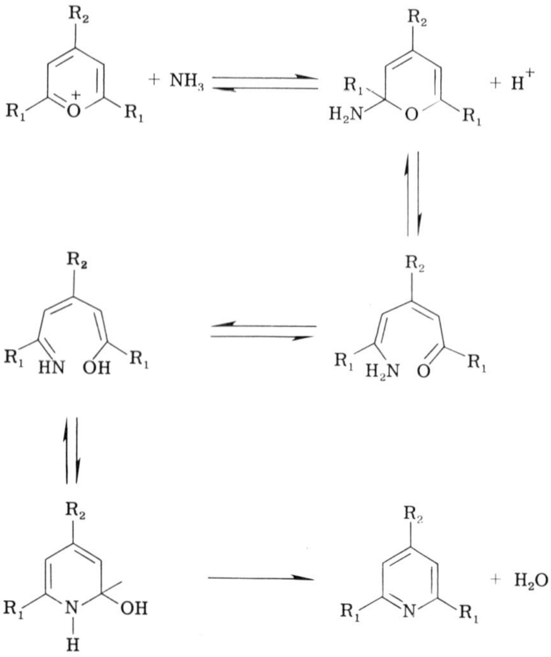

$$\text{(7-158)}$$

process. This step is often written as a nucleophilic addition to a carbonyl group, but the 1,7-hydrogen shift is a very facile process generally and imino enols have been isolated in some cases.[509] The initial addition to the pyrylium ion can give either 1,2 or 1,4-addition products, but 1,2-addition is generally favored.[510] Van der Plas has reviewed the pyridines which have been prepared by this method.[511]

If ammonia is replaced by primary amines either aliphatic or aromatic pyridinium ions are formed.[511] O'Leary has suggested modifying free amino groups in proteins by this procedure.[512] Though obviously less useful preparatively, the ring opening and reclosure of pyridinium ions (Section III,E,1) has found some value for the preparation of N-arylpyridinium ions which are not obtainable directly.[391–393] Generally the use of N-cyano-pyridinium ions[393] has been favored over Zincke's N-2,4-dinitrophenyl-pyridinium salts.

Jutz has enlarged the scope of his electrocyclic–elimination method to permit the preparation of heterocyclics including pyridines, quinolines, iso-quinolines and benzoquinolines of a wide variety of substitution patterns.[513] Some examples are shown here [Eqs. (7-159) and (7-160)]. Other examples

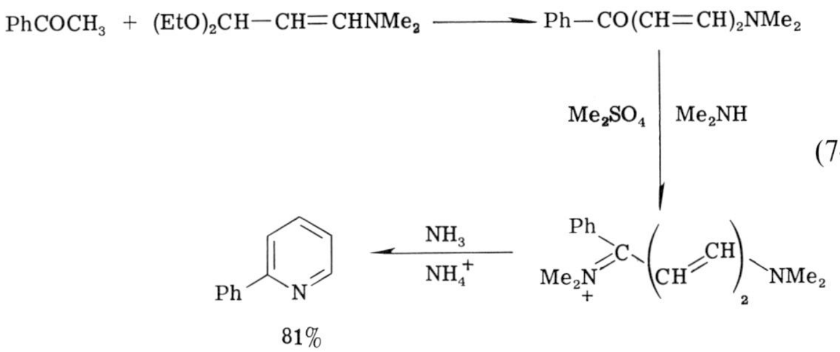

$$PhCOCH_3 \ + \ (EtO)_2CH-CH=CHNMe_2 \longrightarrow Ph-CO(CH=CH)_2NMe_2$$

$$(7\text{-}159)$$

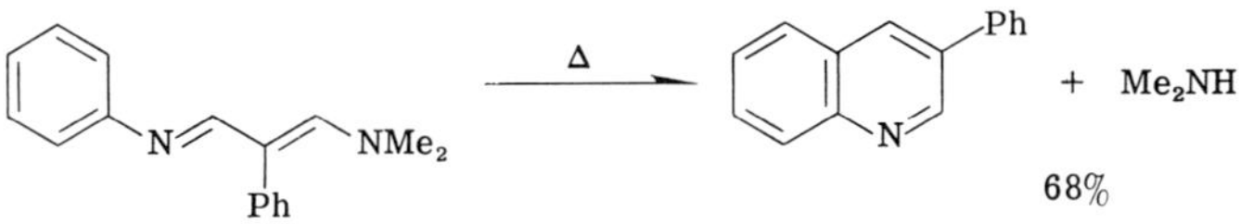

Me$_2$SO$_4$ | Me$_2$NH

81%

+ Me$_2$NH (7-160)

68%

have been observed by Lloyd and Gagan[514] as well as by Acheson and Bolton.[515] A related formation of a pyridine was noted by Harris *et al.*[484] Synthesis of 4-hydroxyquinolines [Eq. (7-161)] probably involves an elec-

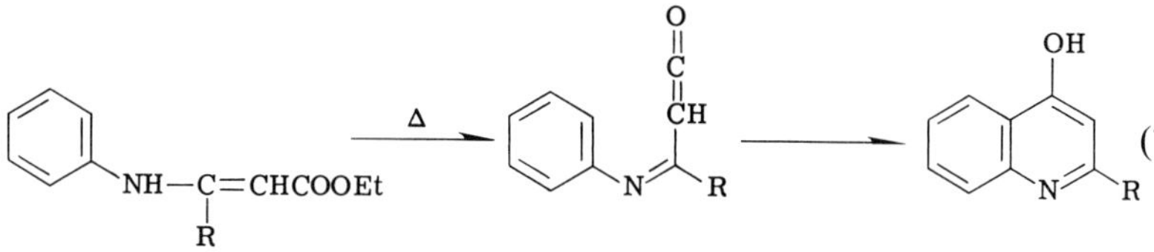

(7-161)

trocyclic reaction, and a run-down on the literature of this reaction has been given by Jutz.[516]

Formation of α-pyridones from α-pyrones may also utilize the electrocyclic ring closure [Eq. (7-162)]. This is a general process which is a very valuable

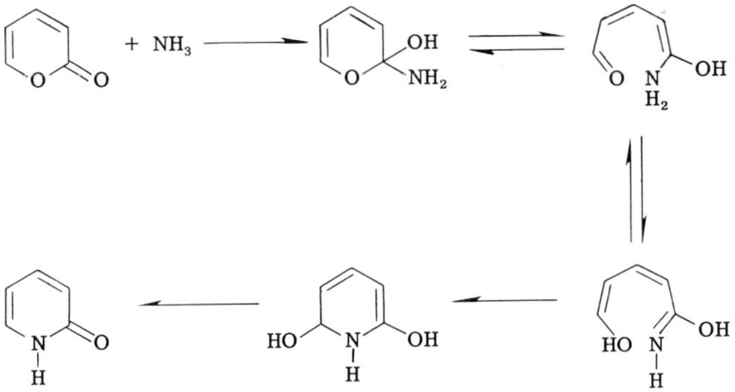

$$(7\text{-}162)$$

preparative reaction.[517,518] Primary amines may replace ammonia thus giving N-alkyl or N-aryl derivatives.[519] Diene isocyanates, obtained via a Curtius rearrangement, cyclize readily to give another route to α-pyridones.[520,521]

Both Oppolzer and Kametani have developed the electrocyclization route to hydroisoquinoline compounds. Kametani has exploited the method for the synthesis of a number of tetracyclic alkaloids and related substances [Eq. (7-163), for example].[522,523] Oppolzer[524,525] obtained some dihydro-

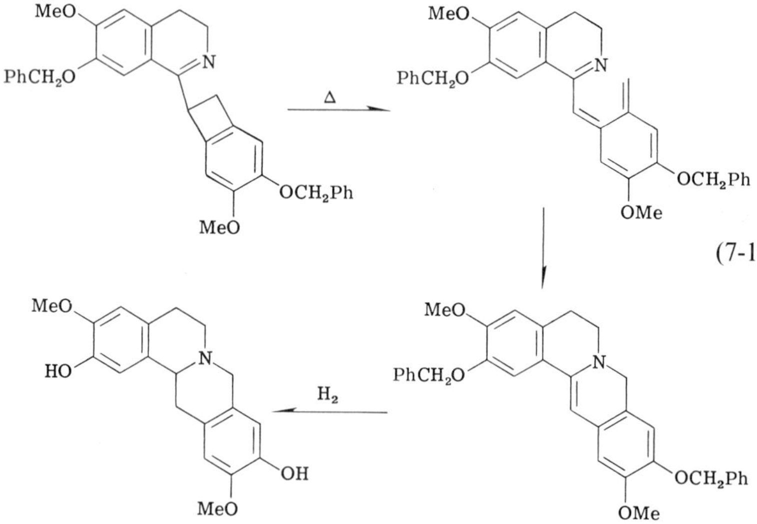

(7-163)

isoquinolines via the same route while attempting to synthesize a steroid derivative [Eq. (7-164)].

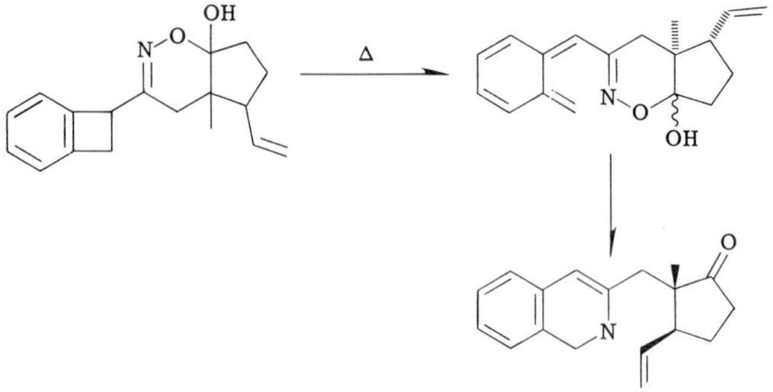

(7-164)

Pyridine N-oxides have been obtained from pyrylium ions by a ring opening and reclosure route which can be written as an electrocyclic–elimination route.[526]

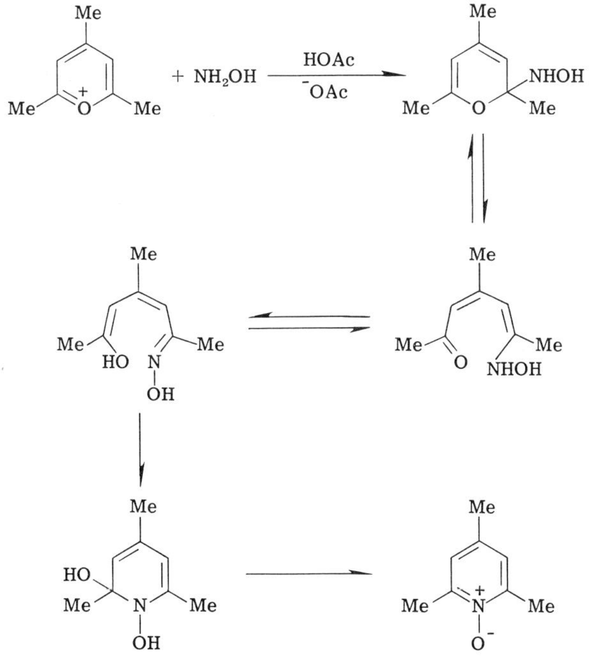

(7-165)

3. One Sulfur Atom

The procedures used to prepare nitrogen heterocycles can in many cases be extended to the preparation of sulfur heterocycles. However, not much has been done to exploit these possibilities. The only widely used reaction is the conversion of pyrylium ions to thiapyrylium ions.[527,528] A rather specialized example is found in the conversion of fluorescein to thiofluorescein.[529]

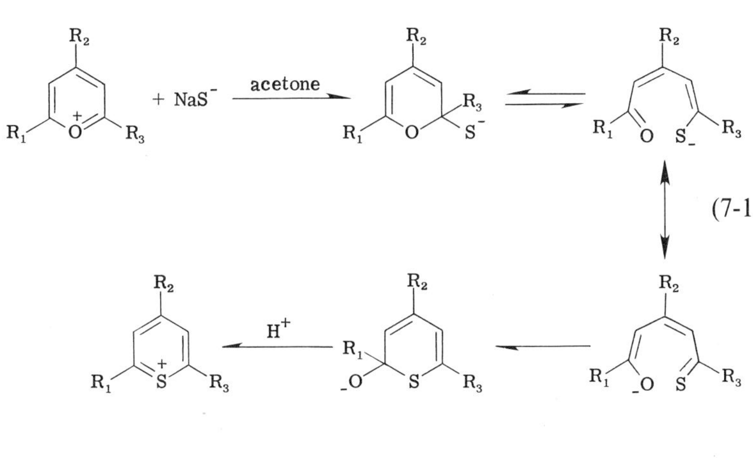

(7-166)

4. Two or More Heteroatoms

Again, synthesis of heterocycles having two or more heteroatoms via electrocyclic processes has been quite common. For the most part these have been applied to nitrogen heterocycles though application to other heteroatoms should be possible. Evidence for mechanistic detail is often lacking and considerable speculation is possible. Examples are illustrated here.

The electrocyclic–elimination scheme has been applied by Jutz to the preparation of a variety of interesting heterocycles having two nitrogen atoms [Eq. (7-167)].[530] Pyrazines can be prepared by a related reaction scheme.[445]

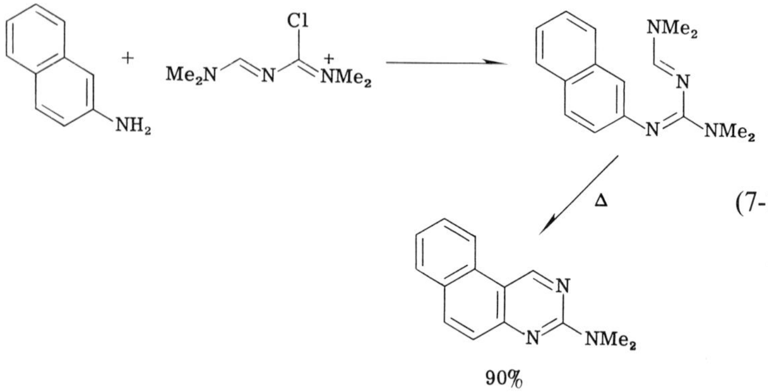

$$(7\text{-}167)$$

90%

Isocyanates and isothiocyanates, prepared *in situ*, cyclize readily to give benzopyrimidones and thiopyrimidones [Eq. (7-130) and (7-131)].[447–449] Some evidence supporting the reaction mechanism has been presented.

Van der Plas has provided strong evidence for the mechanism [Eq. (7-168)]

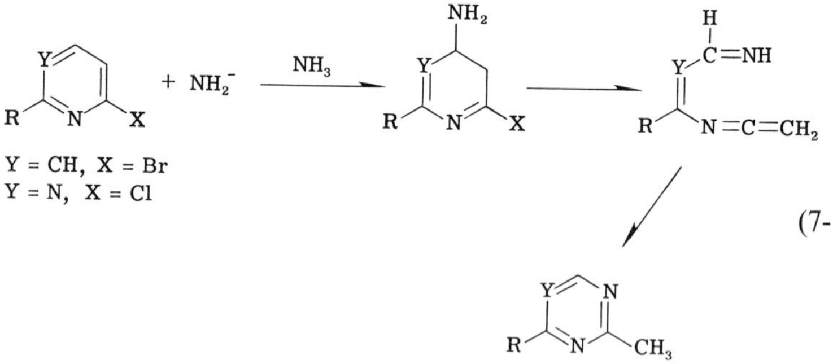

Y = CH, X = Br
Y = N, X = Cl

$$(7\text{-}168)$$

of the conversion of halogenated pyridines to pyrimidines and of halogenated pyrimidines to *S*-triazines by amide ion in liquid ammonia.[531] The process appears to be of broad applicability. Formation of a dihydropyridazine from **169** could reasonably occur via the electrocyclic reaction shown in Eq. (7-169).[532] Other examples involving conversion of pyrimidines to

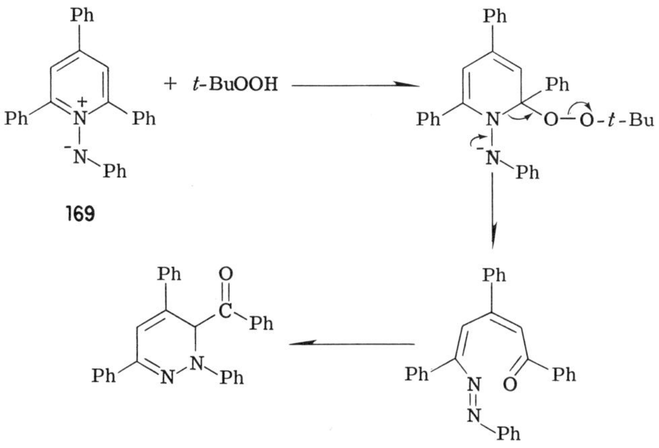

(7-169)

triazines which might utilize an electrocyclic ring closure are known.[533,534]

As would be expected, 3-aza and 3,5-diazapyrylium ions react with am-

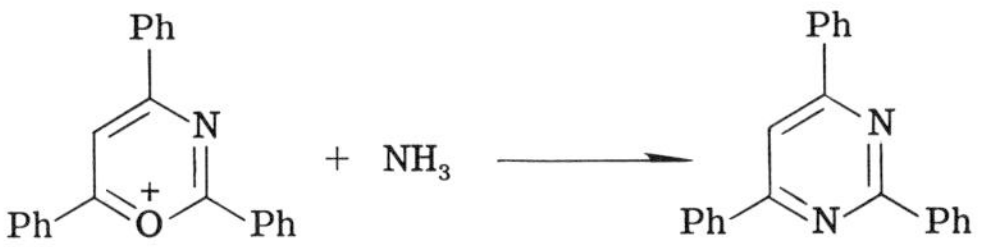

(7-170)

monia, hydrogen sulfide, urea, or thiourea to replace the oxygen by nitrogen or sulfur.[535] The mechanism must be completely analogous to the related reactions of pyrylium salts. An interesting conversion of 2-oxoalkyl-s-triazines to acylaminopyridines could very well use the electrocyclic closure shown in Eq. (7-171) to form the pyrimidine ring.[536,537]

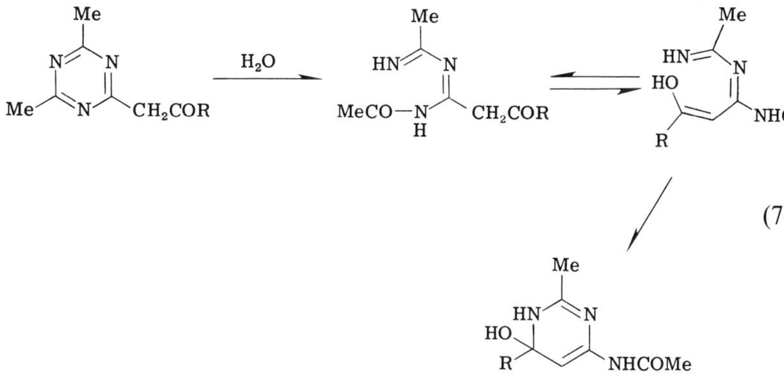

(7-171)

5. Open Chain Products

The ring cleavage of both pyrylium and pyridinium compounds leads initially to acyclic materials which may or may not be stable under the reaction conditions. The earlier parts of Section III,B showed that the unstable acyclics generally recyclized to produce new heterocycles. In some cases, however, the acyclic compounds are isolable and prove useful in their own right. The open chain compounds isolable from pyrylium ions are dienones or dienals and derivatives thereof and these were fully considered in Section II,C. Thus, this section will be restricted to ring opening reactions of pyridinium ions.

Three separate processes play roles in the preparation of acyclic molecules from pyridinium salts; first, the formation of the ion itself from a relevant pyridine, second, the addition of a nucleophile to give a 1,2-dihydropyridine, and third, the retro-electrocyclization. Limitations applying at one or more of these steps make the overall process a very sensitive function of substituents and conditions. Further restrictions often result from the relatively unstable nature of the final products.

Pyridinium ions having a very broad group of N substituents have been used in the cleavage process. These include 2,4-dinitrophenyl,[387–392] cyanide,[393,538] ethyl sulfonate,[539] sulfur trioxide,[540,541] p-nitrophenyl, [538,406] m-nitrophenyl,[406] m-, and p-chlorophenyl,[406] p-tolyl,[406] phenyl,[406] phosphorus pentachloride,[542] 2-amino-6-pyrimidyl,[543] 2,4-dinitronaphthyl,[544] several nitrophenyl groups with another substituent,[545,546] alkoxy, [417–419,547] oxoalkenyl,[548,549] the CSCl group,[550] bispyridiniummethane, [551,552] arylmethanol derivatives,[553] and substituted stilbenes.[554] Of these only a few are really useful in synthesis, and few studies of their limitations exist. König and Bayer[542] have published a list of some of the various N substituents which may lead to cleavage of the pyridine ring, and that list includes a number not noted here.

Zincke's reagent, 2,4-dinitrophenylpyridinium ion, forms readily and gives open chain products which contain the 2,4-dinitrophenyl group when treated with hydroxide or aliphatic amines,[389,390] and with stable anions derived from dicyanomethane or ethyl cyanoacetate.[555] Substitution in the pyridine ring limits the utility of the reaction severely. As would be expected the reaction between 2,4-dinitrochlorobenzene and the pyridine is subject to steric effects which prevent quaternization where substituents are in the 2- or 6- positions,[556] and to reduction of nucleophilicity which also prevents the reaction with 3-Cl, 3-F, 3-NO_2, 3-CN, 3-COOEt, 3-COOH, 4-Cl, 4-Br, 4-CN, and 4-NO_2 substituents.[556,557] Vompe and Turitsyna[558] have indicated that ring cleavage does occur following quaternization when 3-methoxy, 3-methyl, and 3-acetylamino substituents are present. No cleavage was found with 3-amino, 3-hydroxy, 4-acetylamino, or 4-phenylamino substituted pyridines. While a 3-benzyl group apparently prevents ring opening, cleavage does occur with 4-benzylpyridine.[559]

König's reaction, cyanogen bromide and a pyridine, permits cleavage of a wider variety of pyridines being rather less sensitive to substituent effects. Because the reaction is normally carried out without isolation of the pyridinium salt, there is little indication of the specific step which limits ring opening. In addition it appears that the nucleophile may often determine whether cleavage can occur or not with some substituted pyridines. Strell and co-workers[560] using cyanogen bromide and aniline found that glutacondialdehyde derivatives were formed with 3-amino, 3-carboxamido, 3-carboxylic acid, 4-ethyl, 4-amino, 4-hydroxy, 4-benzyl, 2,6-dimethyl, 2-methyl-6-ethyl, and 2-methyl-5-ethyl pyridines. Conversely no reaction occurred with 3-cyano, 2,6-dicarboxylic acid, 2-methyl-6-phenyl, 2,6-diphenyl, 3,5-dibenzyl, or 2,6-dimethyl-3,5-dicarbethoxy pyridines. Further studies with N-methylaniline[561] indicated that the reaction was satisfactory with 2-methyl, 3-methyl, 3-benzyl, and 2-methyl-5-ethylpyridines. A very poor yield was reported with 2,6-lutidine and 2-, 3-, 4-cyano, 2-hydroxymethyl, 4-amino, 4-hydroxy, 2-, 3-, 4-carbonal, and 2-carboxylic acid derivatives did not react. Hafner and Asmus[562] reported that only β- and γ-picolines would react with aniline and polysubstituted pyridines were unreactive. 3-Acetaminopyridine does react with aniline and cyanogen bromide.[563] A relatively thorough study giving yields under controlled conditions showed that aniline and cyanogen bromide will cleave β- and γ-picolines, γ-ethyl, β- and γ-methoxy, γ-phenoxy, γ-thiomethyl, β-acetamino, and γ-chloro pyridines.[564] Low yields were obtained with β-iodo, γ-acetamino and γ-benzoyl pyridines. α-Picoline gave only an 8% yield.

N-methoxypyridinium salts can undergo a ring opening reaction under appropriate conditions. The unsubstituted ion reacts with hydroxide ion[417] or secondary amines to[419] give derivatives of glutacondialdehyde, but other nucleophiles including a number of carbanions either fail to react or give other products.[419,565] Secondary amines, hydroxide ion and a number of carbanions give rise to acyclic products where CN, COOR, COONH$_2$ groups are present on the ring.[422,565,566,421] The sequence CN > COOH > Cl > CONH$_2$ and the order 3 > 2 > 4 was noted for reaction with carbanions from 1,3-dicarbonyl compounds.[565] Either alkyl or chlorine at both C$_2$ and C$_4$ prevent the reaction.

The sulfur trioxide–pyridine complex reacts with hydroxide to give an unstable acyclic product which can be stabilized by treatment with benzoyl chloride.[541] The reaction proceeds with 3-methyl or 3-methoxypyridines, but γ-picoline gave only tar and 3,5-dimethyl or 4-tert-butylpyridines did not react. Kavalek et al.[406] studied the influence of substituents on the phenyl ring of N-phenylpyridinium ion and noted that electron attracting substituents generally enhance the reaction rate ($\rho = +5.05$). The reaction of the 2,4-dinitrophenyl salt is 1×10^{12} times faster than phenyl. A 6% yield of an acyclic product was obtained when 3-cyano-N-methylpyridinium ion was treated with hydroxide, but neither the 2- or 4-cyano derivatives gave any ring cleavage.[567]

In most instances the ring cleavage of pyridinium ions leads to glutacondialdehyde or a derivative thereof, and these have been converted to other products. A number of cyanine dyes have been prepared from the imino derivatives,[561,562,568] and Jutz has used these to prepare polynuclear aromatics.[483] The well-known Ziegler–Hafner synthesis of azulenes utilizes these derivatives in some cases.[569–571] An interesting preparation of a triazole was devised from 3-amino-N-phenylpyridinium ion [Eq. (7-172)].[563]

$$(7\text{-}172)$$

An unexpected extension of the usual ring opening has paved the way for the preparation of a series of azapolyenes. Pyridine with cyanogen bromide and diethylammonium ion gave **170** which could in turn lead to **171** [Eq. (7-173)].[572] The scheme was modified by Dobeneck and Goltzsche,[573]

$$(7\text{-}173)$$

[Eq. (7-174)] but was developed as a repetitive process by Kuhn and Teller

$$(7\text{-}174)$$

[Eq. (7-175)].[553] These latter authors prepared aza and diaza polyenes with several cis double bonds in the chain. The cis double bonds were readily isomerized by iodine to give all trans polyenes. A similar procedure was

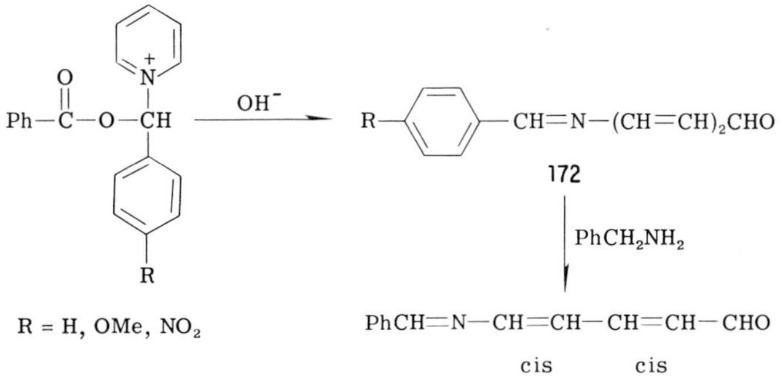

O
Ph—C—O—CH
N
OH⁻
R
R—
—CH=N—(CH=CH)₂CHO
172
PhCH₂NH₂
R = H, OMe, NO₂
PhCH=N—CH=CH—CH=CH—CHO
cis
cis

(7-175)

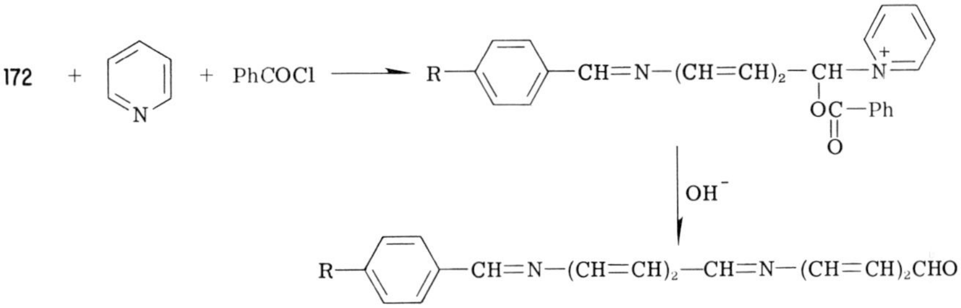

172
+
N
+ PhCOCl
R—
—CH=N—(CH=CH)₂—CH—N
OC—Ph
O
OH⁻
R—
—CH=N—(CH=CH)₂—CH=N—(CH=CH)₂CHO

used by Fischer[549] to prepare longer analogs of glutacondialdehyde [Eq.

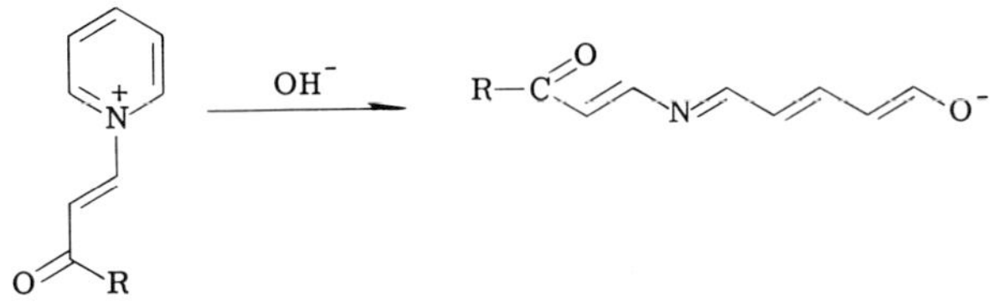

N
O
R
OH⁻
R—C
O
N
O⁻

(7-176)

(7-176)], and other variants on these schemes have been developed.[552,574] A very different approach has been developed by Uwe and Hönig[575] which permits formation of polyene or enynedials of known configuration [Eq. (7-177)].

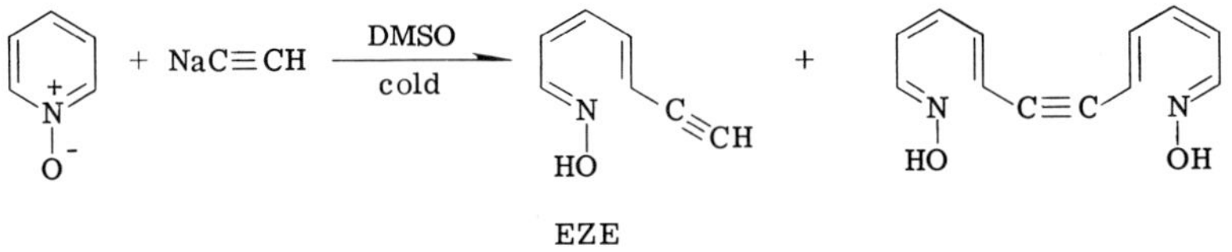

N
O⁻
+ NaC≡CH
DMSO
cold
N
HO
C
CH
EZE
+
N
HO
C≡C
N
OH

(7-178)

362 7. Six Electron–Six Atom Systems

REFERENCES

1. K. E. Lewis and H. Steiner, *J. Chem. Soc.* p. 3080 (1964).
2. M. G. Evans and M. Polanyi, *Trans. Faraday Soc.* **34,** 11 (1938).
3. E. N. Marvell, *Tetrahedron* **29,** 3791 (1973).
4. A. Komornicki and J. W. McIver, Jr., *J. Am. Chem. Soc.* **98,** 5798 (1974).
5. O. S. Tee and K. Yates, *J. Am. Chem. Soc.* **94,** 3074 (1972).
6. N. D. Epiotis, *J. Am. Chem. Soc.* **95,** 1200 (1973).
7. B. K. Carpenter, *Tetrahedron* **34,** 1877 (1978).
8. P. Busse, *Z. Physiol.* **214,** 211 (1933).
9. J. Castello, E. R. H. Jones, G. D. Meakins, and R. W. J. Williams, *J. Chem. Soc.* p. 1159 (1959).
10. L. F. Fieser and M. Fieser, "Steroids," pp. 136–140, Reinhold, New York, 1959.
11. L. Velluz, G. Amiard, and B. Goffinet, *Bull. Soc. Chem. Fr.* **22,** 1341 (1955).
12. E. Havinga and J. L. M. A. Schlattmann, *Tetrahedron* **16,** 146 (1961).
13. E. Vogel, W. Grimme, and E. Dinne, *Tetrahedron Lett.* p. 391 (1965).
14. E. N. Marvell, G. Caple, and B. S. Schatz, *Tetrahedron Lett.* p. 385 (1965).
15. E. N. Marvell, G. Caple, B. Schatz, and W. Pippin, *Tetrahedron* **29,** 3781 (1973).
16. C. Hilton, Ph. D. Thesis, Oregon State Univ., Aug. 1978.
17. A. Padwa, L. Brodsky, and S. Clough, *J. Am. Chem. Soc.* **94,** 6767 (1972).
18. R. Ramage and A. Sattar, *Chem. Commun.* p. 173 (1970).
19. A. S. Cope, A. C. Haven, F. L. Ramp, and E. R. Trumbull, *J. Am. Chem. Soc.* **74,** 4867 (1952).
20. E. Vogel, O. Roos, and K.-H. Disch, *Justus Liebigs Ann. Chem.* **653,** 55 (1962).
21. K. Alder and H. Dortmann, *Chem. Ber.* **87,** 1905 (1954).
22. D. S. Glass, J. W. H. Watthey, and S. Winstein, *Tetrahedron Lett.* p. 377 (1965).
23. P. Radlick and G. Alford, *J. Am. Chem. Soc.* **91,** 6529 (1969).
24. A. G. Anastassiou, V. Orfanos, and J. H. Gebrian, *Tetrahedron Lett.* p. 4491 (1969).
25. G. Boche, H. Böhme, and D. Martens, *Angew. Chem. Int. Edit. Engl.* **8,** 594 (1969).
26. S. Masamune, P. M. Baker, and K. Hojo, *Chem. Commun.* p. 1203 (1969).
27. A. G. Anastassiou and R. P. Cellura, *Tetrahedron Lett.* p. 911 (1970).
28. J. Schwartz, *Chem. Commun.* p. 833 (1969).
29. S. Masamune, K. Hojo, and S. Takada, *Chem. Commun.* p. 1204 (1969).
30. A. G. Anastassiou and J. H. Gebrian, *J. Am. Chem. Soc.* **91,** 4011 (1969).
31. A. G. Anastassiou and J. H. Gebrian, *Tetrahedron Lett.* p. 5239 (1969).
32. A. G. Anastassiou and R. P. Cellura, *Chem. Commun.* p. 903 (1969). This paper assigns the thermal product a trans ring juncture, but this was corrected shortly (refer. 33).
33. A. G. Anastassiou and R. P. Cellura, *Chem. Commun.* p. 1521 (1969).
34. S. Masamune, S. Takada, and R. T. Seidner, *J. Am. Chem. Soc.* **91,** 7769 (1969).
35. A. G. Anastassiou and R. P. Cellura, *Chem. Commun.* p. 484 (1970).
36. V. Boekelheide and J. B. Phillips, *J. Am. Chem. Soc.* **89,** 1695 (1967).
36a. J. M. Ben-Bassat and D. Ginsburg, *Tetrahedron* **30,** 483 (1974).
37. G. B. Kistiakowsky, J. R. Ruhoff, H. A. Smith, and W. E. Vaughn, *J. Am. Chem. Soc.* **58,** 146 (1936).
38. R. B. Turner, B. J. Mallon, M. Tichy, W. von E. Doering, W. R. Roth, and G. Schröder, *J. Am. Chem. Soc.* **95,** 8605 (1973).
39. R. Huisgen, F. Mietzsch, G. Boche, and H. Seidl, *Org. React. Mech. Spec. Publ. Chem. Soc. London* **19,** 3 (1965).
40. R. Huisgen, G. Boche, A. Dahmen, and W. Hechtl, *Tetrahedron Lett.* p. 5215 (1968).
41. S. W. Benson, "Thermochemical Kinetics," p. 179, Wiley, New York, 1968.

42. B. C. Baumann and A. S. Dreiding, *Helv. Chim. Acta* **57**, 1872 (1974).
43. C. W. Spangler and D. L. Boles, *J. Org. Chem.* **37**, 1020 (1972).
44. T. L. Rose, R. J. Seyse, and P. M. Crane, *Int. J. Chem. Kinet.* **6**, 899 (1974).
45. C. W. Spangler, T. P. Jondahl, and B. Spangler, *J. Org. Chem.* **38**, 2478 (1973).
46. E. N. Marvell, G. Caple, C. Delphey, J. Platt, N. Polston, and J. Tashiro, *Tetrahedron* **29**, 3797 (1973).
47. H. M. Prinzbach and E. Druckery, *Tetrahedron Lett.* p. 2959 (1965).
48. G. Moy, Ph. D. Thesis, Oregon State Univ., June 1977.
49. C. Spangler, *Tetrahedron* **32**, 2681 (1976).
50. H. M. Prinzbach, H. Hagemann, J. H. Hartenstein, and R. Kitzing, *Chem. Ber.* **98**, 2201 (1965).
51. H. M. Prinzbch and R. Kaiser, see footnote 7 in refer. 47.
52. J. D. Hobson, M. M. Al Holly, and J. R. Malpass, *Chem. Commun.* p. 764 (1968).
53. K. W. Egger, *Helv. Chim Acta* **51**, 422 (1968).
54. M. Cleary, Ph. D. Thesis, Oregon State Univ., June 1975.
55. C. Delphey, Ph. D. Thesis, Oregon State Univ., June 1977.
56. P. Courtot and R. Rumin, *Bull. Soc. Chim. Fr.* p. 3665 (1969).
57. P. Schiess, R. Seeger, and C. Suter, *Helv. Chim. Acta* **53**, 1713 (1970).
58. C. Hilton and E. N. Marvell, *Abstr. 33rd Am. Chem. Soc. Northwest Regional Meeting, Seattle, 1978, #168.*
59. C. J. Gaasbeek, H. Hogeveen, and N. C. Volger, *Rec. Trav. Chem. Pays-Bas* **91**, 821 (1972).
60. K. R. Huffman, M. Loy, W. A. Henderson, Jr., and E. F. Ullman, *Tetrahedron Lett.* p. 931 (1967).
61. K. R. Huffman, M. Burger, W. A. Henderson, Jr., M. Loy, and E. F. Ullman, *J. Org. Chem.* **34**, 2407 (1969).
62. E. Ciganek, *J. Am. Chem. Soc.* **87**, 1149 (1965).
63. H. Günther, *Tetrahedron Lett.* p. 4085 (1965).
64. M. Görlitz and H. Günther, *Tetrahedron* **25**, 4467 (1969).
65. R. Huisgen, G. Boche, A. Dahmen, and W. Hechtl, *Tetrahedron Lett.* p. 5215 (1968).
66. D. S. Glass, J. Zirner, and S. Winstein, *Proc. Chem. Soc.* p. 276 (1963).
67. R. Huisgen and F. Mietzsch, *Angew. Chem. Int. Edit. Engl.* **3**, 83 (1964).
68. G. Petrowski, Ph. D. dissertation, Univ. of California at Los Angeles, 1969.
69. P. Radlick and G. Alford, *J. Am. Chem. Soc.* **91**, 6529 (1969).
70. M. Neuenschwander and A. Frey, *Chimia* **29**, 212 (1975).
71. E. J. Corey and A. G. Hortmann, *J. Am. Chem. Soc.* **85**, 4033 (1963).
72. J. F. Oth, H. Röttele, and G. Schröder, *Tetrahedron Lett.* p. 61 (1970).
73. H. Röttele, W. Martin, J. F. M. Oth, and G. Schröder, *Chem. Ber.* **102**, 3985 (1969).
74. H.-R. Blattman and W. Schmidt, *Tetrahedron* **26**, 5885 (1970).
75. V. Boekelheide and W. Pepperdine, *J. Am. Chem. Soc.* **92**, 3684 (1970).
76. A. W. Hanson, *Acta Cryst.* **18**, 599 (1965).
77. P. Schiess and P. Fünfschilling, *Tetrahedron Lett.* p. 5191 (1972).
78. P. Schiess and R. Dinkel, *Tetrahedron Lett.* p. 2503 (1975).
79. M. Traetteberg, *Acta Chem. Scand.* **22**, 2294 (1968).
80. W. von E. Doering, V. G. Toscano, and G. H. Beasley, *Tetrahedron* **27**, 299 (1971).
81. H. M. Frey and R. V. Solly, *Trans. Faraday Soc.* **64**, 1858 (1968).
82. M. J. S. Dewar and L. E. Wade, Jr., *J. Am. Chem. Soc.* **99**, 4417 (1977).
83. H. J. Reich, E. Ciganek, and J. D. Roberts, *J. Am. Chem. Soc.* **92**, 5166 (1970).
84. H. Günther, B. D. Tunggal, M. Regitz, H. Scherer, and T. Keller, *Angew. Chem. Int. Edit. Engl.* **10**, 563 (1971).

85. H. Günther, W. Peters, and R. Wehner, *Chem. Ber.* **106**, 3683 (1973).
86. G. Maas and M. Regitz, *Chem. Ber.* **109**, 2039 (1976).
87. W. Betz and J. Daub, *Chem. Ber.* **107**, 2095 (1974).
88. K. M. Rapp and J. Daub, *Tetrahedron Lett.* p. 227 (1977).
89. F.-G. Klärner, *Tetrahedron Lett.* p. 19 (1974).
90. A. J. Bellamy and W. Crilly, *Tetrahedron Lett.* p. 1893 (1973).
91. S. W. Staley, M. A. Fox, and A. Cairncross, *J. Am. Chem. Soc.* **99**, 4524 (1977).
92. H. Dürr, H. Kober, V. Fuchs, and P. Orth, *Chem. Commun.* p. 973 (1972).
93. H. Dürr and H. Kober, *Chem. Ber.* **106**, 1565 (1973).
94. H. Günther, *Angew. Chem.* **77**, 1022 (1965).
95. H. Günther, J. B. Pawliczek, B. D. Tunggal, H. Prinzbach, and R. H. Levin, *Chem. Ber.* **106**, 984 (1973).
96. L. W. Goldblatt and S. Palkin, *J. Am. Chem. Soc.* **63**, 3517 (1941).
97. L. A. Goldblatt and S. Palkin, *J. Am. Chem. Soc.* **66**, 655 (1944).
98. T. R. Savich and L. A. Goldblatt, *J. Am. Chem. Soc.* **67**, 2027 (1945).
99. R. E. Fugitt and J. E. Hawkins, *J. Am. Chem. Soc.* **69**, 319 (1947).
100. G. F. Woods, N. C. Bolgiano, and D. E. Duggan, *J. Am. Chem. Soc.* **77**, 1800 (1955).
101. H. Fleischaker and G. F. Woods, *J. Am. Chem. Soc.* **78**, 3436 (1956).
102. G. F. Woods and A. Viola, *J. Am. Chem. Soc.* **78**, 4380 (1956).
103. K. Alder and H. von Brachel, *Justus Liebigs Ann. Chem.* **608**, 195 (1957).
104. H. Pines and R. Kozlowski, *J. Am. Chem. Soc.* **78**, 3776 (1956).
105. H. Pines and C.-T. Chen, *J. Am. Chem. Soc.* **81**, 928 (1959).
106. K. J. Crowley, *Proc. Chem. Soc.* p. 17 (1964).
107. K. J. Crowley and S. G. Traynor, *Tetrahedron* **34**, 2783 (1978).
108. R. J. De Kock, N. G. Minnaard, and E. Havinga, *Rec. Trav. Chim. Pays-Bas* **79**, 922 (1960).
109. W. G. Dauben and R. M. Coates, *J. Org. Chem.* **29**, 2761 (1964).
110. H. Prinzbach and H. Hagemann, *Angew. Chem. Int. Edit. Engl.* **3**, 653 (1964).
111. L. Skattebøl, J. L. Charleton, and P. de Mayo, *Tetrahedron Lett.* p. 2257 (1966).
112. P. Courtot and R. Rumin, *Tetrahedron Lett.* p. 1849 (1970).
113. P. Courtot and J. Y. Salaün, *Chem. Commun.* p. 124 (1976).
114. F. S. Edmunds and R. A. W. Johnston, *J. Chem. Soc.* p. 2892 (1965).
115. F. S. Edmunds and R. A. W. Johnstone, *J. Chem. Soc.* p. 2898 (1965).
116. F. Näf, R. Decorzant, W. Thommen, B. Wilhalm, and G. Ohloff, *Helv. Chim. Acta* **58**, 1016 (1975).
117. M. Pomerantz, R. N. Wilke, G. W. Gruber, and U. Roy, *J. Am. Chem. Soc.* **94**, 2752 (1972).
118. B. J. Arnold and P. G. Sammes, *Chem. Commun.* p. 1034 (1972).
119. B. J. Arnold, P. G. Sammes, and T. W. Wallace, *J. Chem. Soc. Perkin Trans. 1* p. 415 (1974).
120. W. Sieber, M. Heimgartner, H.-J. Hansen, and H. Schmid, *Helv. Chim. Acta* **55**, 3005 (1972).
121. H. Heimgartner, H.-J. Hansen, and H. Schmid, *Helv. Chim. Acta* **55**, 1385 (1972).
122. P. J. Darcy, R. J. Hart, and H. G. Heller, *J. Chem. Soc. Perkin Trans. 1* p. 571 (1978).
123. G. Quinkert, W. W. Wiersdorf, M. Finke, K. Opitz, and F. G. Von der Haas, *Chem. Ber.* **101**, 2302 (1968).
124. J. Dale and P. O. Kristiansen, *Chem. Commun.* p. 1293 (1968).
125. J. H. Borkent, P. H. F. M. Rouwette, and W. H. Laarhoven, *Tetrahedron* **34**, 2569 (1978).
126. L. Barber, O. L. Chapman, and J. D. Lassila, *J. Am. Chem. Soc.* **90**, 5933 (1968).

127. G. Quinkert, B. Bronstert, P. Michaelis, and U. Krueger, *Angew. Chem. Int. Edit. Engl.* **9,** 240 (1970).

128. J. S. Swenton, E. Saurborn, R. Srinivasan, and F. J. Sonntag, *J. Am. Chem. Soc.* **90,** 2990 (1968).

129. J. C. Floyd, D. A. Plank, and W. H. Starnes, Jr., *Chem. Commun.* p. 1237 (1969).

130. J. Griffiths and H. Hart, *J. Am. Chem. Soc.* **90,** 3297 (1968).

131. H. Hart and R. K. Murray, Jr., *J. Org. Chem.* **35,** 1535 (1970).

132. R. Bastiani and H. Hart, *J. Org. Chem.* **37,** 2830 (1972).

133. M. R. Morris and A. J. Waring, *Chem. Commun.* p. 526 (1969).

134. M. Bellas, D. Bryce-Smith, M. T. Clarke, A. Gilbert, G. Klunkin, S. Krestanovich, C. Manning, and S. Wilson, *J. Chem. Soc. Perkin Trans. 1* p. 2571 (1977).

135. H. Hopf and H. Musso, *Angew. Chem. Int. Edit. Engl.* **8,** 680 (1969).

136. D. A. Ben-Efraim and F. Sondheimer, *Tetrahedron* **25,** 2837 (1969).

137. H. Hopf, *Tetrahedron Lett.* p. 1107 (1970).

138. G. Moy, Ph. D. Thesis, Oregon State Univ., June 1977.

139. R. R. Jones and R. G. Bergman, *J. Am. Chem. Soc.* **94,** 660 (1972).

140. C. W. Spangler and N. Johnson, *Abst. ACS Meeting, April 1968, p. 47.*

141. C. W. Spangler and R. D. Feldt, *Chem. Commun.* p. 709 (1968).

142. S. W. Orchard and B. A. Thrush, *Chem. Commun.* p. 14 (1973).

143. U. R. Nayak, A. H. Kapadi, and S. Dev, *Tetrahedron* **26,** 5071 (1970).

144. G. Frater, *Helv. Chim Acta* **57,** 2446 (1974).

145. E. Buchner, *Chem. Ber.* **21,** 2637 (1888).

146. W. von E. Doering, G. Laber, R. Vonderwahl, N. F. Chamberlain, and R. B. Williams, *J. Am. Chem. Soc.* **78,** 5448 (1956).

147. G. Maier, *Angew. Chem. Int. Edit. Engl.* **6,** 402 (1967).

148. H. Günther, M. Gorlitz, and H.-H. Heinrichs, *Tetrahedron* **24,** 5665 (1968).

149. J. A. Berson, D. R. Hartter, H. Klinger, and P. W. Grubb, *J. Org. Chem.* **33,** 1669 (1968).

150. M. Jones, Jr., *Angew. Chem. Int. Edit. Engl.* **8,** 76 (1969).

151. T. Mukai, H. Kubota, and T. Toda, *Tetrahedron Lett.* p. 3581 (1967).

152. T. Toda, M. Nitta, and T. Mukai, *Tetrahedron Lett.* p. 4401 (1969).

153. M. Jones, Jr. and E. W. Petrillo, *Tetrahedron Lett.* p. 3953 (1969).

154. C. J. Rostock and W. M. Jones, *Tetrahedron Lett.* p. 3957 (1969).

155. H. Günther, *Tetrahedron Lett.* p. 5173 (1970).

156. J. A. Berson, P. W. Grubb, R. A. Clark, D. R. Hartter, and M. R. Willcott III, *J. Am. Chem. Soc.* **89,** 4076 (1967).

157. D. Schönleber, *Chem. Ber.* **102,** 1789 (1969).

158. Y. Kitihara, *Pure Appl. Chem.* **44,** 833 (1975).

159. H. Günther and H. Schmickler, *Pure Appl. Chem.* **44,** 807 (1975).

160. R. Hoffmann, *Tetrahedron Lett.* p. 2907 (1970).

161. L. A. Paquette and L. M. Leichter, *J. Am. Chem. Soc.* **93,** 5128 (1971).

162. G. E. Hall and J. D. Roberts, *J. Am. Chem. Soc.* **93,** 2203 (1971).

163. E. Ciganek, *J. Am. Chem. Soc.* **93,** 2207 (1971).

164. A. Steigl, J. Sauer, D. A. Kleier, and G. Binsch, *J. Am. Chem. Soc.* **94,** 2770 (1972).

165. E. Vogel, *Pure Appl. Chem.* **20,** 237 (1969).

166. G. Günther, H. Schmickler, W. Bremse, F. A. Straube, and E. Vogel, *Angew. Chem. Int. Edit. Engl.* **12,** 570 (1973).

167. G. W. Gruber and M. Pomerantz, *Tetrahedron Lett.* p. 3755 (1970).

168. H. Dürr, H. Kober, I. Halberstadt, U. Neu, T. T. Coburn, T. Mitsuhashi, and W. M. Jones, *J. Am. Chem. Soc.* **95,** 3818 (1973).

169. T. Mitsuhashi and W. M. Jones, *Chem. Commun.* p. 103 (1974).
170. E. Vedejs and W. R. Wilbur, *Tetrahedron Lett.* p. 2679 (1975).
171. H. Dürr, M. Kausch, and H. Kober, *Angew. Chem. Int. Edit. Engl.* **13,** 670 (1974).
172. K.-H. Paul and H. Dürr, *Tetrahedron Lett.* p. 3649 (1976).
173. R. H. Parker and W. M. Jones, *J. Org. Chem.* **43,** 2548 (1978).
174. W.-D. Stohrer, *Chem. Ber.* **106,** 970 (1973).
175. E. Vogel and H. Günther, *Angew. Chem. Int. Edit. Engl.* **6,** 385 (1967).
176. D. M. Jerina, H. Yagi, and J. W. Daly, *Heterocycles* **1,** 267 (1973).
177. H. Günther, *Angew. Chem.* **77,** 1022 (1965).
178. D. R. Boyd, D. M. Jerina, and J. W. Daly, *J. Org. Chem.* **35,** 3170 (1970).
179. L. A. Paquette and J. H. Barrett, *J. Am. Chem. Soc.* **88,** 2590 (1966).
180. L. A. Paquette, J. H. Barrett, and D. E. Kuhla, *J. Am. Chem. Soc.* **91,** 3616 (1969).
181. L. A. Paquette, D. E. Kuhla, and J. H. Barrett, *J. Org. Chem.* **34,** 2879 (1969).
182. L. A. Paquette, D. E. Kuhla, J. H. Barrett, and L. H. Leichter, *J. Org. Chem.* **34,** 2888 (1969).
183. L. A. Paquette, *Angew. Chem. Int. Edit. Engl.* **10,** 11 (1971).
184. J. Rigaudy, C. Izier, and J. Barcelo, *Tetrahedron Lett.* p. 3845 (1975).
185. W. von E. Doering and R. A. Odum, *Tetrahedron* **22,** 81 (1966).
186. R. J. Sundberg, S. R. Suter, and M. Brenner, *J. Am. Chem. Soc.* **94,** 513 (1972).
187. B. A. De Graff, D. W. Gillespie, and R. J. Sundberg, *J. Am. Chem. Soc.* **96,** 7491 (1974).
188. B. Iddon, M. W. Pickering, and H. Suschitzky, *Chem. Commun.* p. 759 (1974).
189. H. Plieninger and D. Wild, *Chem. Ber.* **99,** 3070 (1966).
190. M.-S. Lin and V. Snieckus, *J. Org. Chem.* **36,** 645 (1971).
191. H. Prinzbach, D. Stusche, and R. Kitzing, *Angew. Chem. Int. Edit. Engl.* **9,** 377 (1970).
192. H. Günther, J. B. Pawliczek, B. O. Tunggal, H. Prinzbach, and R. H. Levin, *Chem. Ber.* **106,** 984 (1973).
193. B. A. Hess, Jr., A. S. Bailey, and V. Boekelheide, *J. Am. Chem. Soc.* **89,** 2746 (1967).
194. D. C. Neckers, J. H. Dopper, and H. Wynberg, *Tetrahedron Lett.* p. 2913 (1969).
195. J. H. Dopper and D. C. Neckers, *J. Org. Chem.* **36,** 3755 (1971).
196. U. Eisner and T. Krishnamurthy, *Int. J. Sulfur Chem. B* **6,** 267 (1971).
197. V. Traynelis, *in* "Heterocyclic Compounds, Vol. 26, Seven Membered Ring Compounds Containing Sulfur and Oxygen," Wiley, New York, 1972.
198. D. N. Reinhoudt and C. G. Kouwenhoven, *Chem. Commun.* p. 1232 (1972).
198a. D. N. Reinhoudt and C. G. Kouwenhoven, *Tetrahedron* **30,** 2431 (1974).
199. A. Corvers, Ae. De Groot, and E. F. Godefroi, *Rec. Trav. Chim. Pays-Bas* **92,** 1368 (1973).
200. T. J. Barton, M. D. Martz, and R. G. Zika, *J. Org. Chem.* **37,** 552 (1972).
201. J. M. Hoffman, Jr. and R. H. Schlessinger, *J. Am. Chem. Soc.* **92,** 5263 (1970).
202. D. N. Reinhoudt and C. G. Kouwenhoven, *Chem. Commun.* p. 1233 (1972).
202a. D. N. Reinhoudt and C. G. Kouwenhoven, *Tetrahedron* **30,** 2093 (1974).
203. W. L. Mock, *J. Am. Chem. Soc.* **89,** 1281 (1967).
204. N. Ishibe, K. Hashimoto, and H. Sunami, *J. Org. Chem.* **39,** 103 (1974).
205. T. Q. Mink, L. Christiaens, P. Grandclaudon, and A. Lablache-Combier, *Tetrahedron* **33,** 2225 (1977).
206. J. J. Eisch and J. E. Galle, *J. Am. Chem. Soc.* **97,** 4436 (1975).
207. A. J. Leusink, W. Drenth, J. G. Noltes, and G. J. M. van der Kerk, *Tetrahedron Lett.* p. 1263 (1967); and other references listed in refer. 206.
208. T. J. Barton, R. C. Kippenhan, Jr., and A. J. Nelson, *J. Am. Chem. Soc.* **96,** 2772 (1974).
209. T. J. Barton, N. E. Volz, and J. L. Johnson, *J. Org. Chem.* **36,** 3665 (1971).
210. L. Birkofer and E. Kramer, *Chem. Ber.* **102,** 432 (1969).
211. J. Y. Corey and E. R. Corey, *Tetrahedron Lett.* p. 4669 (1972).

212. J. Streith and J. M. Cassal, *Angew. Chem. Int. Edit. Engl.* **7**, 129 (1968).
213. J. Streith and J. M. Cassal, *Bull. Soc. Chim. Fr.* **6**, 2175 (1969).
214. M. Nastasi, H. Strub, and J. Streith, *Tetrahedron Lett.* p. 4719 (1976).
215. J. Streith, J. P. Luttringer, and M. Nastasi, *J. Org. Chem.* **36**, 2962 (1971).
216. H. Kwart, D. A. Benko, J. Streith, D. J. Harris, and J. L. Shuppiser, *J. Am. Chem. Soc.* **100**, 6501 (1978).
217. H. Kwart, D. A. Benko, J. Streith, and J. L. Shuppiser, *J. Am. Chem. Soc.* **100**, 6502 (1978).
218. M. G. Pleiss and J. A. Moore, *J. Am. Chem. Soc.* **90**, 1369 (1968).
219. G. G. Spence, E. C. Taylor, and O. Buchardt, *Chem. Rev.* **70**, 231 (1970).
220. O. Simonsen, C. Lohse, and O. Buchardt, *Acta Chem. Scand.* **24**, 268 (1970).
221. O. Buchardt, P. L. Kumler, and C. Lohse, *Acta Chem. Scand.* **23**, 2149 (1969).
222. O. Buchardt, C. L. Pedersen, and N. Harrit, *J. Org. Chem.* **37**, 3592 (1972).
223. S. Yamada, M. Ishikawa, and C. Kaneko, *Chem. Commun.* p. 1093 (1972).
224. O. Buchardt and B. Jensen, *Acta Chem. Scand.* **22**, 877 (1968).
225. O. Buchardt, *Tetrahedron Lett.* p. 1911 (1968).
226. P. L. Kumler and O. Buchardt, *J. Am. Chem. Soc.* **90**, 5640 (1968).
227. M. Kröner, *Chem. Ber.* **100**, 3162 (1967).
228. M. Kröner, *Chem. Ber.* **100**, 3172 (1967).
229. M. Brookhart, G. O. Nelson, G. Scholes, and R. A. Watson, *Chem. Commun.* p. 195 (1976).
230. C. R. Graham, G. Scholes, and M. Brookhart, *J. Am. Chem. Soc.* **99**, 1180 (1977).
231. F. A. Cotton and G. Deganello, *J. Am. Chem. Soc.* **95**, 396 (1973).
232. J. L. Kice and T. L. Cantrell, *J. Am. Chem. Soc.* **85**, 2298 (1963).
233. L. A. Paquette, T. Kakihana, J. F. Kelly, and J. R. Malpass, *Tetrahedron Lett.* p. 1455 (1969).
234. M. Brookhart, N. M. Lippman, and E. J. Reardon, *J. Organomet. Chem.* **54**, 247 (1973).
235. R. Criegee, *Angew. Chem. Int. Edit. Engl.* **1**, 519 (1962).
236. G. Maier, U. Heep, M. Wiessler, and M. Strasser, *Chem. Ber.* **102**, 1928 (1969).
237. W. S. Wilson and R. N. Warrener, *Tetrahedron Lett.* p. 4787 (1970).
238. W. S. Wilson and R. N. Warrener, *Tetrahedron Lett.* p. 1837 (1970).
239. E. Vogel, H. Kiefer, and W. Roth, *Angew. Chem.* **76**, 432 (1964).
240. R. Askani, *Chem. Ber.* **102**, 3304 (1969).
241. L. A. Paquette, M. Oku, W. E. Heyd, and R. H. Meisinger, *J. Am. Chem. Soc.* **96**, 5815 (1974).
242. R. Huisgen, W. E. Konz, and G. E. Gream, *J. Am. Chem. Soc.* **92**, 4105 (1970).
243. R. Huisgen and W. E. Konz, *J. Am. Chem. Soc.* **92**, 4102 (1970).
244. W. E. Konz, W. Hechtl, and R. Huisgen, *J. Am. Chem. Soc.* **92**, 4104 (1970).
245. L. A. Paquette and J. C. Phillips, *Chem. Commun.* p. 680 (1969).
246. L. A. Paquette, J. C. Phillips, and R. E. Wingard, Jr., *J. Am. Chem. Soc.* **93**, 4516 (1971).
247. P. J. Collin and W. H. F. Sasse, *Austr. J. Chem.* **24**, 2325 (1971).
248. M. Oda, H. Oikawa, N. Fukazawa, and Y. Kitihara, *Tetrahedron* Lett. p. 4409 (1977).
249. I. W. McCay and R. N. Warrener, *Tetrahedron Lett.* p. 4783 (1970).
250. R. N. Warrener, E. E. Nunn, and M. N. Paddon-Row, *Tetrahedron Lett.* p. 2355 (1976).
251. L. A. Paquette and T. Kakihana, *J. Am. Chem. Soc.* **90**, 3897 (1968).
252. L. A. Paquette and J. C. Phillips, *J. Am. Chem. Soc.* **90**, 3898 (1968).
253. E. H. White, R. L. Stern, T. J. Lobl, S. H. Smallcombe, H. Maskill, and E. W. Friend, Jr., *J. Am. Chem. Soc.* **98**, 3247 (1976).
254. P. Radlich and W. Fenical, *Tetrahedron Lett.* p. 4901 (1967).
255. W. G. Dauben, R. G. Williams, and R. D. McKelvey, *J. Am. Chem. Soc.* **95**, 3932 (1973).

256. W. G. Dauben and M. S. Kellog, *J. Am. Chem. Soc.* **94,** 8951 (1972).
257. L. T. Scott and M. Jones, Jr., *Chem. Rev.* **72,** 181 (1972).
258. E. E. van Tamelen and T. L. Burkoth, *J. Am. Chem. Soc.* **89,** 151 (1967).
259. E. E. van Tamelen and R. H. Greeley, *Chem. Commun.* p. 601 (1971).
260. E. E. van Tamelen, T. L. Burkoth, and R. H. Greeley, *J. Am. Chem. Soc.* **93,** 6120 (1971).
261. S. Masamune and R. H. Seidner, *Chem. Commun.* p. 542 (1969).
262. K. Hojo, R. T. Seidner, and S. Masamune, *J. Am. Chem. Soc.* **92,** 6641 (1970).
263. T. J. Katz, J. J. Cheung, and N. Acton, *J. Am. Chem. Soc.* **92,** 6643 (1970).
264. S. Masamune, C. G. Chin, K. Hojo, and R. T. Seidner, *J. Am. Chem. Soc.* **89,** 4804 (1967).
265. G. Schröder and Th. Martini, *Angew. Chem. Int. Edit. Engl.* **6,** 806 (1967).
266. W. M. Moore, D. D. Morgan, and F. R. Stermitz, *J. Am. Chem. Soc.* **85,** 829 (1963).
267. K. A. Muszkat, D. Gegiou, and E. Fischer, *Chem. Commun.* p. 447 (1965).
267a. K. A. Muszkat and E. Fischer, *J. Chem. Soc. B* p. 662 (1967).
268. R. Korenstein, K. A. Muszkat, and E. Fischer, *Helv. Chim. Acta* **53,** 2102 (1970).
269. K. A. Muszkat and W. Schmidt, *Helv. Chim. Acta* **54,** 1195 (1971).
270. E. V. Blackburn and C. J. Timmons, *Quart. Rev.* **23,** 482 (1969).
271. G. Schröder and W. Martin, *Angew. Chem. Int. Edit. Engl.* **6,** 870 (1967).
272. J. F. M. Oth, G. Antoine, and J. M. Gilles, *Tetrahedron Lett.* p. 6265 (1968).
273. A. Windaus and A. Zomich, *Nachricht. Gesellsch. Wissensch. Göttingen* **11,** 462 (1917).
274. K. von Auwers, *Justus Liebigs Ann. Chem.* **422,** 133 (1921).
275. N. Campbell, in "Chemistry of Carbon Compounds" (E. H. Rodd, ed.), Vol. IVB, p. 810, Elsevier, Amsterdam.
276. W. Dilthey, *J. Prakt. Chem.* **94,** 65 (1916).
277. W. Dilthey, G. Bauriedel, G. Gieselbrecht, A. Seeger, and J. Winkler, *J. Prakt. Chem.* **101,** 177 (1920).
278. J. A. Berson, *J. Am. Chem. Soc.* **74,** 358 (1952).
279. J.-P. Griot, J. Royer, and J. Dreux, *Tetrahedron Lett.* p. 2195 (1969).
280. A. Baeyer and J. Piccard, *Justus Liebigs Ann. Chem.* **384,** 208 (1911).
281. W. Dilthey, *J. Prakt. Chem.* **94,** 53 (1916).
282. G. Büchi and N. C. Yang, *J. Am. Chem. Soc.* **79,** 2318 (1957).
283. R. Gompper and O. Christmann, *Chem. Ber.* **94,** 1784 (1961).
284. A. Hinnen and J. Dreux, *Bull. Soc. Chim. Fr.* p. 1492 (1964).
285. J. Royer and J. Dreux, *Tetrahedron Lett.* p. 5589 (1968).
286. J. Royer and J. Dreux, *Bull. Soc. Chim. Fr.* p. 707 (1972).
287. J.-P. Montillier and J. Dreux, *Bull. Soc. Chim. Fr.* p. 3638 (1969).
288. P. Roullier, D. Gagnaire, and J. Dreux, *Bull. Soc. Chim. Fr.* p. 689 (1966).
289. J. C. Anderson, D. G. Lindsay, and C. B. Reese, *Tetrahedron* **20,** 2091 (1964).
290. C. H. Eugster, C. Garbers, and P. Karrer, *Helv. Chim. Acta* **35,** 1179 (1952).
291. P. Schiess, H. L. Chia, and C. Suter, *Tetrahedron Lett.* p. 5747 (1968).
292. R. C. De Selms and U. T. Kreibich, *J. Am. Chem. Soc.* **91,** 3659 (1969).
293. J. F. Thomas and G. Branch, *J. Am. Chem. Soc.* **75,** 4793 (1953).
294. A. Safieddine, J. Royer, and J. Dreux, *Bull. Soc. Chim. Fr.* p. 703 (1972).
295. A. T. Balaban, G. Mihai, and C. D. Nenitzescu, *Tetrahedron* **18,** 257 (1962).
296. J.-P. LeRoux, G. Letertre, P.-L. Desbene, and J.-J. Basselier, *Bull. Soc. Chim. Fr.* p. 4059 (1971).
297. A. F. Kluge and C. P. Lillya, *J. Org. Chem.* **36,** 1977 (1971).
298. A. de Groot and B. L. M. Jansen, *Tetrahedron Lett.* p. 3407 (1975).
299. S. Bersani, G. Doddi, S. Fornarine, and F. Stegl, *J. Org. Chem.* **43,** 4112 (1978).
300. E. N. Marvell and P. Churchley, *Abst. 145th Meeting American Chemical Society, Los Angeles, Calif., April 1963, p. 45M.*

301. E. N. Marvell, T. Gosink, P. Churchley, and T. H. Li, *J. Org. Chem.* **37**, 2989 (1972).
302. E. N. Marvell, G. Caple, T. A. Gosink, and G. Zimmer, *J. Am. Chem. Soc.* **88**, 619 (1966).
303. E. N. Marvell, T. Chadwick, G. Caple, T. Gosink, and G. Zimmer, *J. Org. Chem.* **37**, 2992 (1972).
304. J. B. Flannery, Jr., *J. Am. Chem. Soc.* **90**, 5660 (1968) see footnote 46.
305. E. N. Marvell and T. Gosink, *J. Org. Chem.* **37**, 3036 (1972).
306. L. A. Burke, J. Elguero, G. Leroy, and M. Sana, *J. Am. Chem. Soc.* **98**, 1685 (1976).
307. A. Williams, *J. Am. Chem. Soc.* **93**, 2733 (1971).
308. R. P. Bell, "The Proton in Chemistry," p. 144, Methuen, London, 1959. It is notable that the material on p. 144 did not relate to data permitting the indicated calculation.
309. See refer. 308, p. 161.
310. S. Benson, "Thermochemical Kinetics," Wiley, New York, 1968.
311. A. Duperrier and J. Dreux, *Tetrahedron Lett.* p. 3127 (1970).
312. P. Rouiller and J. Dreux, *C. R. Acad. Sci., Paris, Ser. C* **258**, 5228 (1964).
313. A. Duperrier, M. Moreau, and J. Dreux, *Bull. Soc. Chim. Fr.* p. 2307 (1975).
314. A. Duperrier, M. Moreau, S. Gelin, and J. Dreux, *Bull. Soc. Chim. Fr.* p. 2207 (1974).
315. G. Köbrich and D. Wunder, *Justus Liebigs Ann. Chem.* **654**, 131 (1962).
316. K. Dimroth and K. H. Wolf, *Angew. Chem.* **72**, 778 (1960).
317. J.-P. Montillier, J. Royer, and J. Dreux, *Bull. Soc. Chim. Fr.* p. 1956 (1970).
318. J. P. Le Roux and C. Goasdoue, *Tetrahedron* **31**, 2761 (1975).
319. M. Trolliet, R. Longeray, and J. Dreux, *Tetrahedron* **30**, 163 (1974).
320. T. Gosink, *J. Org. Chem.* **39**, 1942 (1974).
321. D. G. England and C. G. Krespan, *J. Org. Chem.* **35**, 3300, 3308 (1970).
322. A. Roedig and T. Neukam, *Chem. Ber.* **107**, 3463 (1974).
323. A. Roedig and T. Neukam, *Justus Liebigs Ann. Chem.* p. 240 (1975).
324. A. Roedig, *Angew. Chem. Int. Edit. Engl.* **5**, 680 (1966).
325. W. Surber, V. Theus, L. Colombi, and H. Schinz, *Helv. Chim. Acta* **39**, 1299 (1956).
326. G. Köbrich, *Angew. Chem.* **72**, 348 (1960).
327. K. E. Schulte, J. Reisch, and A. Mock, *Arch. Pharm.* **295**, 627 (1962).
328. S. Sarel and J. Rivlin, *Tetrahedron Lett.* p. 821 (1965).
329. B. Berrang, D. Horton, and J. D. Wander, *J. Org. Chem.* **38**, 187 (1973).
330. R. J. Ferrier and N. Vethaviyasar, *J. Chem. Soc. Perkin Trans. 1* p. 1791 (1973).
331. K. Jankowski and R. Luce, *Tetrahedron Lett.* p. 2069 (1974).
332. W. Anschütz and S. N. Datta, *Justus Liebigs Ann. Chem.* p. 1971 (1975).
333. P. Schiess and H. L. Chia, *Helv. Chim. Acta* **53**, 485 (1970).
334. P. Schiess, R. Seeger, and C. Suter, *Helv. Chim. Acta* **53**, 1713 (1970).
335. G. Maier and M. Wiessler, *Tetrahedron Lett.* p. 4987 (1969).
336. A. Roedig and G. Märkl, *Justus Liebigs Ann. Chem.* **659**, 1 (1962).
357. A. Roedig, G. Märkl, F. Frank, R. Kohlhaupt, and M. Schlosser, *Chem. Ber.* **100**, 2730 (1967).
338. A. Roedig and W. Ruch, *Justus Liebigs Ann. Chem.* **730**, 57 (1969).
339. A. Roedig, G. Märkl, W. Ruch, H. G. Kleppe, R. Kohlhaupt, and H. Schaller, *Justus Liebigs Ann. Chem.* **692**, 83 (1966).
340. A. Roedig, *Angew. Chem.* **76**, 276 (1964).
341. A. Roedig, F. Frank, and G. Roebke, *Justus Liebigs Ann. Chem.* p. 630 (1974).
342. A. Roedig, H.-A. Renk, V. Schaal, and D. Scheutzow, *Chem. Ber.* **107**, 1136 (1974).
343. K. Dimroth and K. H. Wolf, *Angew. Chem.* **72**, 777 (1960).
344. J. Royer, A. Safieddine, and J. Dreux, *C. R. Acad. Sci. Paris, Ser. C* **274**, 1849 (1972).
345. O. Chalvet, C. Decoret, J. Dreux, A. Safieddine, and J. Royer, *Bull. Soc. Chim. Fr.* p. 716 (1972).

346. A. Safieddine, J. Royer, and J. Dreux, *Bull. Soc. Chim. Fr.* p. 2510 (1972).
347. M. Trolliet, J. Royer, R. Longeray, and J. Dreux, *Tetrahedron* **30,** 173 (1974).
348. K. Dimroth, *Angew. Chem. Int. Edit. Engl.* **1,** 462 (1962).
349. N. K. Cuong, F. Fournier, and J. J. Basselier, *Bull. Soc. Chim. Fr.* p. 2117 (1974).
350. B. P. Gusev, E. K. Gorlova, and V. F. Kucherov, *Izv. Akad. Nauk. SSSR Ser. Khim.* p. 1070 (1973).
351. L. A. Rozov, V. Y. Zeifman, N. P. Gambarzan, A. Y. Cheburkov, and I. L. Knunyants, *Izv. Akad. Nauk. SSSR Ser. Khim.* p. 2750 (1976).
352. Z. A. Krasnaya, E. P. Prokof'ev, and V. F. Kucherov, *Izv. Akad. Nauk. SSSR Ser. Khim.* p. 123 (1978).
353. J. H. Day, *Chem. Rev.* **63,** 65 (1965).
354. R. S. Becker and J. Michl, *J. Am. Chem. Soc.* **88,** 5931 (1966).
355. K. R. Huffman and E. F. Ullman, *J. Am. Chem. Soc.* **89,** 5629 (1967).
356. P. Apprion, J. Guillerez, F. Garnier, and R. Guglielmetti, *Helv. Chim. Acta* **58,** 2553, 2563 (1975).
357. T. R. Evans, A. F. Toth, and P. A. Leermakers, *J. Am. Chem. Soc.* **89,** 5060 (1967).
358. J. Zsindely and H. Schmid, *Helv. Chim. Acta* **51,** 1510 (1968).
359. R. Hug, H.-J. Hansen, and H. Schmid, *Helv. Chim. Acta* **55,** 10 (1972).
360. J. Bruhn, J. Zsindely, H. Schmid, and G. Frater, *Helv. Chim. Acta* **61,** 2542 (1978).
361. H. Heaney and J. M. Jablonski, *Chem. Commun.* p. 1139 (1968).
362. H. Heaney and C. T. McCarty, *Chem. Commun.* p. 123 (1970).
363. H. Heaney, J. M. Jablonski, and C. T. McCarty, *J. Chem. Soc. Perkin Trans. 1* p. 2903 (1972).
364. A. T. Bowne and R. H. Levin, *Tetrahedron Lett.* p. 2043 (1974).
365. E. E. Schweizer, D. M. Crouse, and D. L. Dalrymple, *Chem. Commun.* p. 354 (1969).
366. D. W. Hutchinson and J. A. Tomlinson, *Tetrahedron Lett.* p. 5027 (1968).
367. D. G. Clarke, L. Crombie, and D. A. Whiting, *Chem. Commun.* 580, 582 (1973).
368. L. Crombie, P. W. Freeman, and D. A. Whiting, *J. Chem. Soc. Perkin Trans. 1* p. 1277 (1973).
369. D. G. Clarke, L. Crombie, and D. A. Whiting, *J. Chem. Soc. Perkin Trans. 1* p. 1007 (1974).
370. M. J. Begley, L. Crombie, R. W. King, D. A. Slack, and D. A. Whiting, *J. Chem. Soc. Perkin Trans. 1* p. 2393 (1971).
371. W. H. Pirkle, H. Seto, and W. V. Turner, *J. Am. Chem. Soc.* **92,** 6984 (1970).
372. W. H. Pirkle and W. V. Turner, *J. Org. Chem.* **40,** 1617 (1975).
373. J. Ficini, J. Pouliquen, and J.-P. Pauline, *Tetrahedron Lett.* p. 2483 (1971).
374. O. L. Chapman, C. L. McIntosh, and J. Pacansky, *J. Am. Chem. Soc.* **94,** 245 (1973).
375. R. G. S. Pong and J. S. Shirk, *J. Am. Chem. Soc.* **95,** 248 (1973).
376. A. Krantz, *J. Am. Chem. Soc.* **96,** 4992 (1974).
377. A. E. Baydar and G. V. Boyd, *Chem. Commun.* p. 718 (1976).
378. G. Maier and U. Schäfer, *Tetrahedron Lett.* p. 1053 (1977).
379. R. Arad-Yellin, B. S. Green, and K. A. Muszkat, *Chem. Commun.* p. 14 (1976).
380. S. Goldschmidt and H. Wessbecher, *Chem. Ber.* **61,** 372 (1928).
381. J. Baltes and F. Volbert, *Fette Seifen Anstrichmittel* **57,** 660 (1955).
382. M. S. Newman and P. L. Childers, *J. Org. Chem.* **32,** 62 (1967).
383. J. P. Smith and G. B. Schuster, *J. Am. Chem. Soc.* **100,** 2564 (1978).
384. J. Michl, *Photochem. Photobiol.* **25,** 141 (1977).
385. L. Brandsma and D. J. W. Schuijl, *Rec. Trav. Chim. Pays-Bas* **88,** 30 (1969).
386. D. Schuijl-Laros, P. J. W. Schuijl, and L. Brandsma, *Rec. Trav. Chim. Pays-Bas* **91,** 785 (1972).

387. L. Brandsma and H. J. T. Bos, *Rec. Trav. Chim. Pays-Bas* **88**, 732 (1969).
388. L. Brandsma and D. Schuijl-Laros, *Rec. Trav. Chim. Pays-Bas* **89**, 110 (1970).
389. Th. Zincke, G. Heuser, and W. Möller, *Justus Liebigs Ann. Chem.* **333**, 296 (1904).
390. Th. Zincke, *Justus Liebigs Ann. Chem.* **330**, 361 (1903).
391. Th. Zincke and W. Wurker, *Justus Liebigs Ann. Chem.* **338**, 107 (1905).
392. Th. Zincke and W. Wurker, *Justus Liebigs Ann. Chem.* **341**, 365 (1905).
393. W. König, *J. Prakt. Chem.* **69**, 105 (1904).
394. E. N. Marvell, G. Caple, and I. Shahidi, *Tetrahedron Lett.* p. 277 (1967).
395. E. N. Marvell, G. Caple, and I. Shahidi, *J. Am. Chem. Soc.* **92**, 5641 (1970).
396. E. N. Marvell and I. Shahidi, *J. Am. Chem. Soc.* **92**, 5646 (1970).
397. E. N. Marvell, T. H. Li, and C. Paik, *Tetrahedron Lett.* p. 2089 (1973).
398. D. Lu, unpublished results of attempts to obtain evidence for a doubly charged species.
399. W. Hoppe and F. Baumgartner, *Z. Kristallogr. Kristallgeometrie, Kristallphys. Kristallchem.* **108**, 323 (1957).
400. F. Dorn, J. Kotschy, and H. Kausen, *Ber. Bunsenges. Phys. Chem.* **69**, 11 (1965).
401. F. Baumgartner, E. Gunther, and G. Scheibe, *Z. Elektrochem.* **60**, 570 (1956).
402. J. Kavalek and V. Sterba, *Collect. Czech. Chem. Commun.* **38**, 3506 (1973).
403. E. Van den Dunghen, J. Nasielski, and P. Van Laer, *Bull. Soc. Chim. Belges* **66**, 661 (1957).
404. R. Oda and S. Mita, *Bull. Soc. Chem. Japan* **36**, 103 (1963).
405. S. L. Johnson and K. L. Rumon, *Tetrahedron Lett.* p. 1721 (1966).
406. J. Kavalek, J. Polansky, and V. Sterba, *Collect. Czech. Chem. Commun.* **39**, 1049 (1974).
407. J. Kavalek, A. Bartecek, and V. Sterba, *Collect. Czech. Chem. Commun.* **39**, 1717 (1974).
408. T. H. Li, Ph. D. Thesis, Oregon State Univ., 1974.
409. J. Kavalek, A. Lycka, V. Machacek, and V. Sterba, *Collect. Czech. Chem. Commun.* **39**, 2056 (1974).
410. V. Beranek, J. Kavalek, A. Lycka, and V. Sterba, *Collect. Czech. Chem. Commun.* **39**, 2047 (1974).
411. J. Kavalek, A. Lycka, V. Machacek, and V. Sterba, *Collect. Czech. Chem. Commun.* **41**, 1926 (1976).
412. T. Kato and H. Yamanaka, *J. Org. Chem.* **30**, 910 (1965).
413. T. J. van Bergen and R. M. Kellogg, *J. Org. Chem.* **36**, 1705 (1971).
414. P. Schiess and P. Ringele, *Tetrahedron Lett.* p. 311 (1972).
415. P. Schiess. C. Monnier, P. Ringele, and E. Sendi, *Helv. Chim. Acta* **57**, 1676 (1974).
416. P. Schiess, M. L. Chia, and P. Ringele, *Tetrahedron Lett.* p. 313 (1972).
417. R. Eisenthal and A. R. Katritzky, *Tetrahedron* **21**, 2205 (1965).
418. R. Eisenthal, A. R. Katritzky, and E. Lunt, *Tetrahedron* **23**, 2775 (1967).
419. A. R. Katritzky and E. Lunt, *Tetrahedron* **25**, 4291 (1969).
420. T. Kato, H. Yamanaka, T. Adachi, and H. Hiranuma, *J. Org. Chem.* **32**, 3788 (1967).
421. J. Schnekenburger and D. Heber, *Tetrahedron* **30**, 4055 (1974).
422. J. Schnekenburger and D. Heber, *Chem. Ber.* **107**, 3408 (1974).
423. H. Sliwa and A. Tartar, *Tetrahedron Lett.* p. 4717 (1976).
424. Z. Neiman, *J. Chem. Soc. Perkin Trans.* 2 p. 1746 (1962).
425. I. Hassan and F. W. Fowler, *J. Am. Chem. Soc.* **100**, 6696 (1978).
426. T. Kamentani, T. Kato, and K. Fukumoto, *Tetrahedron* **30**, 1043 (1974).
427. R. R. Schmidt, W. Schneider, J. Karg, and U. Burkert, *Chem. Ber.* **105**, 1634 (1972).
428. T. L. Gilchrist, G. E. Gymer, and C. W. Rees, *Chem. Commun.* p. 835 (1973).
429. L. E. Overman and S. Tsuboi, *J. Am. Chem. Soc.* **99**, 2813 (1977).
430. F. C. Schaeffer and W. D. Zimmerman, *J. Org. Chem.* **35**, 2165 (1970).
431. M. P. Cava and L. P. Bravo, *Tetrahedron Lett.* p. 4631 (1970).
432. A. Padwa and P. H. J. Carlsen, *Tetrahedron Lett.* p. 433 (1978).

433. J. L. Asherson and D. W. Young, *Chem. Commun.* p. 916 (1977).
434. W. Pfleiderer and H.-U. Bank, *Angew. Chem. Int. Edit. Engl.* **7,** 535 (1968).
435. A. Pawda, S. Clough, and E. Glazer, *J. Am. Chem. Soc.* **92,** 1778 (1970).
436. A. Padwa and E. Glazer, *Chem. Commun.* p. 838 (1971).
437. A. Padwa, S. Clough, and L. Gehrlein, *Chem. Commun.* p. 74 (1972).
438. A. Padwa and L. Gehrlein, *J. Am. Chem. Soc.* **94,** 4933 (1972).
439. A. Padwa and E. Glazer, *J. Am. Chem. Soc.* **94,** 7788 (1972).
440. A. Padwa, M. Dharan, J. Smolanoff, and S. I. Wetmore, Jr., *J. Am. Chem. Soc.* **95,** 1954 (1973).
441. A. Padwa and S. I. Wetmore, Jr., *Chem. Commun.* p. 1116 (1972).
442. A. Orahovats, H. Heimgartner, H. Schmid, and W. Heintzelmann, *Helv. Chim. Acta* **58,** 2662 (1975).
443. P. Beak and J. L. Miesel, *J. Am. Chem. Soc.* **89,** 2375 (1967).
444. G. McCoy and A. Day, *J. Am. Chem. Soc.* **65,** 2159 (1943).
445. F. Yoneda, M. Higuchi, and M. Kawamura, *Heterocycles* **4,** 1659 (1976).
446. F. Yoneda and M. Higuchi, *J. Chem. Soc. Perkin Trans. 1* p. 1336 (1977).
447. R. C. Shah and M. B. Ichaporia, *J. Chem. Soc.* p. 431 (1936).
448. H. P. Ghadiali and R. C. Shah, *J. Indian Chem. Soc.* **26,** 117 (1949).
449. M. M. Blatter and H. Lukaszewski, *Tetrahedron Lett.* p. 855 (1964).
450. T. L. Gilchrist, C. J. Moody, and C. W. Rees, *J. Chem. Soc. Perkin Trans. 1* p. 1964 (1975).
451. S. Senda, K. Hirota, T. Asao, and Y. Yamada, *Tetrahedron Lett.* p. 2295 (1978).
452. J. E. Baldwin and H. H. Basson, *Chem. Commun.* p. 795 (1969).
453. K. Alder, *Justus Liebigs Ann. Chem.* **585,** 81 (1954).
454. H. C. van der Plas, *Accts. Chem. Res.* **11,** 462 (1978).
455. C. A. H. Rasmussen and H. C. van der Plas, *Tetrahedron Lett.* p. 3841 (1978).
456. K. Adachi, *J. Pharm. Soc. Japan* **77,** 507 (1957).
457. J. Kolc and R. S. Becker, *J. Am. Chem. Soc.* **91,** 6513 (1969).
458. A. R. Katritzsky and J. M. Lagowski, "Chemistry of the Heterocyclic N-oxides," p. 94, Academic Press, New York, 1971.
459. E. N. Marvell and T. H. C. Li, *Synthesis* p. 457 (1973).
460. J. P. Friedrich, M. M. Teeter, J. C. Cowan, and G. E. McManis, *J. Am. Oil Chem. Soc.* **38,** 329 (1961).
461. W. J. De Jarlais and H. M. Teeter, *J. Am. Oil Chem. Soc.* **39,** 421 (1962).
462. J. P. Friederich and R. E. Beal, *J. Am. Oil Chem. Soc.* **39,** 528 (1962).
463. R. A. Eisenhauer, R. E. Beal, and E. L. Griffin, *J. Am. Oil Chem. Soc.* **40,** 129 (1963).
464. L. T. Black and R. A. Eisenhauer, *J. Am. Oil Chem. Soc.* **40,** 272 (1963).
465. R. A. Eisenhauer, R. E. Beal, and E. L. Griffin, *J. Am. Oil Chem. Soc.* **41,** 60 (1964).
466. M. Yoshida, H. Sugihara, S. Tsushima, and T. Miki, *Chem. Commun.* p. 1223 (1969).
467. M. M. Radcliffe and W. P. Weber, *J. Org. Chem.* **42,** 297 (1977).
468. B. I. Rosen and W. P. Weber, *Tetrahedron Lett.* p. 151 (1977).
469. B. I. Rosen and W. P. Weber, *J. Org. Chem.* **42,** 47 (1977).
470. G. Frater, *Helv. Chim. Acta* **60,** 515 (1977).
471. F. R. Jensen and W. E. Coleman, *J. Am. Chem. Soc.* **80,** 6149 (1950).
472. M. R. DeCamp, R. H. Levin, and M. Jones, Jr., *Tetrahedron Lett.* p. 3575 (1974).
473. J. Rigaudy and P. Capdevielle, *Tetrahedron* **33,** 767 (1977).
474. H. Straub and J. Hambrecht, *Synthesis* p. 425 (1975).
475. H. G. Heller and K. Salisbury, *J. Chem. Soc. C* p. 399 (1970).
476. G. Märkl and H. Hauptmann, *Tetrahedron* **32,** 2131 (1976).

477. J. Druey, E. F. Jenney, K. Schenker, and R. B. Woodward, *Helv. Chim. Acta* **45,** 600 (1962).
478. E. W. Neuse and B. R. Green, *Justus Liebigs Ann. Chem.* p. 1534 (1974).
479. R. Huisgen and H. Mayr, *Chem. Commun.* p. 55, 57 (1976).
480. R. Brelow, M. Oda, and J. Pecoraro, *Tetrahedron Lett.* p. 4415 (1972).
481. R. F. C. Brown and G. L. McMullen, *Austr. J. Chem.* **27,** 2385, 2605 (1974).
482. C. Jutz and R. M. Wagner, *Angew. Chem. Int. Edit. Engl.* **11,** 315 (1972).
483. J. C. Jutz, *Topics Curr. Chem.* **73,** 125 (1978).
484. R. L. N. Harris, J. L. Huppatz, and J. N. Phillips, *Angew. Chem. Int. Edit. Engl.* **15,** 498 (1976).
485. K. Elbs, *Chem. Ber.* **17,** 2847 (1884).
486. O. Diels and K. Alder, *Chem. Ber.* **60,** 716 (1927).
487. A. Balaban and C. Nenitzescu, *Justus Liebigs Ann. Chem.* **625,** 74 (1959).
488. K. Dimroth, G. Braüniger, and G. Neubauer, *Chem. Ber.* **90,** 1634 (1957).
489. K. Dimroth, G. Neubauer, H. Möllenkamp, and G. Oosterloo, *Chem. Ber.* **90,** 1688 (1957).
490. K. Dimroth and G. Neubauer, *Chem. Ber.* **92,** 2042 (1959).
491. R. Gompper and O. Christman, *Angew. Chem.* **71,** 32 (1959).
492. R. Gompper and O. Christman, *Chem. Ber.* **94,** 1795 (1961).
493. G. Märkl, *Angew. Chem. Int. Edit. Engl.* **1,** 511 (1962).
494. G. Märkl and H. Baier, *Tetrahedron Lett.* p. 4379 (1968).
495. G. A. Reynolds and J. A. Van Allan, *J. Org. Chem.* **34,** 2736 (1969).
496. K. Dimroth, *Angew. Chem.* **72,** 331 (1960).
497. R. S. Sagitullin, S. P. Gromov, and A. N. Kost, *Dokl. Akad. Nauk. SSSR* **236,** 634 (1977).
498. R. S. Sagitullin, S. P. Gromov, and A. N. Kost, *Tetrahedron* **34,** 2213 (1978).
499. L. H. Knox, E. Verlarde, and A. D. Cross, *J. Am. Chem. Soc.* **85,** 2533 (1963).
500. R. Darms, T. Threlfall, M. Pesaro, and A. Eschenmoser, *Helv. Chim. Acta* **46,** 2893 (1963).
501. E. Vogel, British Chem. Soc. Special Publication No. 21 pp. 113–147 (1967).
502. W. C. Agosta and W. W. Lowrance, Jr., *J. Org. Chem.* **35,** 3851 (1970).
503. H. C. van der Plas, "Ring Transformations of Heterocycles," Vol. 2, Academic Press, New York, 1973.
504. W. Schroth and G. Fischer, *Angew. Chem. Int. Edit. Engl.* **2,** 394 (1963).
505. M. J. Begley, L. Crombie, R. W. King, D. A. Slack, and D. A. Whiting, *Chem. Commun.* p. 138 (1976).
506. L. Crombie, D. A. Slack, and D. A. Whiting, *Chem. Commun.* p. 139 (1976).
507. M. J. Begley, L. Crombie, D. A. Slack, and D. A. Whiting, *Chem. Commun.* p. 140 (1976).
508. J. W. Huffman and T. M. Hsu, *Tetrahedron Lett.* p. 141 (1972).
509. A. T. Balaban and C. Toma, *Tetrahedron Suppl.* **7,** 1 (1966).
510. A. T. Balaban, G. R. Bedford, and A. R. Katritzky, *J. Chem. Soc.* p. 1646 (1964).
511. Cf. refer. 503 pp. 18–25.
512. M. H. O'Leary and G. A. Samberg, *J. Am. Chem. Soc.* **93,** 3530 (1971).
513. Cf. refer. 483 pp. 194–208.
514. D. Lloyd and J. M. F. Gagan, *J. Chem. Soc. C.* p. 2488 (1970).
515. R. M. Acheson and R. G. Bolton, *Tetrahedron Lett.* p. 2821 (1973).
516. Cf. refer. 483 p. 211.
517. Cf. refer. 503, pp. 34–35.
518. H. Meislich, "Pyridine and its Derivatives" (E. Klingsberg, ed.), pp. 509–866, Wiley (Interscience), New York, 1962.

519. A. S. Afridi, A. R. Katritzky, and C. A. Ramsden, *J. Chem. Soc. Perkin Trans. 1*, p. 1436 (1977).

520. F. Eloy and A. Deryckere, *J. Heterocycl. Chem.* **7**, 1191 (1970).

521. J. H. McMillan and S. S. Washburn, *J. Org. Chem.* **38**, 2982 (1973).

522. T. Kametani, T. Takahashi, K. Ogasawara, and K. Fukumoto, *Tetrahedron* **30**, 1047 (1974).

523. T. Kametani, M. Takemura, K. Ogasawara, and K. Fukumoto, *J. Heterocycl. Chem.* **11**, 179 (1974).

524. W. Oppolzer, *Angew. Chem. Int. Edit. Engl.* **16**, 10 (1977).

525. W. Oppolzer, M. Petrzilka, and K. Bättig, *Helv. Chim. Acta* **60**, 2964 (1977).

526. C. L. Pedersen, N. Harrit, and O. Buchardt, *Acta Chem. Scand.* **24**, 3435 (1970).

527. K. Kanai, M. Umehara, H. Kitano, and K. Fukui, *Nippon Kagaku Zasshi* **84**, 432 (1963).

528. R. Wizinger and P. Ulrich, *Helv. Chim. Acta* **39**, 207, 217 (1956).

529. R. Meyer and J. Szanecki, *Chem. Ber.* **33**, 2577 (1900).

530. Cf. refer. 483 p. 208, 216.

531. Cf. refer. 503 p. 87, 137.

532. V. Snieckus and G. Kan, *Tetrahedron Lett.* p. 2267 (1970).

533. E. C. Taylor and C. W. Jefford, *J. Am. Chem. Soc.* **84**, 3744 (1962).

534. W. Pfeiderer and F. E. Kempter, *Chem. Ber.* **103**, 908 (1970).

535. Cf. refer. 483 pp. 196–197, 208–209.

536. D. R. Osborne and R. Levine, *J. Org. Chem.* **28**, 2933 (1963).

537. O. R. Osborne, W. T. Wieder, and R. Levine, *J. Heterocycl. Chem.* **1**, 145 (1964).

538. W. König, *J. Prakt. Chem.* **70**, 19 (1904).

539. P. Baumgarten, *Chem. Ber.* **57**, 1622 (1924).

540. P. Baumgarten, *Chem. Ber.* **59**, 1166 (1926).

541. J. Becher, N. Haunsø, and T. Pedersen, *Acta. Chem. Scand. B* **29**, 124 (1975).

542. W. König and R. Bayer, *J. Prakt. Chem.* **83**, 325 (1910).

543. E. P. Lira, *J. Heterocycl. Chem.* **9**, 713 (1972).

544. Th. Zincke and F. Krollpfeiffer, *Justus Liebigs Ann. Chem.* **408**, 285 (1915).

545. B. Lipke, *Z. Chem.* **10**, 463 (1970).

546. B. Lipke, *Z. Chem.* **11**, 150 (1971).

547. H. Sliwa and A. Tartar, *Tetrahedron* **35**, 341 (1979).

548. G. W. Fischer, *Z. Chem.* **8**, 379 (1968).

549. G. W. Fischer, *Chem. Ber.* **103**, 3489 (1970).

550. R. Hull, *J. Chem. Soc. C* p. 1777 (1968).

551. R. A. Olofson and D. M. Zimmerman, *J. Am. Chem. Soc.* **89**, 5057 (1967).

552. R. J. Molyneux and R. Y. Wong, *Tetrahedron* **33**, 1431 (1977).

553. R. Kuhn and E. Teller, *Justus Liebigs Ann. Chem.* **715**, 106 (1968).

554. H. Albrecht and F. Krohnke, *Justus Liebigs Ann. Chem.* **701**, 126 (1967).

555. Y. Tamura, K. Sumoto, M. Mano, and T. Masui, *Yakugaku Zasshi* **92**, 371 (1972).

556. Y. Tamura, Y. Miki, T. Honda, and M. Ikeda, *J. Heterocycl. Chem.* **9**, 865 (1972).

557. J. Epztajn, E. Lunt, and A. R. Katritzky, *Tetrahedron* **26**, 1665 (1970).

558. A. F. Vompe and N. F. Turitsyna, *Zhur. Obshch. Khim.* **28**, 2864 (1958).

559. J. F. Tilney-Bassett and W. Waters, *J. Chem. Soc.* p. 2123 (1959).

560. M. Strell, W. B. Braunbruck, W. F. Fuhler, and O. Huber, *Justus Liebigs Ann. Chem.* **587**, 177 (1954).

561. M. Strell and K. Rost, *Chem. Ber.* **90**, 1905 (1957).

562. K. Hafner and K.-D. Asmus, *Justus Liebigs Ann. Chem.* **671**, 31 (1964).

563. W. König, M. Coenen, W. Lorenz, F. Bahr, and A. Bassl, *J. Prakt. Chem.* **30**, 96 (1965).

564. A. F. Vompe, I. I. Loevkoev, N. F. Turitsyna, V. V. Durmashkina, and L. V. Ivanova, *Zhur. Obshch. Khim.* **34,** 1758 (1964).

565. J. Schnekenburger, D. Heber, and E. Heber-Brunschweiger, *Justus Liebigs Ann. Chem.* p. 1799 (1976).

566. J. Schnekenburger, D. Heber, and E. Heber-Brunschweiger, *Tetrahedron* **33,** 457 (1977).

567. E. H. Kosower and J. W. Patton, *Tetrahedron* **22,** 2081 (1966).

568. F. Hamer, "Cyanine Dyes and Related Compounds," pp. 244–269, Wiley (Interscience), New York, 1964.

569. K. Hafner, *Justus Liebigs Ann. Chem.* **606,** 79 (1957).

570. K. Hafner and H. Kaiser, *Justus Liebigs Ann. Chem.* **618,** 140 (1958).

571. K. Hafner, *Angew. Chem.* **70,** 419 (1958).

572. G. Schwartzenbach and R. Weber, *Helv. Chim. Acta* **25,** 1628 (1942).

573. H. von Dobeneck and W. Goltzsche, *Chem. Ber.* **95,** 1484 (1962).

574. T. Tamura, N. Tsujimoto, and H. Yumi, *Yakugaku Zasshi* **92,** 546 (1972).

575. F. Uwe and S. Hünig, *Justus Liebigs Ann. Chem.* p. 1407 (1974).

576. C. W. Spangler, B. Keys, and D. C. Bookbinder, *J. Chem. Soc. Perkin Trans. 2* p. 810 (1979).

577. C. W. Spangler, S. Ibrahim, D. C. Bookbinder, and S. Ahmad, *J. Chem. Soc. Perkin Trans. 2* p. 717 (1979).

578. C. Jutz, R.-M. Wagner, A. Kraatz, and H.-G. Löbering, *Justus Liebigs Ann. Chem.* p. 874 (1975).

579. O. Crescente, H. G. Heller, and S. Oliver, *J. Chem. Soc. Perkin Trans. 1* p. 150 (1979).

580. H. G. Heller, S. Oliver, and H. Shawe, *J. Chem. Soc. Perkin Trans. 1* p. 154 (1979).

581. R. H. Bradbury, T. L. Gilchrist, and C. W. Rees, *Chem. Commun.* p. 528 (1979).

582. H. Bross, R. Schneider, and H. Hopf, *Tetrahedron Lett.* p. 2129 (1979).

583. B. M. Trost, P. H. Scudder, R. M. Cory, N. J. Turro, V. Ramamurthy, and T. J. Katz, *J. Org. Chem.* **44,** 1264 (1979).

584. W. Schroth, F. Billig, and G. Reinhold, *Angew. Chem.* **79,** 685 (1967).

585. W. Schroth, F. Billig, and H. Langguth, *Angew. Chem.* **77,** 919 (1965).

586. W. Schroth, F. Billig, and R. Langguth, *Z. Chem.* **5,** 352 (1964).

587. G. J. Baxter and R. F. C. Brown, *Australian J. Chem.* **28,** 1551 (1975).

CHAPTER 8
CONJUGATED SYSTEMS WITH
EIGHT OR MORE Π ELECTRONS

Experimental difficulties in the preparation of appropriate steroisomeric reactants and the inherent instability of large polyenes combined with the obviously established predictability of the orbital conservation theory have reduced the studies of the electrocyclic reactions of longer π systems to a very modest number. The majority of these are of the eight electron–eight atom type.

I. EIGHT ELECTRON SYSTEMS

There have been no detailed theoretical studies of the eight electron electrocyclization process, so the only theoretical work has been the general treatment of electrocyclic processes already described in Chapter 2.

A. Eight Electron–Seven Atom Systems

Two general groups of molecules fit this classification, heptatrienyl anions and their heteroatom electrologs such as 5-aza-1,3,6-heptatrienes.

1. Heptatrienyl Anions

Staley[1] has reviewed the work of electrocyclizations of heptatrienyl anions. Present knowledge can be summarized briefly as follows. Heptatrienyl anions which are not specially stabilized generally cyclize to cycloheptadienyl anions. The stereochemistry, predicted to be conrotatory, has not been cleanly confirmed experimentally. Stabilized anions such as the enolate of hexadienal do not cyclize and may be generated from the cyclic isomers such as dihydrooxepins by treatment with potassamide.

2. Heteroheptatrienes

There appear to have been no examples of cyclization of any heteroatom systems of the type illustrated in Eq. (8-1). Potential examples of Eq. (8-2)

$$(8\text{-}1)$$

$$(8\text{-}2)$$

$$(8\text{-}3)$$

type have been observed[2,3] but close via the six electron–five atom system. Apparently formation of the requisite all cis conformation is too slow to compete with the five atom cyclization. However, when the requisite cis geometry is attained the eight electron closure proceeds nicely. Mukai and Sukawa[4] suggested the scheme of Eq. (8-4) to account for their unexpected

$$(8\text{-}4)$$

results, but little evidence was available to support the suggestion. Competition between a six electron and an eight electron cyclization occurs when a cis double bond is present but the final double bond is part of an aromatic ring [Eq. (8-5)].[5,6,6a,19,93] The requirement for a cis double bond is clearly

$$(8\text{-}5)$$

illustrated by the reactions of Eqs. (8-6) and (8-7).[7] Irradiation of a similar

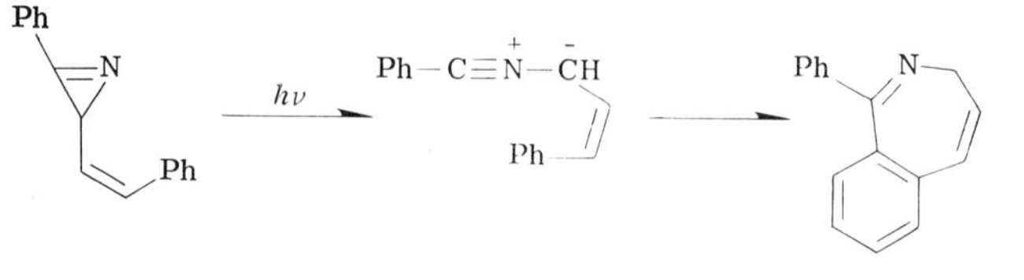

(8-6)

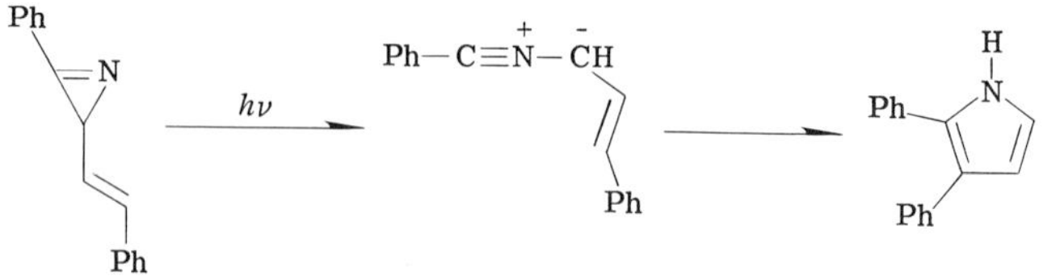

(8-7)

azirine, however, led to a very different eight electron cyclization which apparently requires a trans → cis double bond mutation [Eq. (8-8)].[8] The

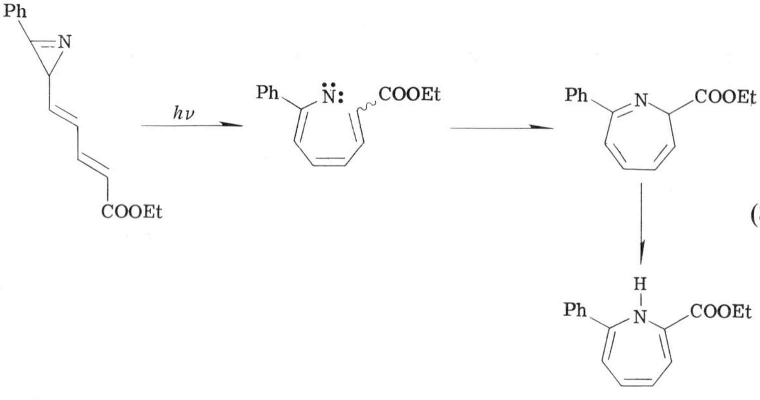

(8-8)

ring closure step is a thermal process since nitrenes can be trapped. Nitrilimines also can undergo the eight electron reaction [Eq. (8-9)].[9]

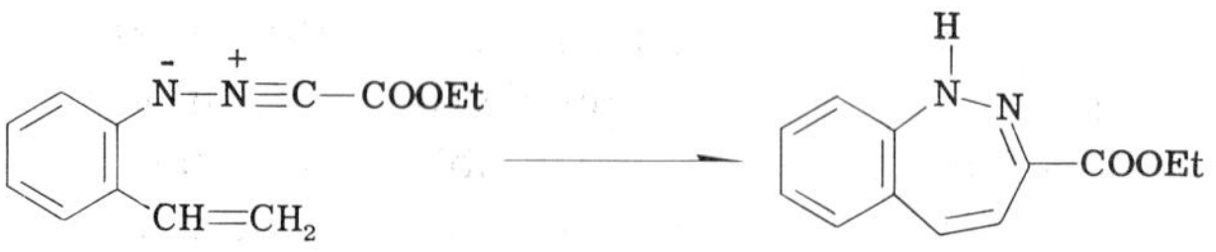

(8-9)

The geometric requirement for a cis double bond persists in cases of Eq. (8-3) type. One possible example gave only the six electron reaction, the double bond being trans.[10]

Retro-electrocyclic reactions have been observed in some instances. When **173** is heated to 450° several products result from an apparent ring cleavage and a six electron electrocyclization [Eq. (8-10)].[51] A similar sequence has

$$(8\text{-}10)$$

also been observed with a diazaoxepine [Eq. (8-11)].[52]

$$(8\text{-}11)$$

B. Eight Electron–Eight Atom Systems

When the Woodward–Hoffmann papers first appeared the stereochemical results for the electrocyclic reactions of cyclobutenes and hexatrienes were established. Consequently the real experimental test of the theoretical extention to $4n + 2$ and $4n$ systems was the electrocyclic reaction of octatetraenes.

1. Stereochemistry

The initial determination of the dominant stereochemistry of ring closure of 2,4,6,8-decatetraenes was done indirectly.[11] A pure sample of 2,8-di-*trans*-deca-4,6-diyne-2,8-diene was hydrogenated over Lindlar catalyst. Hydrogenation is not totally specific but does not alter the stereochemistry of the terminal double bonds. The 7,8-dimethylbicyclo[4.2.0]octadiene was isolated and its stereochemistry was ascertained as shown in Eq. (8-12). The

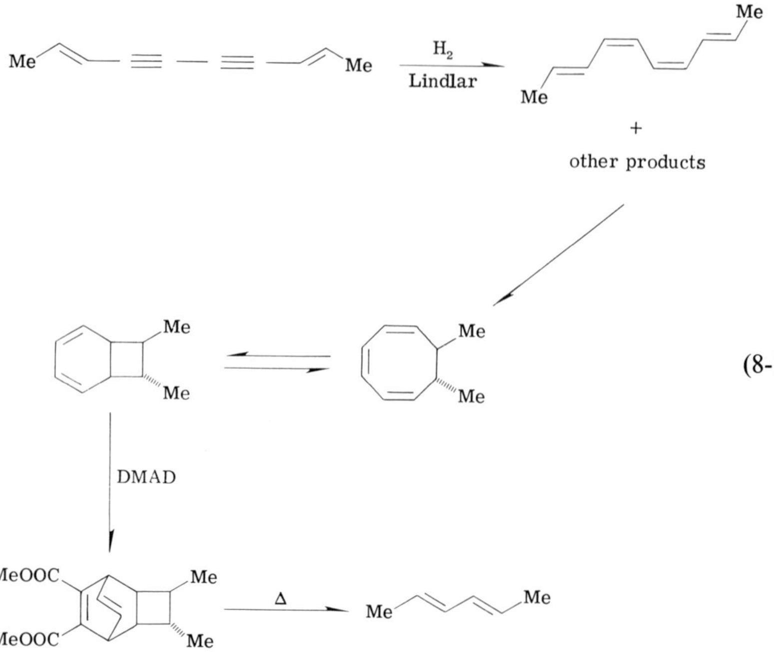

Me —⎯ ≡ — ≡ —⎯ Me $\xrightarrow[\text{Lindlar}]{H_2}$

+

other products

(8-12)

DMAD

Δ

course of the reaction was followed by completely hydrogenating each sample prior to analysis. A similar study with *trans*-2-*cis*-8-deca-4,6-diyne-2,8-diene gave mainly *cis*-7,8-dimethylbicyclo[4.2.0]octadiene. Thus the conrotatory route was shown to be the main course for the reaction, but no measure of the stereoselectivity was possible.

Shortly after publication of the above Huisgen published the results of his elegant studies of the stereoisomeric decatetraenes which were isolated, identified, and allowed to react separately.[12,13] All underwent conrotatory

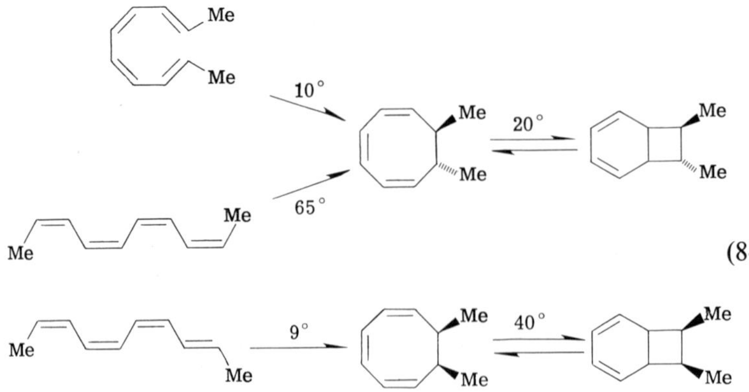

10°   20°

65°

(8-13)

9°   40°

reaction [Eq. (8-13)], and these reactions were stereoselective within the limits of experimental error (99.5–99.95%).[12] Conrotatory cyclization is therefore a general phenomenon in the eight electron system and may be expected to occur with high stereoselectivity.

2. The Disrotatory Reaction

Great interest still is attached to the problem of identifying an authentic case of a "forbidden" electrocyclization and in determining the energy differential between the "allowed" and "forbidden" reactions. While no certain example of a "forbidden" process has been identified there have appeared two measures of the energy differential. Dahmen and Huisgen[14] have exploited the reversibility (see Chapter 8, Section I,B,3) of the electrocyclization of *trans,cis,cis,trans*-2,4,6,8-decatetraene to obtain a measure of the enthalpy difference between the rates of formation of the trans and *cis*-7,8-dimethylcyclooctatrienes. Reaction was carried out in a gas chromatograph at temperatures varying from 145° to 205°, and the $\Delta\Delta H^{\ddagger}$ obtained ran from 10.9 to 11.8 kcal/mol. Various routes for formation of

$$\text{(8-14)}$$

the cis product were considered, but perhaps the most reasonable route via the known[15,16] equilibrium between 1,3,5- and 1,3,6-cyclooctatrienes was not eliminated. Since the unsubstituted cyclooctatrienes are in equilibrium at 100° this route needs to be eliminated before the ca. 11 kcal/mol differential can be considered as more than a lower bound.

A second estimate of $\Delta\Delta H^{\ddagger}$ has been made by Staley and Henry.[17,18] The basis for their estimate was the observation that **174** is in mobile equi-

$$\text{(8-15)}$$

174

librium with *trans*-bicyclo[6.2.0]decatriene [Eq. (8-15)].[17] From nmr studies of the rate they estimated $\Delta G^{\ddagger}$ as ca. 15 kcal/mol. When *cis*-bicyclo[6.2.0]-decatriene is heated it gives **176** accompanied by a small amount of **177**

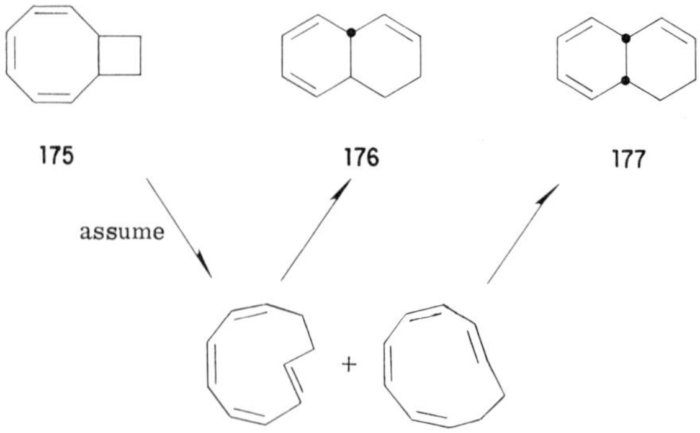

175   176   177

assume   (8-16)

[Eq. (8-16)]. The measured $\Delta H^{\ddagger}$ for formation of **176** is 32 kcal/mol, and the corresponding value for **177** is ca. 4 kcal/mol higher. Assuming the mechanism shown in Eq. (8-16) the authors suggest the $\Delta\Delta H^{\ddagger}$ between "allowed" and "forbidden" routes is ca. 18–20 kcal/mol. The validity of this estimate depends on the requirement that the strain energy change from **174** to its transition state does not decrease more than the strain energy change from **175** to its non-allowed transition state.

3. Equilibria and Rates

The work by Reppe and his colleagues on the preparation and reactions of cyclooctatetraene (COT) has evoked a long series of studies on COT and its partially hydrogenated derivatives.[20] Although 1,3,5-cyclooctatriene is a stable molecule, addition of a Grignard reagent to **178** gave an acyclic product [Eq. (8-17)].[21] Meister[22] showed that the ring closure reaction does

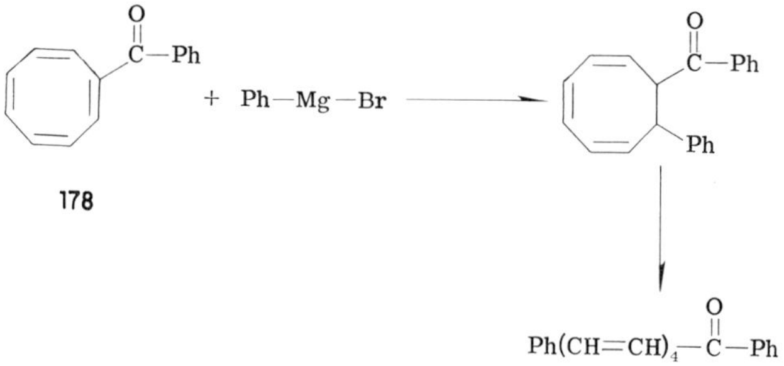

178

(8-17)

$$Ph(CH{=}CH)_4{-}\overset{O}{\overset{\|}{C}}{-}Ph$$

proceed when he heated all *cis*-1,8-dimethoxyoctatetraene (130°) and obtained 7,8-dimethoxybicyclo[4.2.0]octa-2,4-diene. The facile ring closure of di-*cis*-1,3,5,7-octatetraene was observed by Ziegenbein,[23] though he was

$$
\begin{array}{c}
\text{—C}\equiv\text{C—C}\equiv\text{C—} \xrightarrow[\text{Lindlar}]{H_2} \left[\text{（tetraene）} \right]
\end{array}
$$

$$\downarrow 25° \qquad\qquad (8\text{-}18)$$

80%

not able to isolate the acyclic tetraene. The rate of this reaction was measured by preparing the tetraene from cyclooctatriene by flash photolysis and identifying it spectrally at 20°K.[24,25] At 25° the half-life of the tetraene is 23 sec and $E_a = 17.0$ kcal/mol, log $A = 11.0$.

Apparently retro-electrocyclization of the dianion of cyclooctatrien-7,8-diol must be as rapid as the above ring closure.[26] When **179** is treated with

$$
\begin{array}{c}
\text{（bicyclic diacetate）} + \text{LiAlH}_4 \xrightarrow[25°]{\text{Et}_2\text{O}} \xrightarrow{H^+} \text{OHC}-(\text{CH}=\text{CH})_3\text{CHO}
\end{array}
$$
all trans

179

$$\downarrow \qquad\qquad \Big\downarrow H^+ \qquad (8\text{-}19)$$

$$
\text{（bicyclic dianion）} \rightleftharpoons \text{（cyclic dianion）} \longrightarrow {}^-\text{O}-(\text{CH}=\text{CH})_4-\text{O}^-
$$

lithium aluminum hydride for two minutes at 25° octatrienedial is obtained [Eq. (8-19)]. The diacetate **179** is relatively stable and exists as the bicyclic compound predominantly.[27] Other substituted cyclooctatrienes which do not revert to acyclic valence tautomers are noted in Chapter 7, Section I,F, Table 7-6.

The most comprehensive study of rates and equilibria was made with the isomeric decatetraenes.[13] The relevant data are listed in Table 8-1. No information about the equilibrium between the all cis decatetraene and *trans*-7,8-dimethylcyclooctatriene was obtained since under the ring closure conditions the main product was 7,8-dimethylbicyclo[4.2.0]octadiene. Dimethyl *trans,cis,cis,trans*-2,4,6,8-decatetraenedioate also cyclizes to the bicyclic tautomer [Eq. (8-18)], and at 40° the rate is 3.0×10^{-5} sec^{-1}.[28] The equilibrium constant is 16.0 at 50°. Unfortunately it is not possible to ascertain whether the rate determining step is the eight electron or the six

TABLE 8-1

Rates and Equilibria for Some Octatetraene–Cyclooctatriene Compounds

Compound	$\Delta H^{\ddagger}$	$\Delta S^{\ddagger}$	$K \dfrac{\text{cycl.}}{\text{acy.}}$	$T\ (^\circ)$
Decatetrane				
trans cis cis trans	15.5	−17	5.67	0
trans cis cis cis	19.4	−12	9.0	10
cis cis cis cis	21.7	−12		

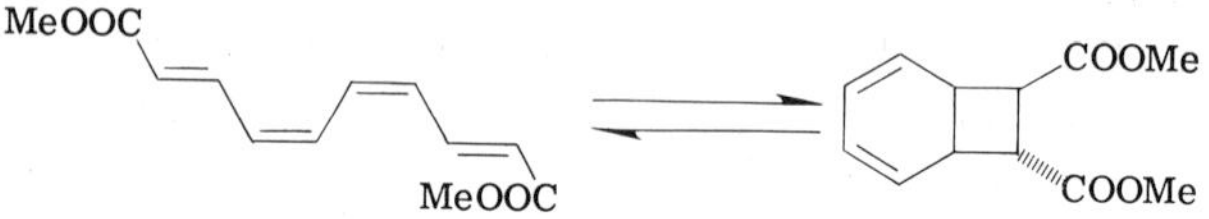

(8-20)

electron electrocyclization. The cis,cis,cis,trans ester is reported to be in equilibrium with the bicyclic ester at 24°.[29] Staley and Henry[17] studied the reaction of Eq. (8-15) and found $K = 1.38$ at 76°. The coalescence temperature was 48° which gives $\Delta G^{\ddagger} \simeq 15$ kcal/mol.

The most rapid eight electron cyclization reported is that of **180** which is formed photochemically from cyclooctatrienone at −196°. The ketene

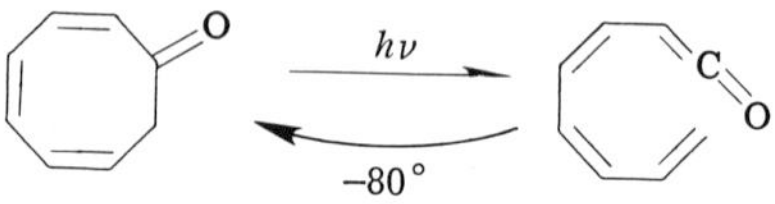

(8-21)

180

reverts to the cyclic ketone at temperatures above −80° but it can be trapped by reaction with the solvent when the photolysis is carried out in methanol.[49]

Two additional ring closure reactions have been reported both proceeding rapidly at room temperature. Preparation of **181** by partial hydrogenation of the diyne gave only cyclized product,[28] cyclization being too rapid to

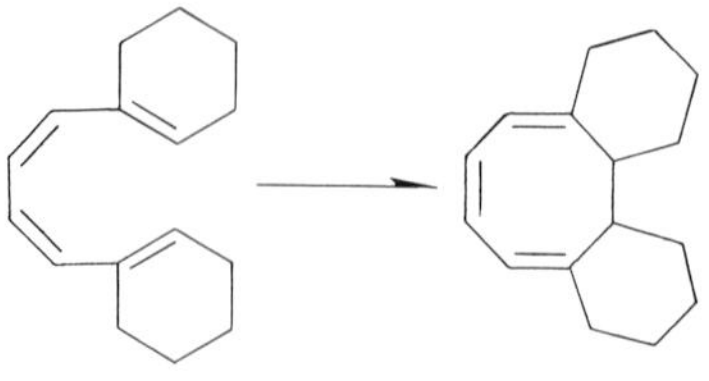

(8-22)

181

permit isolation of the tetraene. A very interesting modification was found by Mitchell and Sondheimer[30] who noted that 3,5-octadien-1,7-diyne forms

(8-23)

the dimer of benzocyclobutadiene in 85% yield. The reaction has a half-life of 10 min at 25°.

All of the data show that the eight electron electrocyclic reaction is a rapid process more facile than either four or six electron reactions and comparable to the cis-dienone-pyran reaction. The geometry of the transition state, which must resemble a single turn helix, is clearly very favorable. In particular, this geometry favors the double bond migration process, since a high degree of overlap is maintained at the transition state. Substituents larger than hydrogen which are cis on the terminal double bonds show the expected steric interference raising the enthalpy of activation about 3 kcal/mol per methyl group. The equilibrium data show the ΔG° must be small for the unsubstituted molecules, though the cyclic tautomer is more stable. However, minor alteration by substitution can tip the balance in favor of the acyclic compound.

4. Retroelectrocyclization and Its Use in Synthesis

The tenuous free energy difference between acyclic and cyclic tautomers raises problems and also provides a useful route to synthesis of a number of substituted polyenes. When one or the other of the valence isomers is the sole isolable form under normal conditions, no thermal reaction is apparent until either can undergo a different reaction. For example, when 1,3,5-cyclooctatriene is heated to 450° in a flow system there appear a variety of products derived from all three valence tautomers, acyclic, monocyclic, and bicyclic

$$(8\text{-}24)$$

[Eq. (8-24)].[33] Ziegenbein obtained mainly 1-vinyl-1,3-cyclohexadiene from pyrolysis of 1,2-diacetoxy-5-cyclooctene at 500°.[34] Conversely when the diphenyltetraene (182) was heated to 175° six products were obtained [Eq. (8-25)].[28] A rationale for the routes to their formation was provided.

$$(8\text{-}25)$$

In neither of these cases could any of the less stable valence isomer be found. Cyclooctatetraene dianion reacts with a variety of carbonyl compounds to give in each case a series of products including in most cases one derived from retro-electrocyclization of a cyclooctatriene.[35-39] An example is shown in Eq. (8-26) where ring opening is attributed to steric effects.[36] The

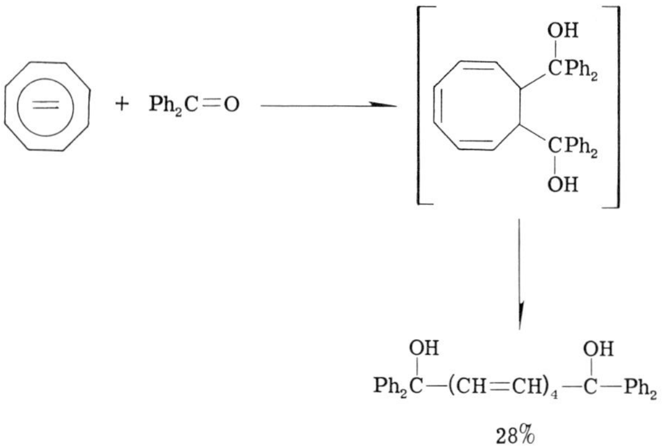

$$\text{(cyclooctatetraene)} + Ph_2C=O \longrightarrow \left[\begin{array}{c} \text{OH} \\ \overset{|}{C}Ph_2 \\[1ex] \overset{|}{C}Ph_2 \\ \text{OH} \end{array} \right]$$

$$(8\text{-}26)$$

$$\underset{\text{OH}}{Ph_2\overset{|}{C}}-(CH=CH)_4-\underset{\text{OH}}{\overset{|}{C}}-Ph_2$$

28%

trans,cis,cis,trans form of 2,4,6,8-decatetraenedioic acid can be prepared by carbonation of the dianion.[39]

Cyclooctatetraene provides the starting point for preparation of a group of polyenes via cyclooctatrienone,[20] 7-bromo-1,3,5-cyclooctatriene,[40] and trans-7,8-dibromobicyclo[4.2.0]octadiene (183).[20,26,31] Thus, Anet prepared octatrienedial [Eq. (8-19)] from the diacetate obtainable from 183. Shortly thereafter Hoever[31] prepared 1,8-dicyanooctatetraene [probably all trans,

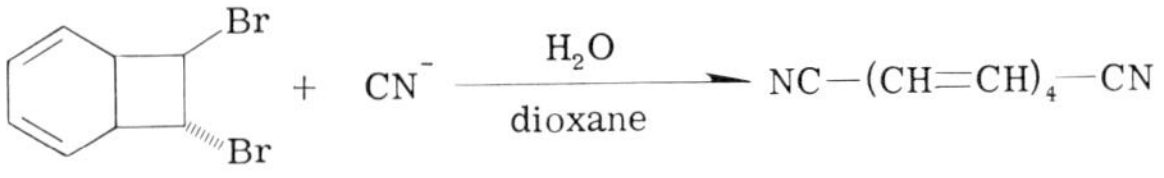

$$\text{(183)} + CN^- \xrightarrow[\text{dioxane}]{H_2O} NC-(CH=CH)_4-CN \qquad (8\text{-}27)$$

183

Eq. (8-27)]. Exploitation of this route has continued with the reaction of 183 with Grignard reagents being particularly interesting.[41,42] The disulfide

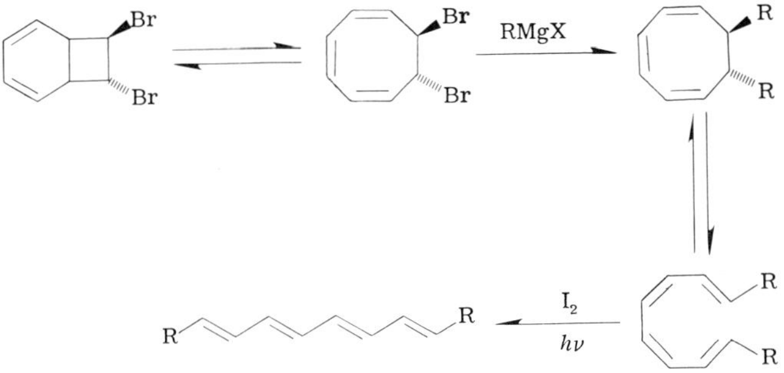

$$(8\text{-}28)$$

$$R'\diagdown\!=\!\diagup\!=\!\diagdown\!=\!\diagup R \xleftarrow[h\nu]{I_2} \text{(cyclooctatetraene-R)}$$

R = PhC≡C, Me_3C—C≡C, Ph, Me—C≡C, CH_2=C—C≡C
                                              |
                                             Me

184 was prepared by a similar route,[43] while several silylated polyacetylenic tetraenes were obtained analogously.[44]

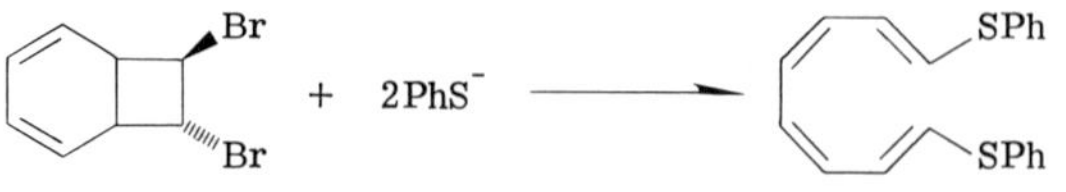

$$(8\text{-}29)$$

184

A series of 1,3,5-cyclooctatrienols prepared from cyclooctatrienone by reduction or Grignard addition can be converted thermally to trienones

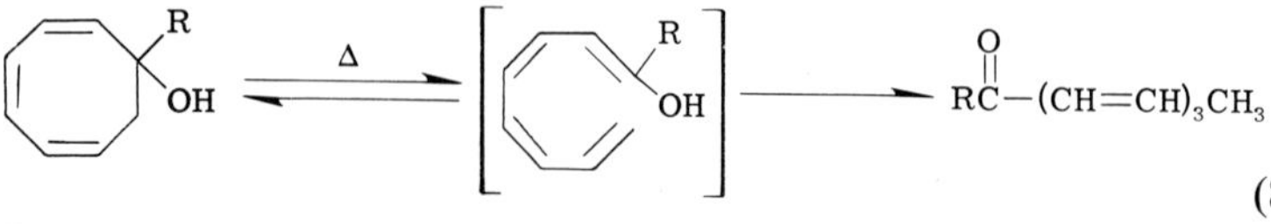

$$(8\text{-}30)$$

R = H, Me, Ph, Et

[Eq. (8-30)].[45,46] Rates of the thermal reaction appear to increase in the order H < Me ≪ Ph. Acetylenic Grignard reagents react with 7-bromocyclo-octatriene giving mixtures of acyclic, monocyclic and bicyclic products.[47]

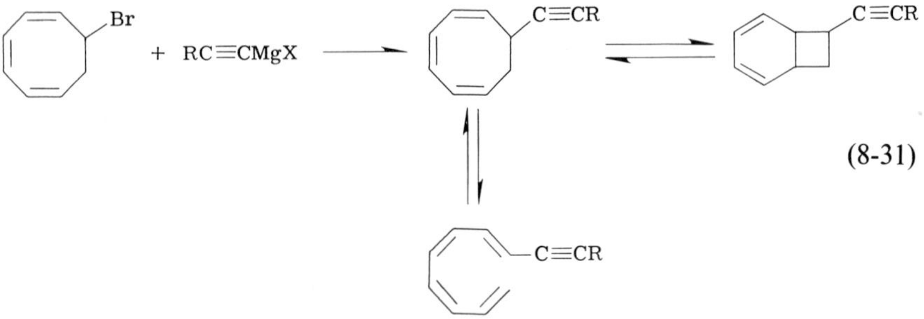

$$(8\text{-}31)$$

The bromide can be converted to an azide which is converted thermally to 2-butadienylpyrrole.[40] Cyclooctatrienone undergoes 1,8- addition of piperidine and the initial adduct produces a trienone [Eq. (8-32)].[48] Some acetophenone is formed from the trienone by the electrocyclic elimination sequence.

$$(8\text{-}32)$$

Bicyclo[4.2.0]octadienes can be obtained by photoaddition to benzene and some of its derivatives.[29,32] For example, addition of trimethylethylene to benzonitrile gave an adduct thermally convertible to an unstable nitrile

$$(8\text{-}33)$$

$$Me_2C=\overset{\underset{\displaystyle CN}{|}}{C}-(CH=CH)_3Me$$

[Eq. (8-33)].[32] Dimethyl-*trans,cis,cis,trans*-decatetraenedioate was obtained in the same way from benzene.[29]

Several nitrogen heterocycles also undergo the reverse electrocyclization and that initial product then reacts further. Thus, N-carbethoxy-1,2-dihydroazocine gives 2-vinyl-N-carbethoxypyridine [Eq. (8-34)].[50]

$$(8\text{-}34)$$

A general reaction which may have some value in synthesis, but is much more notable for the mechanistic problem it poses, can be considered here for it may proceed via an eight electron retro-electrocyclization to a cyclononatetraene.[53] In accord with a mechanism proceeding via all-*cis*-cyclonon-

$$(8\text{-}35)$$

atetraene, the 9,9-dideuterio derivative gives 1,1-dideuterio-*cis*-dihydro-indene.[54] Also in reasonable agreement with the retro-electrocyclization

$$(8\text{-}36)$$

$$(8\text{-}37)$$

route is the suggestion of Staley and Henry[55,56] that the cis-anti compound gives *cis*-dihydroindene [Eq. (8-36)] and the cis-cis gives *trans*-dihydroindene [Eq. (8-37)]. This idea received further support from the studies of Anastassiou and Griffith.[57,58] The influence of substituents on the rate of reaction indicates that the 1,8-bond is being broken in the rate determining step.[59] Finally it has been shown that a cyclononatetraene is a permissable intermediate.[60-63] Despite this there has been no clear evidence that the retro-cyclic reaction actually does occur. A different mechanism has been suggested,[64] and that paper can be consulted for added references to this problem.

5. Electrocyclization in Synthesis

Little work has appeared which exploits the cyclization process in synthesis. However, it appears that this may change as more natural products with a bicyclo[3.3.0]octane skeleton are found. Kaiser and Hafner[65,66] were the first to develop approaches to that skeleton which was important for preparation of pentalenes. The basic reaction of Eq. (8-38) offers a very

$$(8\text{-}38)$$

nice potential route to pentalene and some of its derivatives. However, the isolated products are hydrogen shift isomers of the direct cyclization product. The reaction rate is remarkably enhanced when $R_1 = R_3 = $ Me or $R_2 = $ Me [Eq. (8-39)].

$$(8-39)$$

The hydrogen shift problem has proved to be a real obstacle to study of the stereochemistry of this cyclization. The direct cyclization product of Eq. (8-40) could not be observed under any conditions.[67] A reasonable

$$(8-40)$$

candidate for stereochemical study has been examined, but the stereochemical result was not noted [Eq. (8-41)].[68]

$$(8\text{-}41)$$

II. TEN ELECTRON SYSTEMS

Very little is known mechanistically about ten electron electrocyclic reactions. The predicted disrotatory stereochemistry has not been fully verified experimentally. Some rather tenuous evidence for disrotatory closure can be adduced from the reaction sequence of Eq. (8-42).[84] If one assumes that

$$(8\text{-}42)$$

the conrotatory retroelectrocyclization of the benzocyclobutene moiety takes place with the methyl rotating outward, then the sole product isolated results from an allowed disrotatory ten (or 14) electron cyclization.[85] The stereochemistry was deduced from the coupling ($J = 4$ Hz) of the significant protons. The majority of the examples published are related to the well known Ziegler–Hafner synthesis of azulenes [Eq. (8-43)].[69,70] When

$$(8\text{-}43)$$

monosubstituted cyclopentadienes are used in the synthesis 1-substituted azulenes are obtained,[70] and both 5- and 6-methylazulenes have been prepared from β- and γ-picolines.[71] The generality of the ring closure has been discussed.[72] Both 6-dimethylamino and 6-chloroazulenes have been prepared by this route.[73] Prinzbach and Herr[74] showed that ring closure can be used to form the five membered as well as the seven-membered ring

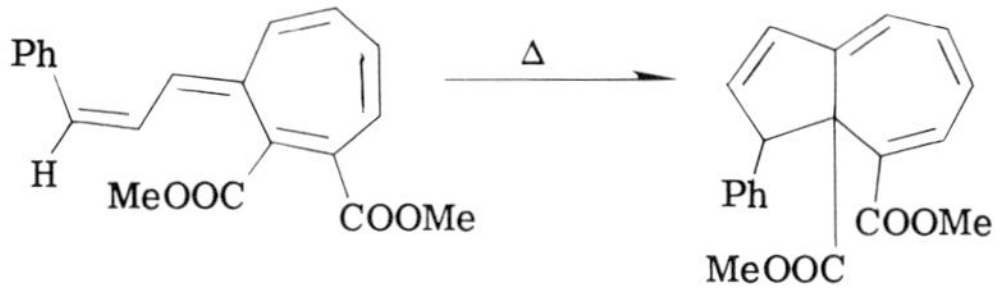

(8-44)

[Eq. (8-44)]. When the terminal double bond bears two phenyls ring closure goes to the unsubstituted side of the ring.

Jutz and co-workers have adapted this synthesis for polycyclic systems.[75-77] An example is shown in Eq. (8-45) which may raise questions

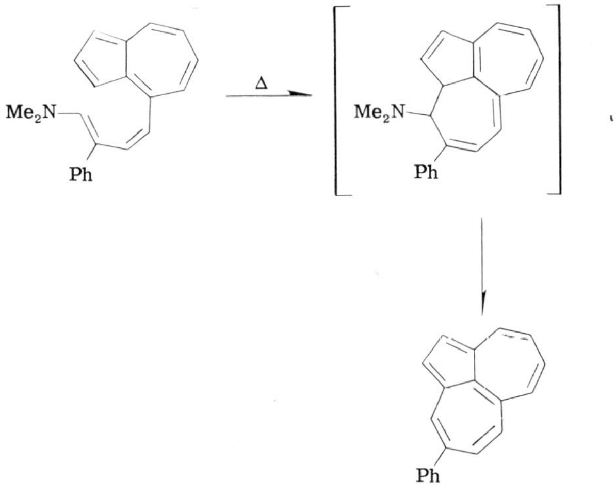

(8-45)

about the number of electrons involved. For convenience this has been rather arbitrarily defined here as the shortest conjugated system which makes the sigma bonding atoms the termini of that system. In this example ten electrons are involved, and the reaction has proceeded exclusively via ten rather than an alternative six electron reaction. Hafner[70] notes having carried out an equivalent reaction. An interesting double cyclization was employed in one case [Eq. (8-46)], and cyclization at a different position was used in

$$(8\text{-}46)$$

$$(8\text{-}47)$$

another case [Eq. (8-47)]. Jutz has reviewed his extensive work on reactions of the electrocyclic-elimination route to aromatic compounds.[78] Nitrogen atoms in the conjugate chain do not influence the reaction adversely [Eq. (8-48)].[78]

$$(8\text{-}48)$$

An interesting reversible ten electron electrocyclic process has been studied with some care [Eq. (8-49)].[79] The stable form is sensitive to solvent

$$(8\text{-}49)$$

effects and to protonation or deprotonation. A fascinating set of unusual electrocyclic reactions, two of the ten electron type, account for both of the

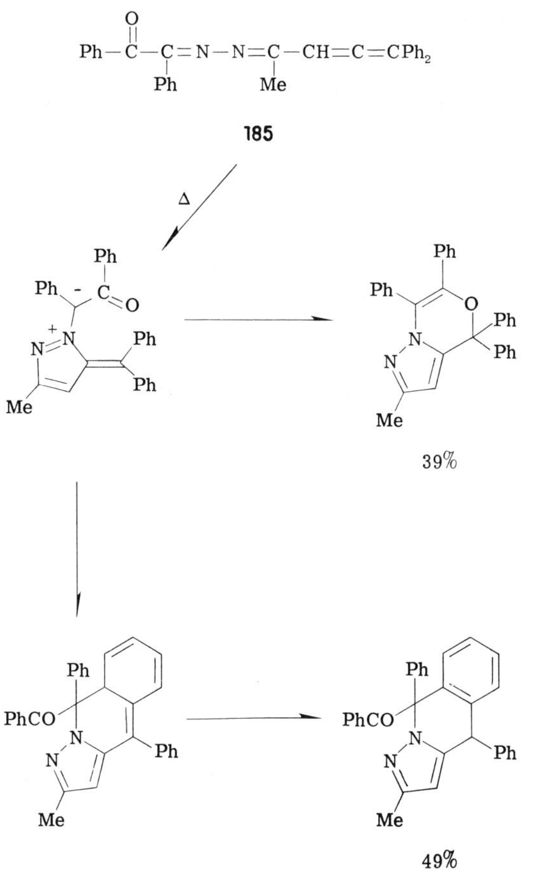

$$\underset{\underset{\text{Ph}}{|}}{\overset{\overset{\text{O}}{\|}}{\text{Ph}-\text{C}}}-\text{C}=\text{N}-\text{N}=\underset{\underset{\text{Me}}{|}}{\text{C}}-\text{CH}=\text{C}=\text{CPh}_2$$

185

Δ

39%   (8-50)

49%

isolable products of thermolysis of **185**.[94]

III. LONG CONJUGATE SYSTEMS

Experimental study of long acyclic conjugate systems is difficult, if not impossible, because of the problems of preparing specific poly-cis isomers, rapid cyclizations of sections of the system and extremely negative entropies of activation. The sole solution has been to incorporate portions of the conjugate chain into one or more rings. Even so, problems associated with synthesis, instability of reactant and/or product and identification of the stereochemistries of reactants and products has thoroughly limited studies in this area.

A. Twelve Electron Systems

Twelve electron conrotatory electrocyclization leading to a six-membered ring is a facile process [Eq. (8-51)].[80,81] Fortunately the 1,5-hydrogen shifts

$$(8\text{-}51)$$

are slower than cyclization and isolation of the initial product was possible. The stereochemistry of the tricyclic product was ascertained by a combination of cycloaddition (DMAD) and nmr studies.[81] No evidence for any disrotatory product was obtained. The reaction is rapid, energetically as favorable as the eight electron cyclization and entropically more advantageous, i.e., $\Delta H^{\ddagger} = 19.4$ kcal/mol and $\Delta S^{\ddagger} = -9$ eu.

This electrocyclization can be useful in synthesis [Eq. (8-52)],[82] but few examples have appeared.

$$(8\text{-}52)$$

B. Fourteen Electron Systems

A particularly interesting situation has developed in this case, since it appears that the first authentic example of a concerted but symmetry forbidden electrocyclization has been uncovered [Eq. (8-53)].[83] At 80° the ring

$$(8\text{-}53)$$

$J = 10$ Hz

closure is relatively rapid ($t_{1/2} = 12$ min), but the isolated product has undergone hydrogen shifts to generate an aromatic compound. However, in the presence of a very large excess of DMAD a 1:1 adduct, presumably derived from the initial cyclization product, is obtained. Evidence from coupling constants supports the conrotatory stereochemistry of the cyclization, and the activation parameters, $\Delta H^{\ddagger} = 23.3$ kcal/mol, $\Delta S^{\ddagger} = +6.3$ eu, bolster the suggestion that the cyclization is concerted. Steric problems obviously inhibit the disrotatory reaction which was not observed. It is unfortunate that the electrocyclization reaction itself could not be directly observed. An example which can be considered as an allowed fourteen electron reaction was treated in Section II.[84]

C. Sixteen Electron Systems

While no completely clear cut examples of this type have appeared, a reaction in which such an electrocyclization is a reasonable participant has been uncovered.[85] The authors show the stereochemistry given in Eq. (8-54),

(8-54)

but no basis for this assignment was made. Prinzbach *et al.*[83] has stated that there is "little doubt that if it undergoes a sixteen electron cyclization, it does so in allowed conrotatory fashion." This appears to be a safe bet since both orbital symmetry and steric factors lead in that direction.

D. Eighteen Electron Systems

An elongated equivalent of the norcaradiene-cycloheptatriene reaction has been proposed to account reasonably for some reactions which convert porphins to homoporphins.[86,87,88] The first case observed gave an azahomoporphin [Eq. (8-55)],[86] while the second gave a tetraphenylhomoporphin.[87]

This latter reaction was shown to be reversible.[88]

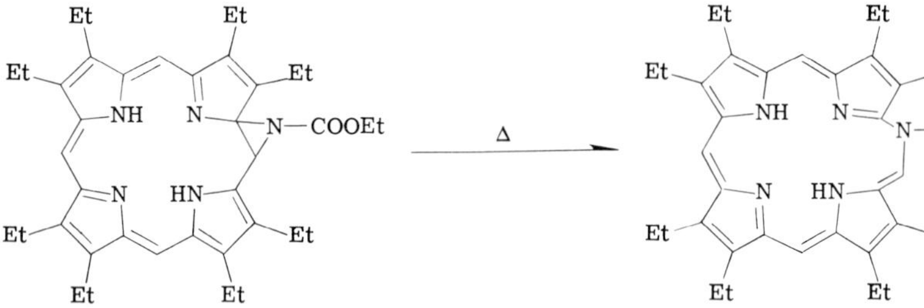

(8-55)

E. Twenty-Two Electron Systems

Whether or not an example of this class has been observed appears problematical. The reaction was initially invoked to rationalize the observations of Eq. (8-56).[89] An allowed twenty-two electron cyclization would produce

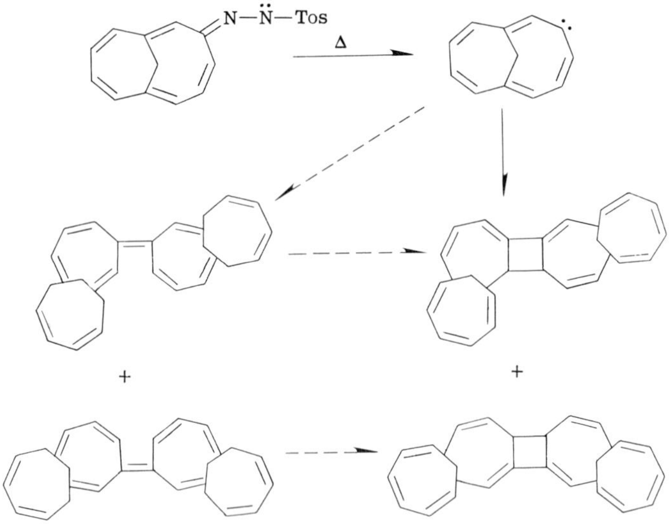

(8-56)

via disrotatory closure a molecule with the two cyclobutane hydrogens in a cis configuration. When it was demonstrated in elegant fashion that the hydrogens in question were trans oriented, the orbital symmetry rule was invoked to eliminate this route.[90] However, this raises some very interesting questions about application of the rule in cases of large conjugated systems. For annulenes the last aromatic compound is apparently 22-annulene, by calculation[91] and also by nmr study of isolated annulenes.[92] Though the theoretical basis for the effect with orbital symmetry differs (i.e., HOMO–

LUMO separation), the result will be similar and the stereoselectivity based on orbital symmetry should decrease with length of the polyene chain. The results with the fourteen electron system suggest that application of the rule in the present case is not on very sound ground.

REFERENCES

1. S. W. Staley, *in* "Pericyclic Reactions" (A. P. Marchand and R. E. Lehr, eds.), Vol. I, Academic Press, New York, 1977, pp. 230–236.
2. J. C. Pommelet, N. Manisse, and J. Chuche, *Tetrahedron* **28**, 3929 (1972).
3. W. Eberbach and U. Trostmann, *Tetrahedron Lett.* p. 3569 (1977).
4. T. Mukai and H. Sukawa, *Tetrahedron Lett.* p. 1835 (1973).
5. J. Dingwal and J. T. Sharp, *Chem. Commun.* p. 128 (1975).
6. J. T. Sharp, R. H. Findlay, and P. B. Thorogood, *Chem. Commun.* p. 909 (1970).
6a. J. T. Sharp, R. H. Findlay, and P. B. Thorogood, *J. Chem. Soc. Perkin Trans. 1* p. 102 (1975).
7. A. Padwa, J. Smolanoff, and A. Tremper, *J. Am. Chem. Soc.* **97**, 4682 (1975).
8. A. Padwa, J. Smolanoff, and A. Tremper, *J. Org. Chem.* **41**, 543 (1976).
9. L. Garanti and G. Zecchi, *J. Chem. Soc. Perkin Trans. 1* p. 2092 (1977).
10. A. Kakehi, S. Ito, T. Maeda, R. Takeda, M. Nishimura, M. Tamashima, and T. Yamaguchi, *J. Org. Chem.* **43**, 4837 (1978).
11. E. N. Marvell and J. Seubert, *J. Am. Chem. Soc.* **89**, 3377 (1967).
12. R. Huisgen, A. Dahmen, and H. Huber, *J. Am. Chem. Soc.* **89**, 7130 (1967).
13. R. Huisgen, A. Dahmen, and H. Huber, *Tetrahedron Lett.* p. 1461 (1969).
14. A. Dahmen and R. Huisgen, *Tetrahedron Lett.* p. 1465 (1969).
15. D. S. Glass, J. Zirner, and S. Winstein, *Proc. Chem. Soc.* p. 276 (1963).
16. W. R. Roth, *Justus Liebigs Ann. Chem.* **671**, 25 (1964).
17. S. W. Staley and T. J. Henry, *J. Am. Chem. Soc.* **92**, 7612 (1970).
18. S. W. Staley and T. J. Henry, *J. Am. Chem. Soc.* **93**, 1292 (1971).
19. A. A. Reid, J. T. Sharp, H. R. Sood, and P. B. Thorogood, *J. Chem. Soc. Perkin Trans. 1* p. 2543 (1973).
20. G. Schröder, "Cyclooctatetraen" Verlag Chemie, Weinheim, Germany, 1965.
21. A. C. Cope and D. J. Marshall, *J. Am. Chem. Soc.* **75**, 3208 (1953).
22. H. Meister, *Chem. Ber.* **96**, 1688 (1963).
23. W. Ziegenbien, *Chem. Ber.* **98**, 1427 (1965).
24. T. D. Goldfarb and L. Lindquist, *J. Am. Chem. Soc.* **89**, 4588 (1967).
25. P. Datta, T. D. Goldfarb, and R. S. Boikess, *J. Am. Chem. Soc.* **91**, 5429 (1969).
26. R. Anet, *Tetrahedron Lett.* p. 720 (1961).
27. R. Huisgen, A. Dahmen, G. Boche, and W. Hechtl, *Tetrahedron Lett.* p. 5215 (1968).
28. E. N. Marvell, J. Seubert, G. Vogt, G. Zimmer, G. Moy, and J. R. Siegmann, *Tetrahedron* **34**, 1323 (1978).
29. G. Kaupp and E. Jostkleigrewe, *Angew. Chem.* **88**, 812 (1976).
30. G. H. Mitchell and F. Sondheimer, *J. Am. Chem. Soc.* **91**, 7520 (1969).
31. H. Hoever, *Tetrahedron Lett.* p. 255 (1962).
32. J. G. Atkinson, D. E. Ayer, G. Büchi, and E. W. Robb, *J. Am. Chem. Soc.* **85**, 2257 (1963).
33. W. R. Roth and B. Pelzer, *Justus Liebigs Ann. Chem.* **685**, 56 (1965).
34. W. Ziegenbein, *Angew. Chem.* **77**, 42 (1965).
35. T. S. Cantrell and H. Schechter, *J. Am. Chem. Soc.* **85**, 3300 (1963).
36. T. S. Cantrell and H. Schechter, *J. Am. Chem. Soc.* **87**, 136 (1965).

37. T. S. Cantrell and H. Schechter, *J. Am. Chem. Soc.* **89,** 5868, 5877 (1967).
38. H. Schechter, *J. Am. Chem. Soc.* **94,** 5361 (1972).
39. T. S. Cantrell, *Tetrahedron Lett.* p. 5635 (1968).
40. M. Kroener, *Chem. Ber.* **100,** 3162 (1967).
41. E. Müller, H. Straub, and J. M. Rao, *Tetrahedron Lett.* p. 773 (1970).
42. M. Straub, J. M. Rao, and E. Müller, *Justus Liebigs Ann. Chem.* p. 1339 (1973).
43. L. Moegel, W. Schroth, and B. Werner, *Chem. Commun.* p. 57 (1978).
44. S. J. Harris and D. M. R. Walton, *Tetrahedron* **34,** 1037 (1978).
45. M. Kroener, *Chem. Ber.* **100,** 3172 (1967).
46. M. Ogawa, M. Takagi, and T. Matsuda, *Tetrahedron* **29,** 3813 (1973).
47. H. Straub, J. M. Rao, and E. Müller, *Justus Liebigs Ann. Chem.* p. 1352 (1973).
48. M. Ogawa and T. Matsuda, *Chem. Lett.* p. 47 (1975).
49. L. L. Barber, O. L. Chapman, and J. D. Lassila, *J. Am. Chem. Soc.* **91,** 531 (1969).
50. W. H. Okamura, *Tetrahedron Lett.* p. 4717 (1969).
51. H. Sukawa, O. Seshimoto, T. Tezuka, and T. Mukai, *Chem. Commun.* p. 696 (1974).
52. P. L. Kumler and O. Buchardt, *J. Am. Chem. Soc.* **90,** 5640 (1968).
53. K. F. Bangert and V. Boekelheide, *J. Am. Chem. Soc.* **86,** 905 (1964).
54. W. Grimme, *Chem. Ber.* **100,** 113 (1967).
55. S. W. Staley and T. J. Henry, *J. Am. Chem. Soc.* **91,** 1239 (1969).
56. S. W. Staley and T. J. Henry, *J. Am. Chem. Soc.* **91,** 7787 (1969).
57. A. G. Anastassiou and R. C. Griffith, *J. Am. Chem. Soc.* **95,** 2379 (1973).
58. A. G. Anastassiou and R. C. Griffith, *Chem. Commun.* p. 399 (1972).
59. G. Boche and G. Schneider, *Tetrahedron Lett.* p. 2449 (1974).
60. P. Radlick and W. Fenical, *J. Am. Chem. Soc.* **91,** 1560 (1969).
61. P. Radlick and G. Alford, *J. Am. Chem. Soc.* **91,** 6529 (1969).
62. A. G. Anastassiou, V. Orfanos, and J. H. Gebrian, *Tetrahedron Lett.* p. 4491 (1969).
63. M. Neuenschwander and A. Frey, *Chimia* **29,** 212 (1975).
64. G. D. Ewing, S. V. Ley, and L. A. Paquette, *J. Am. Chem. Soc.* **100,** 2909 (1978).
65. R. Kaiser and K. Hafner, *Angew. Chem. Int. Edit. Engl.* **9,** 892 (1970).
66. R. Kaiser and K. Hafner, *Angew. Chem. Int. Edit. Engl.* **12,** 335 (1973).
67. J. J. Gajewski and C. J. Cavender, *Tetrahedron Lett.* p. 1057 (1971).
68. B. M. Trost and L. S. Melvin, Jr., *Tetrahedron Lett.* p. 2675 (1975).
69. K. Ziegler and K. Hafner, *Angew. Chem.* **67,** 301 (1955).
70. K. Hafner, *Justus Liebigs Ann. Chem.* **606,** 79 (1957).
71. K. Hafner and K.-O. Asmus, *Justus Liebigs Ann. Chem.* **671,** 31 (1964).
72. K. Hafner, *Pure Applied Chem.* **28,** 153 (1971).
73. W. Bauer and U. Müller-Westerhoff, *Tetrahedron Lett.* p. 1021 (1972).
74. H. Prinzbach and H.-J. Herr, *Angew. Chem. Int. Edit. Engl.* **11,** 135 (1972).
75. C. Jutz and E. Schweiger, *Chem. Ber.* **107,** 2383 (1974).
76. C. Jutz, E. Schweiger, H.-G. Löbering, A. Kraatz, and W. Kosbahn, *Chem. Ber.* **107,** 2956 (1974).
77. C. Jutz and E. Schweiger, *Synthesis* p. 193 (1974).
78. J. C. Jutz, *Top. Current Chem.* **73,** 125 (1978).
79. C. Temple, Jr., C. L. Kussner, and J. A. Montgomery, *J. Org. Chem.* **32,** 2241 (1967).
80. H. Sauter and H. Prinzbach, *Angew. Chem. Int. Edit. Engl.* **11,** 296 (1972).
81. H. Sauter, B. Gallenkamp, and H. Prinzbach, *Chem. Ber.* **110,** 1382 (1977).
82. K. Hafner, K. H. Hafner, C. König, M. Kreuder, G. Ploss, G. Schulz, E. Sturm, K. H. Vöpel, *Angew. Chem. Int. Edit. Engl.* **2,** 123 (1963).
83. H. Prinzbach, H. Babsch, and D. Hunkler, *Tetrahedron Lett.* p. 649 (1978).
84. M. A. O'Leary and D. Wege, *Tetrahedron Lett.* p. 2811 (1978).

85. S. Kuroda, T. Asao, M. Funamizu, H. Kurihara, and Y. Kitahara, *Tetrahedron Lett.* p. 251 (1976).

86. R. Grigg, *J. Chem. Soc. C* p. 3664 (1971).

87. H. J. Callot and Th. Tschamber, *Tetrahedron Lett.* p. 3155 (1974).

88. H. J. Callot and Th. Tschamber, *Tetrahedron Lett.* p. 3159 (1974).

89. P. H. Gebert, R. W. King, R. A. LaBar, and W. M. Jones, *J. Am. Chem. Soc.* **95,** 2357 (1973).

90. R. A. LaBar and W. M. Jones, *J. Am. Chem. Soc.* **96,** 3645 (1974).

91. M. J. S. Dewar and G. J. Gleicher, *J. Am. Chem. Soc.* **87,** 685 (1965).

92. F. Sondheimer, *Accts. Chem. Res.* **5,** 81 (1972).

93. K. L. M. Stanley, J. Dingwall, J. T. Sharp, and T. W. Naisby, *J. Chem. Soc., Perkin Trans. 1* p. 1433 (1979).

94. E. E. Schweizer and S. Evans, *J. Org. Chem.* **43,** 4328 (1978).

CHAPTER 9
ODD ELECTRON SYSTEMS

Conjugated systems with an odd number of electrons constitute a special case in the conservation of orbital symmetry. In one sense this arises because both Hückel and Möbius aromaticities are defined in terms of even numbers of electrons. Woodward and Hoffmann pointed out that the frontier orbital procedure indicates that a radical should follow the same rotational path for electrocyclic reactions as the entity having one more electron in the conjugate system. As was noted in Chapter 2, Longuet-Higgins and Abrahamson[1] demonstrated that correlation diagrams allow neither conrotatory nor disrotatory routes. Thus, to make predictions it is necessary to carry out more sophisticated calculations.

I. QUANTITATIVE CALCULATIONS

The year 1971 must rate as the year of the radical calculation since Clark and Adams,[2] Dewar and Kirschner,[3] Haselbach,[4] and Boche and Szeimies[5,6] all showed that semi-empirical molecular orbital calculations made the disrotatory path more favorable than the conrotatory for the cyclopropyl to allyl radical reaction. Later studies[7,8,9] generally concurred in this result, but the energy difference in favor of the disrotatory reaction has not been subject to that sort of agreement. The difference varies from 0.2,[9] 7.4,[6] 12,[7] 14,[4] 18,[2] 20 (see below),[7] to 27.6.[3] Generally both routes show the patterns expected of forbidden processes, i.e., the bond breaking and rotational motions are not synchronous. It is interesting that of the three disrotatory routes, that one is preferred in which the hydrogen on the methine group moves out of the plane towards the pair of hydrogens which are rotating outwards.[4,6,7] The disrotation with the hydrogen moving toward the groups rotating inward has been found less favorable than the conrotatory route.[6] One calculation indicated a planar disrotatory process was less favorable than the non-planar by ca. 8 kcal/mol.[7] Perhaps most surprising is that the most recent MINDO/3 calculation leaves the situation in much the same condition it was when Longuet-Higgins left it, that is indeterminate.

All of the calculations above were made explicitly for the cyclopropyl to allyl radical reaction, which leaves uncertain how these results can be generalized. Would [4n + 3] types all be disrotatory and (4n + 1) conrotatory? Haselbach makes the note that [4n + 3] neutral perimeters, as well as [4n + 1], are slightly aromatic, hence of Hückel type and disrotatory. This would imply that all radical reactions of uncharged species would be disrotatory. Theoretical studies of the pentadienyl radical would be helpful.

Two quite different radical systems are available with even atom groupings to which an electron can be added to give a radical anion or from which an electron can be abstracted to give a radical cation. In general, the extra electron of a radical anion will enter an anti-bonding orbital, while the odd electron of the radical cation will be in a bonding orbital, so the two types may behave very differently. Few detailed studies of either type of charged radical have appeared, but Bauld and his students have made a general appraisal of the anion radical problem.[10,11,25,26] One generalization was proposed in early work[25]—electrocyclization will be allowed for the same rotational mode preferred by the neutral precursor only when the orbitals occupied by the odd electron also correlate. Thus, benzocyclobutene-o-xylylene radical anions undergo allowed electrocyclization in the conrotatory mode. INDO/MO calculations gave E_a(conrot.) = 0.5 and E_a (disrot.) = 23.7 kcal/mol.[10,26] In general, neither route is allowed, and a survey of the problem in such cases has been made by Bauld and Cessac.[11] They divide the analysis into two parts—orbital topology and molecular symmetry effects. Orbital topology effects can be summarized readily— neutral odd atomic species show no preference, both anion and cation radicals of the same species show the same preference, C_4 (conrot.), C_6 (disrot.), C_8 (conrot.), with a decreasing selectivity as the molecular size increases. Since the selectivity imposed by orbital topology is relatively slight molecular symmetry must also be taken into account. The general principle that correlation can be made only for electrocyclic partners of the same symmetry, hence no allowed (by correlation) route can exist between partners of differing symmetry, is used in the following sense. If orbital topology provides a preference for a conrotatory course and the partners both have C_2 molecular symmetry, the two effects bolster, but if the reactant has C_s symmetry the two effects are opposed and indeed molecular symmetry can dominate. Quantitative calculations of some hypothetical examples were used to support this analysis.

Aside from the orbital topology analysis of Bauld and Cessac,[11] there have appeared no theoretical studies of the rotational modes for electrocyclic reactions of specific radical cations. However, in a paper describing a PMO approach to the interpretation of mass spectra, Dougherty[12] suggests that for fragmentations which give rise to significant metastable peaks the rules for thermal "electrocyclic" (means pericyclic) reactions apply. Further, he states that Dewar's rule that these reactions take place by aromatic transition states can be applied to radical cations via the PMO method. This does not define whether the reaction will proceed better by one or the other of the rotational modes, thus leaving the questions of preferred routes unanswered in any complete sense.

II. EXPERIMENTAL RESULTS

The great difficulties associated with obtaining direct information about a reaction which generally converts one transient entity into another has left us with only modest experimental results in this entire area.

A. Neutral Odd-Atomic Radicals

The cyclopropyl–allyl radical reaction has been examined with some care, principally by Ruchardt and co-workers. Cyclopropyl radical does convert to the allyl radical in the gas phase, and in accord with its non-allowed nature the reaction has an activation energy of 20–22 kcal/mol.[13,13a] Bicyclobutanyl radical apparently reacts rapidly at −90° to give cyclobutenyl radical, but at −170° a second radical reasonably assumed to be the bicyclobutyl radical is relatively stable.[14] Presumably this ring opening could be considered a disrotatory process facilitated by the high ring strain. Monocyclic cyclopropyl radicals are not prone to ring scission, though this can be facilitated by the presence of substituents such as phenyl which will stabilize the allyl radical.[15,16]

The activating influence of phenyl groups has been used in studies of the stereochemistry of the ring opening. Both **186** and **187** were generated by heating *tert*-butyl peresters at 80°–130° in various aromatic solvents. The allyl radicals were stabilized by dimerization or by reaction with benzyl radicals formed by hydrogen abstraction from the solvent by the *tert*-butoxy radicals.[17,18,19] It is difficult to see how anything except qualitative information could be derived from this since the stereochemical information would be derived from the double bond geometry [Eq. (9-1)]. Interpretation was simplified when it was found that both **186** and **187** gave the same mixture of meso and dl trans dimers.[17] The interconversion of the allylic radicals was thus faster than their capture.

A clever attempt was made to use kinetic information to answer the stereochemical problems. If the ring cleavage is disrotatory **187** should *a priori* react faster than **186**, but **186** is favored by a statistical factor of two

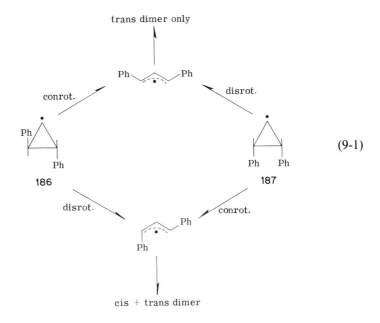

$$(9\text{-}1)$$

which leaves some ambiguity as to which has a greater rate overall. If the process is conrotatory **187** is favored by the statistical factor but the *a priori* rates are difficult to assess since the less stable **187** goes to the less stable allylic radical. In both cases then one would find it desirable to have additional quantitative information about the relative stabilities of **186** and **187** and of the two allyl radicals. The rate determining step for formation of allylic products precedes the ring opening so kinetic information about that important step had to be obtained indirectly.[18,19] Ring opening was competitive with cage recombination, and if all recombinations are assumed to proceed at the same rate the ratio of allylic to recombination products will permit the relative rates of ring opening for **186** and **187** to be determined. With this assumption ring opening for **186** is twice as fast as **187**.[18] The statistical factor was considered dominant and this was interpreted as evidence in favor of the disrotatory path. Later results showed that the recombination rates were subject to steric hindrance and thus the basic assumption underlying the experimental rate determination was not valid.[19] Though the authors still favored the disrotatory route, the evidence does not permit any unequivocal statement about the stereochemistry of ring opening.

That disrotatory ring opening apparently can occur was demonstrated further by the study of bicyclic molecules.[20] Thus for example **188** gives an

$$
\text{188} \qquad \xrightarrow[-35°]{h\nu} \qquad \left[\quad \right] \qquad \longrightarrow \qquad \tag{9-2}
$$

esr spectrum for the ring opened radical at $-35°$.[20] However, **189** (R = Ph)

$$
\text{189} \qquad \xrightarrow[\text{PhEt}]{130°} \qquad \begin{array}{c} \text{CH—Ph} \\ | \\ \text{Me} \end{array} \quad + \quad \begin{array}{c} \text{products with} \\ \text{unaltered} \\ \text{skeleton} \end{array} \tag{9-3}
$$

gave products both from ring cleavage and ring retention, and an example where ring opening did not occur was observed.[20] Compound **189** (R = Me, as the diacyl peroxide) forms a monocyclic radical which dimerizes.[46] Since the diacyl peroxide of 2,2,3,3-tetramethylcyclopropane does not give acyclic products via thermolysis, this disparity of behavior for the two tetraalkyl-cyclopropyl radicals gives some evidence that the disrotatory process is more favorable than the conrotatory.[46]

Aziridyl radicals appear to react in much the same fashion as their carbo-cyclic analogs.[22] Thus, **190** undergoes ring opening [Eq. (9-4)], but *trans-*

$$
\text{190} \qquad \xrightarrow[\text{PhEt}]{130°} \quad (\text{PhCH}=\text{N}-\underset{\underset{\text{Ph}}{|}}{\text{CH}}-)_2 \quad + \quad \text{PhCH}=\text{N} \atop \underset{\underset{\text{Me}-\text{CH}-\text{Ph}}{|}}{\text{CHPh}} \tag{9-4}
$$

dimethylaziridinyl radical does not undergo ring opening at $40°$.[22]

Cyclization of a pentadienyl radical has been demonstrated to occur, but it is not a facile process [Eq. (9-5)].[24] The reaction gives mainly the product

$$
\xrightarrow{>210°} \tag{9-5}
$$

of conrotatory closure though evidence that a small amount of cis product could be present was found. Whether steric or electronic factors dominate this result is not evident. A retro-electrocyclization which must occur via a

$$
\xrightarrow{-50°} \tag{9-6}
$$

disrotatory mode was found by esr methods using an adamantane matrix [Eq. (9-6)].[21] The spectrum for the cyclohexadienyl radical cannot be observed in solution. The $\Delta G^{\ddagger}$ for ring opening is 14.5 kcal/mol at $-50°$. Quite obviously both conrotatory and disrotatory reactions can occur.

B. Radical Anions

The cyclobutene–butadiene radical anion system has been treated theoretically and predicted to be non-allowed in both rotatory modes but with an energetic preference for the conrotatory route.[10,11] An experimental test of this projection was made using cis and trans isomers of **191** [Eq. (9-7)].[10,26]

The radical anion **191** from the cis isomer is stable at $-78°$ and converts to a ring opened dianion at $0°$. No radical anion can be observed for the trans

(9-7)

191

isomer even at $-78°$, the dianion being the first species observed. The dianions are not configurationally stable even at $-78°$, but two pieces of evidence support the thesis that ring opening is predominantly conrotatory. First the rates are trans $\gg$ cis, a common result for conrotatory processes.

(9-8)

Second, a pentacyclic product is obtained from the cis isomer after quenching, but not from the trans, a reaction explicable by the electrocyclization of Eq. (9-8).

Benzocyclobutene-o-xylylene represents an allowed reaction for radical anions.[10,25] However, the early results on this system appeared contradictory in that Rieke and co-workers[27] found no evidence for retro-electrocyclization, while Bauld and Farr[28] reported rapid ring opening at $-80°$. Later results convinced the latter of an error in their interpretation and it now seems clear that the radical anion of benzocyclobutene does not undergo this valence isomerization.[29] The radical anion of naphtho-[2.3]cyclobutene also fails to open the cyclobutene ring.[30] The expected reaction was finally observed with 1,2-diphenylbenzocyclobutene.[25,10] Both cis and trans isomers gave no observable anion radical at $-78°$ but produced only the dianion of diphenyl-o-xylylene [Eq. (9-9)]. The lack of stereospecificity was attributed to the quenching process.

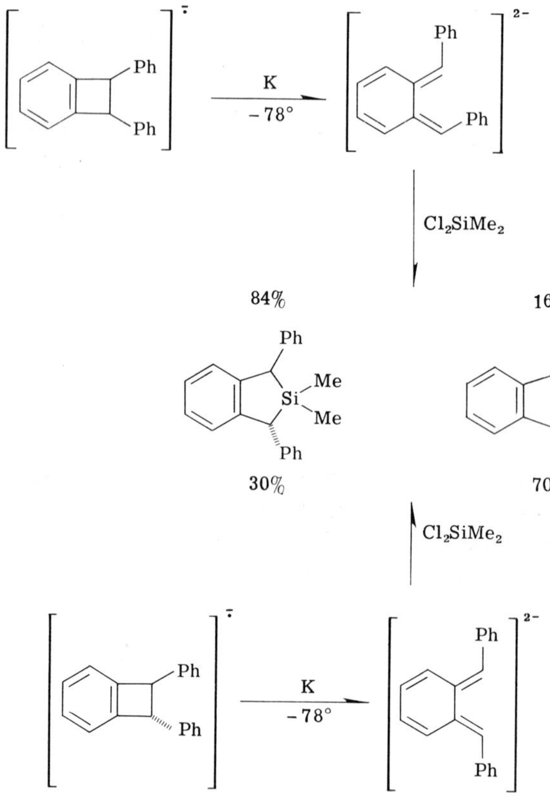

(9-9)

Bauld and Hudson[31] have investigated another allowed (disrot.) anion radical reaction [Eq. (9-10)]. Though the product was not conformationally

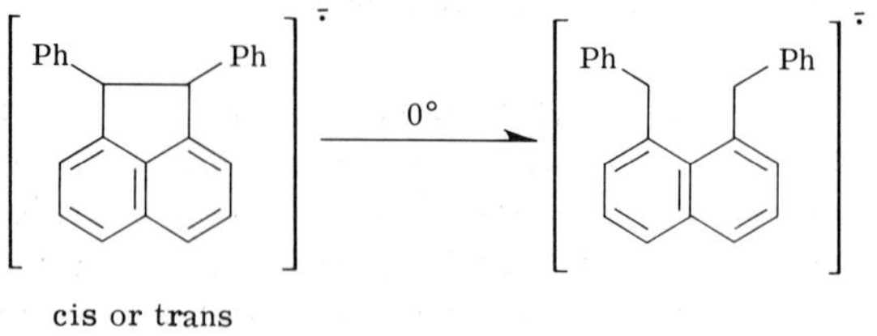

cis or trans

(9-10)

stable, arguments for a preferred disrotatory reaction were presented. A related case gave only a dihydro product [Eq. (9-11)], but the authors felt

$$ \xrightarrow{-80°} \qquad\qquad (9\text{-}11) $$

cis or trans

that ring cleavage followed by rapid reduction was the most reasonable mechanism.[32] The reaction would follow a disrotatory path. Retro-electro-cyclization of **192** was not observed.[33]

$$ \xrightarrow{\;\;\Delta\;\;}\!\!\!\!\!\!/ \qquad\qquad (9\text{-}12) $$

An eight atom radical anion system has received some study.[34-37] *cis*-Bicyclo[6.1.0]nonatriene radical anion is stable at low temperature but converts to the ring opened dianion at higher temperatures.[34,35] Conversely

$$ \xrightarrow[\text{DME}]{\text{K}} \qquad\qquad (9\text{-}13) $$

the trans isomer does not open at $-60°$ where it decomposes.[36,37] Bauld and Cessac[11] predicted the reaction should be conrotatory whereas HOMO predicts disrotatory. Though it appears to follow a disrotatory path, Bauld and Cessac argue that the reaction is dominated by molecular symmetry. It has also been noted that the dianion might open in a disrotatory mode and that it may be a dianion reaction which is observed.[37]

One ten atom radical anion, that of dimethylenecyclooctatriene, does not cyclize.[38]

C. Radical Cations

All of the experimental work with radical cations has been done with the mass spectrometer, and as one might suspect it has been difficult to provide convincing evidence of stereochemical modes of electrocyclic processes. One early attempt to relate mass spectral results to electrocyclic reactions

used similarities between the spectra of *cis*- and *trans*-cyclobutenedicar-
boxylic acids and the muconic acids to establish relations between these
isomers.[39] The authors indicated that reasonable correlations between the
cis-cyclobutenedicarboxylic acid and *cis,cis*- and *trans,trans*-muconic acids
or between the trans cyclobutene isomer and *cis,trans*-muconic acid existed.
Thus, the radical cations of cyclobutene were following a disrotatory course.
Observation of some metastable peaks suggests these reactions belong to
Dougherty's class I and should follow thermal rules.[12] According to Bauld
and Cessac[11] this would be a conrotatory reaction for radical cations.

This result was interpreted as indicating an excited state interrelation exists
for radical cations in the mass spectrometer. Further evidence for the same
correlation was found from ion kinetic energy spectra of tetramethylcyclo-
butenes and dimethyl-2,4-hexadienes.[40] On the other hand no evidence for
any correlation was obtained with dimethyl cyclobutenedicarboxylates and
the dimethyl muconates.[41]

Johnstone and Ward[42,43] used a rather different approach to the same
problem. They assume that two adjacent hydrogens having a *cis* arrange-
ment will be lost as a hydrogen molecule, but in a trans array fragmentation
will proceed by consecutive loss of two hydrogen atoms. Some of their

$$\text{Ph(CH=CH)}_2\text{Ph}^{+\cdot} \longrightarrow \qquad \text{loss of H + H}$$

results are listed above and these led them to conclude that reactions followed excited state rules. Bishop and Fleming[44] criticized the interpretation
suggesting that the results were indicative of the general rule that cations lose
molecules and cation radicals lose radicals. They conclude that the Woodward–Hoffmann rules are a poor probe into energy levels of ions fragment-
ing in the mass spectrometer.

Recently Gross and Russell[45] developed some evidence that cyclobutene radical cation converts rapidly to butadiene radical cation in the mass spectrometer. They estimated E_a as about 7 kcal/mol for this reaction. The final note on these problems has certainly not been sounded.

REFERENCES

1. H. C. Longuet-Higgins and E. W. Abrahamson, *J. Am. Chem. Soc.* **87**, 2045 (1965).
2. D. T. Clark and D. B. Adams, *Nature (London) Phys. Sci.* **233**, 121 (1971).
3. M. J. S. Dewar and S. Kirschner, *J. Am. Chem. Soc.* **93**, 4290, 4291 (1971).
4. E. Haselbach, *Helv. Chim. Acta* **54**, 2257 (1971).
5. G. Boche and G. Szeimies, *Angew. Chem. Int. Edit. Engl.* **10**, 911 (1971).
6. G. Boche and G. Szeimies, *Angew. Chem. Int. Edit. Engl.* **10**, 912 (1971).
7. P. Merlet, S. D. Peyerimhoff, R. J. Buenker, and S. Shih, *J. Am. Chem. Soc.* **96**, 959 (1974).
8. L. Farnell and W. G. Richards, *Chem. Commun.* p. 334 (1973).
9. M. J. S. Dewar noted in footnote in refer. 19.
10. N. L. Bauld, J. Cessac, C.-S. Chang, F. R. Farr, and R. Holloway, *J. Am. Chem. Soc.* **98**, 4561 (1976).
11. N. L. Bauld and J. Cessac, *J. Am. Chem. Soc.* **99**, 23 (1977).
12. R. C. Dougherty, *J. Am. Chem. Soc.* **90**, 5780 (1968).
13. G. Greig and J. C. J. Thynne, *Trans. Faraday Soc.* **62**, 3338 (1966).
13a. G. Greig and J. C. J. Thynne, *Trans. Faraday Soc.* **63**, 1369 (1967).
14. P. J. Krusic, J. P. Jesson, and J. K. Kochi, *J. Am. Chem. Soc.* **91**, 4566 (1969).
15. H. M. Walborsky and J.-C. Chen, *J. Am. Chem. Soc.* **92**, 7573 (1970).
16. H. M. Walborsky and P. C. Collins, *J. Org. Chem.* **41**, 940 (1976).
17. S. Sustmann, C. Ruchardt, A. Bieberbach, and G. Boche, *Tetrahedron Lett.* p. 4759 (1972).
18. S. Sustmann and C. Ruchardt, *Tetrahedron Lett.* p. 4765 (1972).
19. S. Sustmann and C. Ruchardt, *Chem. Ber.* **108**, 3043 (1975).
20. A. Barmetler, C. Ruchardt, R. Sustmann, S. Sustmann, and R. Verhulsdonk, *Tetrahedron Lett.* p. 4389 (1974).
21. R. Sustmann and F. Lübbe, *J. Am. Chem. Soc.* **98**, 4867 (1976).
22. S. Sustmann, R. Sustmann, and C. Ruchardt, *Chem. Ber.* **108**, 1527 (1975).
23. R. Sustmann and F. Lübbe, *Tetrahedron Lett.* p. 2831 (1974).
24. R. E. Lehr, J. M. Wilson, J. W. Harder, and P. T. Cohenour, *J. Am. Chem. Soc.* **98**, 4867 (1976).
25. W. L. Bauld, C.-S. Chang, and F. R. Farr, *J. Am. Chem. Soc.* **94**, 7164 (1972).
26. N. L. Bauld and J. Cessac, *J. Am. Chem. Soc.* **97**, 2284 (1975).
27. R. D. Rieke, S. E. Bales, P. M. Hudnall and C. F. Meares, *J. Am. Chem. Soc.* **92**, 1418 (1970).
28. N. L. Bauld and F. Farr, *J. Am. Chem. Soc.* **91**, 2788 (1969).
29. N. L. Bauld, F. Farr, and G. R. Stevenson, *Tetrahedron Lett.* p. 625 (1970).
30. R. D. Rieke, C. F. Meares, and L. I. Rieke, *Tetrahedron Lett.* p. 5275 (1968).
31. N. L. Bauld and C. E. Hudson, *Tetrahedron Lett.* p. 3147 (1974).
32. S. F. Nelson and J. P. Gillespie, *J. Org. Chem.* **38**, 3592 (1973).
33. R. Allendoerfer, L. L. Miller, M. E. Larscheid, and R. Chang, *J. Org. Chem.* **40**, 97 (1975).

34. R. Rieke, M. Ogliaruso, R. McClung, and S. Winstein, *J. Am. Chem. Soc.* **88,** 4729 (1966).
35. T. J. Katz and C. Talcott, *J. Am. Chem. Soc.* **88,** 4732 (1966).
36. G. Moshuk, G. Petrowski, and S. Winstein, *J. Am. Chem. Soc.* **90,** 2179 (1968).
37. S. Winstein, G. Moshuk, R. Rieke, and M. Ogliaruso, *J. Am. Chem. Soc.* **95,** 2624 (1973).
38. R. L. Blankespoor and C. M. Snavely, *J. Org. Chem.* **41,** 2071 (1976).
39. M. K. Hoffman, M. M. Bursey, and R. E. K. Winter, *J. Am. Chem. Soc.* **92,** 727 (1970).
40. M. E. Rennekamp and M. K. Hoffman, *Org. Mass Spectrom.* **10,** 1067 (1975).
41. A. Mandelbaum, S. Winstein, E. Gil-Av, and J. H. Leftin, *Org. Mass Spectrom.* **10,** 842 (1975).
42. R. A. W. Johnstone and S. D. Ward, *J. Chem. Soc. C* p. 1805 (1968).
43. R. A. Johnstone and S. D. Ward, *J. Chem. Soc. C* p. 2540 (1968).
44. M. J. Bishop and I. Fleming, *J. Chem. Soc. C* p. 1712 (1969).
45. M. L. Gross and D. H. Russell, *J. Am. Chem. Soc.* **101,** 2082 (1979).
46. R. P. Corbally, M. J. Perkins, and A. P. Elnitski, *J. Chem. Soc. Perkin Trans. I* p. 793 (1979).

INDEX

A

Alkenes
  hindered
    synthesis of, 112–114
  medium ring
    optically active, 51
    stereochemistry, 49
    synthesis of, 49, 50
  poly
    synthesis of, 113, 114
Allenes
  asymmetric synthesis of, 81
  formation from alkylidenecyclopropanes, 47
  oxidation of, 57
  synthesis of, 47, 67
    table, 79
  vinyl
    synthesis of, 189
Allene oxides
  formation of, 57, 58
  rearrangement, 58
Allowed reaction, definition of, 9
Allyl
  anions
    configuration interconversion, 95
  cations
    from cyclopropyl cations, 23
    ring closure of, 24, 33
  radicals
    formation from cyclopropyl radical, 402, 404–405
    stereochemical studies, 404–405
    theory, 402–403
Amarine, 222
Anions (*see* under individual ions)
Annulenes, ring closure of, 304–305
Antarafacial, definition of, 5

Arene oxides
  equilibrium with oxepins, 290–291
  theory, 290–291
Aromaticity
  Hückel, 13
  Möbius, 13, 14
  of transition states, 13
Aromatic rings, synthesis from pyrylium ions, 348
Azaallenium ion, 81
2-Azabicyclo[3,2,0]heptadienes, ring opening of, 169
2-Azabutadienes, ring closure of, 193
Azacyclooctatrienes, synthesis of, 187
1-Azahexatrienes, ring closure of, 323–333
2-Azahexatrienes
  ring closure of, 333–335
    stereochemistry, 334
    theory, 334–335
3-Azahexatrienes,
  ring closure of, 336–341
    theory, 336
Azahomoporphin, 398
Azaoxyallyl, 60, 61
Azanorcaradienes, 290–292
Azapolyenes, synthesis of, 360
Azepines
  ring closure of, 290–293
    theory, 290–291
Azetines
  ring opening, 192–196
    2-alkoxy, 193–194
    bicyclic, 195–196
    theory, 192
Azetinium ions, 196
Azides
  vinyl
    azirines from, 90, 240
    theory of ring closure, 241

ORGANIC CHEMISTRY

A SERIES OF MONOGRAPHS

EDITOR

HARRY H. WASSERMAN

Department of Chemistry
Yale University
New Haven, Connecticut

1. Wolfgang Kirmse. CARBENE CHEMISTRY, 1964; 2nd Edition, 1971

2. Brandes H. Smith. BRIDGED AROMATIC COMPOUNDS, 1964

3. Michael Hanack. CONFORMATION THEORY, 1965

4. Donald J. Cram. FUNDAMENTALS OF CARBANION CHEMISTRY, 1965

5. Kenneth B. Wiberg (Editor). OXIDATION IN ORGANIC CHEMISTRY, PART A, 1965; Walter S. Trahanovsky (Editor). OXIDATION IN ORGANIC CHEMISTRY, PART B, 1973; PART C, 1978

6. R. F. Hudson. STRUCTURE AND MECHANISM IN ORGANO-PHOSPHORUS CHEMISTRY, 1965

7. A. William Johnson. YLID CHEMISTRY, 1966

8. Jan Hamer (Editor). 1,4-CYCLOADDITION REACTIONS, 1967

9. Henri Ulrich. CYCLOADDITION REACTIONS OF HETEROCUMULENES, 1967

10. M. P. Cava and M. J. Mitchell. CYCLOBUTADIENE AND RELATED COMPOUNDS, 1967

11. Reinhard W. Hoffmann. DEHYDROBENZENE AND CYCLOALKYNES, 1967

12. Stanley R. Sandler and Wolf Karo. ORGANIC FUNCTIONAL GROUP PREPARATIONS, VOLUME I, 1968; VOLUME II, 1971; VOLUME III, 1972

13. Robert J. Cotter and Markus Matzner. RING-FORMING POLYMERIZATIONS, PART A, 1969; PART B, 1; B, 2, 1972

14. R. H. DeWolfe, CARBOXYLIC ORTHO ACID DERIVATIVES, 1970

15. R. Foster. ORGANIC CHARGE-TRANSFER COMPLEXES, 1969

16. James P. Snyder (Editor). NONBENZENOID AROMATICS, VOLUME I, 1969; VOLUME II, 1971

17. C. H. Rochester. ACIDITY FUNCTIONS, 1970

18. Richard J. Sundberg. THE CHEMISTRY OF INDOLES, 1970

19. A. R. Katritzky and J. M. Lagowski. CHEMISTRY OF THE HETEROCYCLIC N-OXIDES, 1970

20. Ivar Ugi (Editor). ISONITRILE CHEMISTRY, 1971

ORGANIC CHEMISTRY